Advances in Delivery Science and Technology

Series Editor
Michael J. Rathbone

For further volumes:
http://www.springer.com/series/8875

Kenneth A. Howard
Editor

RNA Interference from Biology to Therapeutics

Springer

Editor
Kenneth A. Howard
Department of Molecular Biology and Genetics,
Interdisciplinary Nanoscience Center
Aarhus University
Aarhus, Denmark

ISSN 2192-6204 ISSN 2192-6212 (electronic)
ISBN 978-1-4614-4743-6 ISBN 978-1-4614-4744-3 (eBook)
DOI 10.1007/978-1-4614-4744-3
Springer New York Heidelberg Dordrecht London

Library of Congress Control Number: 2012948979

Printed on acid-free paper

Springer is part of Springer Science+Business Media (www.springer.com)

I dedicate this book to my Mother,
Joan Howard.

Preface

The 2006 Nobel Prize in Medicine or Physiology was awarded to Andrew Fire and David Mello for their 1998 discovery of double-stranded RNA-mediated gene silencing by the process of RNA interference (RNAi). The capability to control and study cellular gene expression has opened up completely new research areas, shed exciting new light on existing fields, and provided researches with an unprecedented tool for investigating functional genomics and the potential to harness the RNAi mechanism as a potent therapeutic. This has resulted in an explosion of activity in both academia and industry.

Understanding the molecular mechanisms of RNAi is crucial for its transformation into a therapeutic modality. This dependency is the focus of "RNA Interference from Biology to Therapeutics", a concept applied to a Controlled Release Society (CRS) Educational Workshop at the 2009 CRS Annual meeting organised by the Editor.

The volume is structured to introduce the reader to the biological principles of RNAi followed by therapeutic delivery and disease treatment; however, integration of these aspects is a common thread running throughout. Education was an important consideration in the book preparation; therefore, the text provides sufficient background of the subject matter to allow utilisation as a learning tool for students.

The opening chapter gives an overview of RNAi pathways and the ground rules for therapeutic exploitation using synthetic small interfering RNA (siRNA) and vector-based approaches highlighted in subsequent chapters. SiRNA design towards RNAi pathway engagement is continued into chapter 2, focused on the development of Dicer-substrate therapeutics. The rapid emergence of the microRNA (miRNA) field, fuelled by its inherent role in regulation of cellular processes in normal and disease states, is highlighted in chapters 3, 13, and 14 dedicated to this subject. Chapter 3, for example, describes miRNA biology in tissue development and repair. In keeping with the "biology-therapeutic" link, the second part of this chapter describes its therapeutic exploitation in tissue regenerative medicine.

The clinical translation of RNAi therapeutics is dependent on enabling technologies to overcome both extracellular and intracellular delivery requirements; this is the focus of chapters 4–10. A number of delivery solutions and RNAi applications are covered that include nanoparticle-based systems composed of polymer, lipid, or exosomes. Systemic and mucosal routes of administration are addressed as well as stealth and targeting strategies.

Chapters 11–14 focus on RNAi treatment for specific disease types. Chapters 11 and 12 discuss target site selection within the viral RNA genome, viral escape, and solutions such as targeting cellular host factors in the treatment of HIV and influenza. Chapters 13 and 14 describe the application of miRNA in cancer such as deregulated miRNA expression for the identification of novel diagnostic and prognostic biomarkers and novel therapeutic targets as well as a description of miRNA-based anticancer therapies.

The clinical translation of RNAi-based treatments is an ultimate goal. Chapter 15 describes Alnylam's clinical development of a siRNA therapeutic for Respiratory Synitial Virus (RSV) lung infections. Preclinical steps including siRNA screen, antiviral efficacy, toxicology, and immune studies are presented before an overview of recent Phase 1 and 2 clinical trials.

This book is highly relevant for experts in, or at the interface of, RNAi, delivery science, and medicine from a personal field perspective as well as opening up new interdisciplinary research possibilities. An attractive feature is the "Future Perspectives" section ending each chapter that gives global experts the opportunity to express a personal view on where the field is going, offering potential new research directives to the reader.

"*RNA Interference from Biology to Therapeutics*" is a comprehensive and truly unique text for those involved or interested in this extremely exciting, important, and high impact field.

Aarhus, Denmark Kenneth A. Howard

Acknowledgements

I would like to express my sincere gratitude to all the authors for the time and effort they have given in the preparation of the excellent chapters contained in the book. I really appreciate your willingness to take on board editorial suggestions. I am grateful to the CRS and especially Series Editor Michael Rathbone for giving me the opportunity to put together a book that I believe was needed in the field. Cheers Mike My thanks go to Springer, Carolyn Honour, and Renata Hutter, and Renata, thank you for your great work and always being so polite.

Contents

Contributors

Katharina Ahrens Max Planck Institute for Infection Biology, Berlin, Germany

Samir El Andaloussi Department of Physiology, Anatomy and Genetics, University of Oxford, Oxford, UK

Morten Østergaard Andersen Department of Molecular Biology and Genetics, Interdisciplinary Nanoscience Center, Aarhus University, Aarhus, Denmark

Mark A. Behlke Integrated DNA Technologies, Inc., Coralville, IA, USA

Ben Berkhout Laboratory of Experimental Virology, Department of Medical Microbiology, Center for Infection and Immunity Amsterdam, Academic Medical Center, University of Amsterdam, Amsterdam, The Netherlands

Jesper B. Bramsen Department of Molecular Biology and Genetics, Interdisciplinary Nanoscience Center, Aarhus University, Aarhus, Denmark

Jeff Cehelsky Alnylam Pharmaceuticals, Cambridge, MA, USA

Philipp Dillschneider Department of Molecular Biology and Genetics, Interdisciplinary Nanoscience Center, Aarhus University, Aarhus, Denmark

Julia J.M. Eekels Laboratory of Experimental Virology, Department of Medical Microbiology, Center for Infection and Immunity Amsterdam, Academic Medical Center, University of Amsterdam, Amsterdam, The Netherlands

Michael Goldberg Department of Cancer Immunology and AIDS, Dana-Farber Cancer Institute, Harvard Medical School, Boston, MA, USA

Jared Gollob Alnylam Pharmaceuticals, Cambridge, MA, USA

Borja Ballarín González Department of Molecular Biology and Genetics, Interdisciplinary Nanoscience Center, Aarhus University, Aarhus, Denmark

Christa Haldrup Department of Molecular Medicine, Aarhus University Hospital, Aarhus, Denmark

Thomas B. Hansen Department of Molecular Biology and Genetics, Interdisciplinary Nanoscience Center, Aarhus University, Aarhus, Denmark

Kenneth A. Howard Department of Molecular Biology and Genetics, Interdisciplinary Nanoscience Center, Aarhus University, Aarhus, Denmark

Dennis K. Jeppesen Department of Molecular Medicine, Aarhus University Hospital, Aarhus, Denmark

Alexander Karlas Max Planck Institute for Infection Biology, Berlin, Germany

Verena Karsten Alnylam Pharmaceuticals, Cambridge, MA, USA

Kazunori Kataoka Division of Clinical Biotechnology, Center for Disease Biology and Integrative Medicine, Graduate School of Medicine, The University of Tokyo, Tokyo, Japan

Jørgen Kjems Department of Molecular Biology and Genetics, Interdisciplinary Nanoscience Center, Aarhus University, Aarhus, Denmark

Nobuyoshi Kosaka Division of Molecular and Cellular Medicine, National Cancer Center Research Institute, Tokyo, Japan

Helle Kristensen Department of Molecular Medicine, Aarhus University Hospital, Aarhus, Denmark

Samira Lakhal Department of Physiology, Anatomy and Genetics, University of Oxford, Oxford, UK

Jing Li Department of Pharmaceutical Sciences, Wayne State University, Detroit, MI, USA

Jinghuan Li Department of Physiology, Anatomy and Genetics, University of Oxford, Oxford, UK

Rachel Meyers Alnylam Pharmaceuticals, Cambridge, MA, USA

Ebbe Bech Nielsen Department of Molecular Biology and Genetics, Interdisciplinary Nanoscience Center, Aarhus University, Aarhus, Denmark

Sara Nochur Alnylam Pharmaceuticals, Cambridge, MA, USA

Takahiro Ochiya Division of Molecular and Cellular Medicine, National Cancer Center Research Institute, Tokyo, Japan

Aisling J. O'Loughlin Department of Physiology, Anatomy and Genetics, University of Oxford, Oxford, UK

Torben F. Ørntoft Department of Molecular Medicine, Aarhus University Hospital, Aarhus, Denmark

Marie S. Ostenfeld Department of Molecular Medicine, Aarhus University Hospital, Aarhus, Denmark

David Oupický Department of Pharmaceutical Sciences, Wayne State University, Detroit, MI, USA

Dan Peer Laboratory of Nanomedicine, Department of Cell Research and Immunology, George S. Wise Faculty of Life Science, Tel Aviv University, Tel Aviv, Israel

Center for Nanoscience and Nanotechnology, Tel Aviv University, Tel Aviv, Israel

Frederico Pittella Division of Clinical Biotechnology, Center for Disease Biology and Integrative Medicine, Graduate School of Medicine, The University of Tokyo, Tokyo, Japan

John J. Rossi Irell and Manella Graduate School of Biological Sciences, Division of Molecular and Cellular Biology, Beckman Research Institute of City of Hope, City of Hope, Duarte, CA, USA

Amy R. Simon Alnylam Pharmaceuticals, Cambridge, MA, USA

Karina D. Sørensen Department of Molecular Medicine, Aarhus University Hospital, Aarhus, Denmark

Scott D. Rose Integrated DNA Technologies, Inc., Coralville, IA, USA

Fumitaka Takeshita Division of Molecular and Cellular Medicine, National Cancer Center Research Institute, Tokyo, Japan

Troels Bo Thomsen Department of Molecular Biology and Genetics, Interdisciplinary Nanoscience Center, Aarhus University, Aarhus, Denmark

Akshay Vaishnaw Alnylam Pharmaceuticals, Cambridge, MA, USA

Matthew M.J. Wood Department of Physiology, Anatomy and Genetics, University of Oxford, Oxford, UK

Yusuke Yoshioka Division of Molecular and Cellular Medicine, National Cancer Center Research Institute, Tokyo, Japan

Jiehua Zhou Division of Molecular and Cellular Biology, Beckman Research Institute of City of Hope, Duarte, CA, USA

Yu Zhu Department of Pharmaceutical Sciences, Wayne State University, Detroit, MI, USA

Chapter 1
RNA Interference Pathways and Therapeutic Exploitation

Jesper B. Bramsen and Thomas B. Hansen

Abstract RNA interference (RNAi) is currently the method of choice to experimentally silence endogenous gene expression for functional genomics studies, while ongoing clinical trials point to its great therapeutic potential. Harnessing endogenous RNAi pathways to effectuate gene silencing by introducing artificial RNAi substrates or inhibitors translates into effective silencing efficiencies with high predictability and reliability but also has the potential to disturb endogenous gene regulation by the native inhabitants of the RNAi pathway, the microRNAs (miRNAs). A wealth of RNAi strategies have been developed over the last decade to produce optimal experimental triggers of RNAi entering at all levels in the RNAi pathway. Here we provide an overview of RNAi silencing pathways and its transformation into a therapeutic mode using vector-based approaches and chemically optimized, small interfering RNAs (siRNAs).

1.1 Entering the Age of RNA Interference

Nobel Prizes are rarely awarded based on discoveries <10 years old. Yet, in 2006, Craig Mellow and Andrew Fire were awarded the Nobel Prize in medicine or physiology for their 1998 discovery of double-stranded RNA-mediated gene silencing by a process of RNA interference (RNAi) [1]. Here a decade later, the impact of this discovery has been immense; RNAi has opened up completely new research areas, shed exciting new light on existing research fields, and provided researches with an unprecedented powerful tool for functional genomics investigations and potential diagnostic and therapeutic agents.

J.B. Bramsen (✉) • T.B. Hansen
Department of Molecular Biology and Genetics, Interdisciplinary Nanoscience Center, Aarhus University, C.F. Mollers Alle, 8000 Aarhus C, Denmark
e-mail: jebb@mb.au.dk

K.A. Howard (ed.), *RNA Interference from Biology to Therapeutics*, Advances in Delivery Science and Technology, DOI 10.1007/978-1-4614-4744-3_1,

Fire and Mellow showed that the introduction of exogenous long double-stranded RNA (dsRNA) into the nematode worm *Caenorhabditis elegans* triggers potent silencing of genes that share perfect sequence complementary to the introduced dsRNA, a silencing effect far superior to what was typically seen using single-stranded antisense RNA, the prevalent gene silencing tool at the time.

Several important observations quickly followed; the exogenous long dsRNA was found to be enzymatically processed into shorter 20–25-base pair (bp) dsRNAs acknowledged as the true effectuators of RNAi [2–4]. The subsequent observation that exogenous, synthetic 21-bp dsRNAs could trigger highly effective and sequence-specific gene silencing upon introduction in the cytoplasm in almost all investigated eukaryotic organisms provided researches with an extremely potent gene silencing tool [5]. These small dsRNA species, coined small interfering RNAs (siRNAs), marked the birth of a completely new silencing mechanism and the entry into the RNAi era of molecular biology and medicine. Besides being established as an ubiquitous tool in basic research, the number of RNAi-based preclinical and clinical trials has increased rapidly over recent years providing optimism for successful clinical translation [6, 7], a high achievement for such a fledgling technology.

1.2 RNAi Pathways

1.2.1 The Natural Biological Functions of RNAi Interference

As presented throughout this chapter, the unprecedented power of RNAi as a gene silencing tool originates from harnessing natural gene silencing pathways to effectuate efficient degradation of any chosen target mRNA. All key enzymes in the RNAi pathway are broadly conserved throughout the eukaryotic clade, suggesting the existence of a minimized version of the RNAi machinery in the last common ancestor of eukaryotes [8]. Here, RNAi is hypothesized to have served primarily as an ancient defense mechanism that functions *in cis* to endonucleolytically cleave and degrade exogenous long cytoplasmic dsRNA originating from viral infection [9–11], aberrantly expressed transgenes [10, 12–14], mobile genetic elements [15, 16], or aberrant processing of endogenous mRNA [17]. In higher eukaryotes, however, the RNAi machinery has evolved to be a key aspect in the general regulation of endogenous gene expression. Here, a class of endogenous 20–25 nucleotide single-stranded RNAs, known as microRNAs (miRNAs), are found to regulate the stability and translation of mRNAs *in trans* upon directly base pairing to their 3′untranslated regions (3′UTRs). The first example of a miRNA, a noncoding gene termed *lin-4* encoding a small RNA species of ~22 nt, was identified in the nematode worm *C. elegans* in 1993 [18]. It was hypothesized that *lin-4* represses translation of the *lin-14* gene by base pairing partially with putative target sites positioned in the 3′UTR of the *lin-14* mRNA. It took almost 7 years before the second miRNA, *let-7*, was shown to regulate the *lin-41* mRNA through a similar mechanism. Interestingly, *let-7* was found to be evolutionary conserved in a wide range of animal species indicating a more general role of miRNA in gene regulation [19–21].

Accordingly, thousands of miRNAs have since been identified, a task greatly enhanced by next generation sequencing techniques (see http://www.mirbase.org), and the RNAi machinery has indeed proved widespread and important in controlling gene expression in nearly all aspects of cellular metabolism.

1.2.2 RNAi Pathways: Step by Step

Considering the essential roles for RNAi pathways in organism homeostasis, it is clear that the development of safe RNAi-based therapeutics requires a thorough understanding of the RNAi pathways by which eukaryotic organisms regulate their transcriptome (Fig. 1.1). In essence, any artificial RNAi trigger employed, either introduced into the cytoplasm as synthetic siRNAs or transcribed in the cell nucleus from artificial RNAi vectors, will have to be carefully engineered to structurally mimic endogenous RNA intermediates in the RNAi pathway in order not to interfere with normal miRNA function. This section will, therefore, highlight key aspects of the maturation and function of miRNAs in order to establish the ground rules under which RNAi-harnessing approaches must comply.

Overall, the RNAi pathway has the potential to respond to a variety of RNA substrates of both endogenous origin, that is, primary miRNA precursors (pri-miRNAs) transcribed in the cell nucleus and exogenous origin such as cytoplasmic, foreign dsRNAs, e.g., viral RNA, siRNAs, etc. The terms "miRNA pathway" and "siRNA pathway" are often encountered in the literature to distinguish RNA substrates of endogenous and exogenous origin, however, in essence all mature RNAi triggers (miRNA and siRNA produced by the RNAi proteins from nuclear miRNA precursor and cytoplasmic dsRNA, respectively) are functionally identical once incorporated into the effector protein complex RISC in the cytoplasm [22, 23]. Concurrently, multiple entry routes into the RNAi pathway can be exploited for gene silencing therapeutics.

Focusing first on endogenous RNAi substrates, the majority of miRNAs are transcribed as primary transcripts (pri-miRNA) in the nucleus by RNA polymerase II and are situated either intronically in coding mRNA or as separate transcription units [24, 25]. The pri-miRNA is hairpin structured and is initially recognized by the bipartite complex, the microprocessor, composed of the endonucleolytic enzyme Drosha and its cofactor DGCR8 [26, 27]. The microprocessor crops the hairpin structure out of its flanking sequence, resulting in a 50–70-nt hairpin RNA, termed the precursor miRNA (pre-miRNA). Drosha is composed of dual RNase III domains responsible for the endonucleolytic cleavage, and the resulting product consists of a 5′ phosphate and 3′ hydroxyl group. Moreover, from a structural point of view, the Drosha-processed pre-miRNA typically has two unpaired nucleotides in the 3′ end, referred to as a 2-nt 3′ overhang. Basically, microprocessor products (with a few exceptions) serve as *bona fide* substrates in the downstream biogenesis pathway; therefore, the specific pri-miRNA recognition by the microprocessor is a defining step in miRNA biogenesis although the exact determinants of microprocessor recognition are still debated.

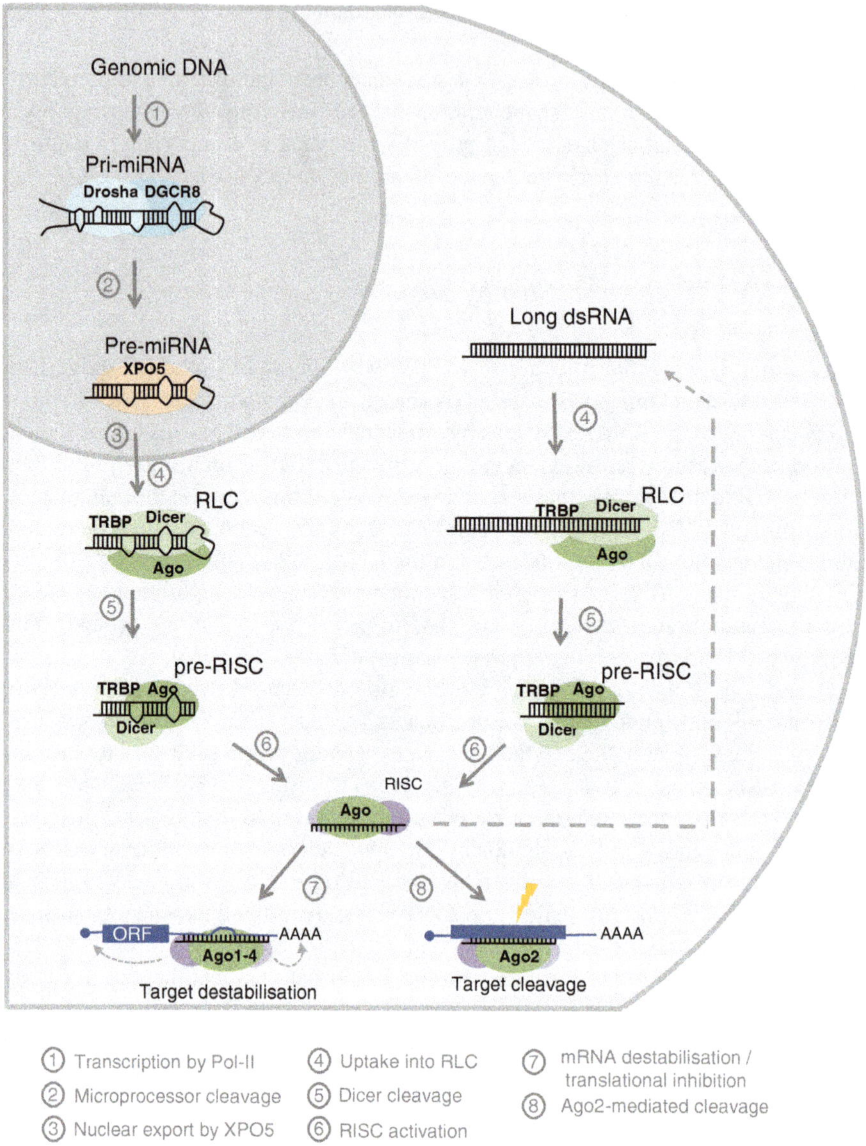

Fig. 1.1 Overview of the RNAi pathways. Natural miRNAs are transcribed by pol II as part of long pri-miRNAs ①, which are processed in the nucleus (*gray circle*) by the microprocessor (containing DGCR8 and Drosha) into pre-miRNAs ②. These are subsequently exported to the cytoplasm by XPO5 ③ and loaded into a RISC-loading complex (RLC) containing, e.g., Dicer and TRBP ④. In the cytoplasm, Dicer cleaves the pre-miRNA (*left*) and cytoplasmic long dsRNA (*right*) into a miRNA or siRNA duplex, respectively ⑤, thereby generating pre-RISC in which one strand is selectively loaded into active RISC ⑥. RISC can now direct mRNA destabilization/translational inhibition of target sharing only partial sequence complementarity to the guiding strand ⑦ (typical for miRNAs) or RNA cleavage of targets sharing perfect sequence complementarity to guide strand ⑧ (typical for siRNAs). SiRNAs generated from cytoplasmic long dsRNA can, due to their perfect sequence identity, target their original source, e.g., viral dsRNA for degradation (*dashed line*). Refer to text for more detailed description

With the initial step in miRNA biogenesis being compartmentalized to the nucleus, a translocation of the pre-miRNA across the nuclear membrane through the nucleopore complex (NPC) is a required step in the pathway. Pre-miRNA export is dependent on Exportin-5 (XPO5) [28], which is believed to be the rate-limiting step in miRNA biogenesis [29]. XPO5 has high affinity toward hairpin-like RNA with a structured stem exceeding 18 bp in length and with 3′ rather than 5′ overhangs [30], consistent with microprocessor products serving as canonical XPO5 substrates.

After export to the cytoplasm, the pre-miRNA is subjected to further processing by the so-called RISC-loading complex (RLC) [31–33] in which Dicer, another RNase III enzyme, cleaves off the pre-miRNA loop to produce a 21–22-nt miRNA duplex [3]. The pre-miRNA structure is, in part, recognized by the terminal 2-nt 3′ overhang, and consequently Dicer is typically regarded as an exonuclease. Like Drosha, Dicer-cleavage produces 2-nt 3′ overhangs; thus, the resulting miRNA duplex displays overhangs at both termini. Here, at the levels of Dicer cleavage, the miRNA and siRNA pathways meet (Fig. 1.1); in addition to processing pre-miRNA, Dicer also consecutively cleaves long dsRNA, such as foreign cytoplasmic dsRNA, from the terminus into smaller 21–22-nt siRNA duplexes [34]. Subsequently, miRNA and siRNA duplexes are incorporated into a pre-RISC [35], a complex composed primarily of one of four mammalian Argonaute proteins (Ago 1–4) as the core component. One strand in the miRNA or siRNA duplex (the passenger, non-guiding strand) is subsequently removed during RISC activation, resulting in single-stranded RNA in association with the activated RISC (the guide strand). This process is referred to as strand selection. The selection and incorporation of guide strand is at least, in part, determined by the molecular thermodynamics of the miRNA/siRNA duplex [36, 37]: Here, the least stable 5′ end in terms of base pairing strength is predominantly observed as the guiding strand. An examination of RISC activation in more detail reveals two distinct mechanisms: a RISC-mediated cleavage of the passenger strand, restricted to near-perfect RNA duplexes, such as siRNAs [38, 39], or a helicase-dependent mechanism responsible for unwinding the duplex [38, 40]; in both cases, strand selection is complying with the thermodynamics-based rules above.

Once incorporated into RISC, the single-stranded miRNA base pairs with target sites typically positioned in the 3′UTRs of mRNA, whereas siRNA target sites are located according to their origin or original design, typically within the open reading frame (ORF). Two possible effector mechanisms occur depending solely on the nature of Argonaute protein found in RISC and the level of complementarity between the guiding strand and the target mRNA: (1) in case of full complementary between guide strand and target, which is typical to siRNAs guide strands; Ago2 catalyzes the endonucleolytic cleavage, termed slicing, of target mRNAs, at the position located opposite the 9–10th nucleotide position counting from the guide strand 5′ end [41, 42]. This subjects the target mRNA to rapid exonucleolytic decay [22, 43] that is the intended and prevalent process in most typical siRNA-mediated mRNA "knockdown" experiments. (2) In case of more imperfect base pairing, essentially requiring only complementarity between the seed sequence of the miRNA (position 2–7/8) and the 3′UTR (seed match), cleavage is not observed, and

RISC instead destabilizes mRNA or represses its translation [23, 44, 45] in a process much less potent than siRNA-mediated slicing. Of the four closely related Ago proteins, Ago1–4, only Argonaute2 has endonucleolytic mRNA cleavage potential, whereas the three other closely related Ago proteins (Ago1, 3, and 4) are devoid of endonucleolytic activity [41]. The vast majority of characterized miRNA-mediated effects are facilitated by seed match target sites in the mRNA 3′UTR leading to mRNA destabilization or translational repression by mechanisms that are still debated but likely involve recruitment of a deadenylation complex (CAF1:CCR4:NOT1) and mRNA destabilization through deadenylation. Due to their functional identity, however, siRNAs also behave as miRNAs to destabilize hundreds of RNAs in addition to the intended target upon base pairing of the siRNA seed region to 3′UTRs, a process known as off-targeting (discussed in Sect. 1.4.6).

1.3 Harnessing RNAi Pathways for Gene Silencing Therapeutics

1.3.1 The Benefits of RNAi-Based Therapeutics

The breakthrough for RNAi as a potential human therapeutic was the observation of potent, sequence-specific, and seemingly safe knockdown of endogenous gene function in cell culture upon introduction of synthetic 21mer siRNAs with perfect sequence complementarity toward its target [5]. The importance of this observation for RNAi therapeutics cannot be underestimated, and synthetic siRNAs are the preferred gene silencing tool in vitro, a success that emanates from their general consistency, high efficiency, and ease of use. Today, it is a simple matter to order designed commercial synthetic siRNAs to obtain the desired gene knockdown in short-term cell culture experiments using commercial transfection reagents. Hitherto, synthetic siRNAs have provided the bedrock for the successful RNAi-based therapeutics: The first knockdown of an endogenous gene, apolipoprotein B (ApoB) using a clinically relevant formulation and administration route, was observed in mouse livers after standard intravenous injections of a chemically modified, but naked (non-formulated), siRNA in 2004 [46]. Also, the first successful knockdown of a cancer target gene in humans, the M2 subunit of ribonucleotide reductase (RRM2), was achieved in a clinical phase I trial in tumors from melanoma patients upon systemic delivery of siRNA nanoparticles [7] (for more details on ongoing clinical trials, refer to http://clinicaltrials.gov).

There are several contributing components to the great success of RNAi-based gene silencing as compared to the competing antisense oligonucleotide (ASOs) technology developed in the 1970–1980s [47]. Similarly to ASO technology, the high predictability and specificity of nucleic acid base pairing provides fully programmable and specific gene silencing which in practice renders all genes "druggable": Researchers need only to produce short dsRNA species with a structure recognizable to RNAi proteins and perfect sequence complementarity to a particular mRNA target to achieve gene knockdown. Yet, RNAi-based silencing strategies hold additional benefits; exogenous RNAi substrates, such as synthetic siRNAs, are rapidly incorporated into RNAi protein complexes to protect them from nuclease degradation [48, 49] and, very importantly, to orchestrate their transport to their target mRNA in the cytoplasm, thereby effectuating multiple rounds of target cleavage [10]. Furthermore, the structural mimicry of endogenous miRNA species by artificial siRNA triggers allows them, at least in theory, to remain undetected to cellular sensors of foreign dsRNA, thereby preventing triggering of innate and adaptive immune responses.

1.3.2 Therapeutic Entry into the RNAi Pathways

Despite the overwhelming popularity of synthetic siRNAs, a wealth of strategies for harnessing endogenous RNAi pathway as a gene silencing tool have been developed to best suit the given experimental system or organism. Collectively, these strategies exploit the fact that protein complexes in the RNAi pathways respond to the structure of their RNA substrates rather than their sequences and that miRNA pathways and siRNA pathways intersect in the cytoplasm at the level of RISC loading and shares degradation pathways (Fig. 1.1). Therefore, any desired guide strand can be loaded into Ago2-RISC to effectuate mRNA target cleavage once embedded within a suitably structured RNAi substrate mimic, e.g., primary miRNA transcripts, precursor miRNAs, miRNA duplexes, dsRNA shorter than 30 bp, or most popularly 21mer siRNAs (Fig. 1.2).

As discussed in greater detail in Sect. 1.4, most experimenters typically use synthetic 21mer siRNAs as triggers of RNAi: These have perfect structural identity to natural Dicer cleavage products (typically two 21-nt RNA strands annealed to form a 19-bp dsRNA duplex stem and 2-nt 3′ overhangs at both ends) and are upon introduction into the cell cytoplasm loaded into RISC by RLC to facilitate transient gene knockdown. The siRNA sequence is typically designed to target the mRNA ORF, and silencing effect persists for 2–7 days in typical cell-cultured siRNA. As described in Sect. 1.4, numerous successful siRNA designs utilizing shorter of longer RNA backbones and chemical modification have been developed to enhance silencing potency, specificity, and longevity. Still, the lack of efficient means to achieve cytoplasmic delivery in vivo is the major bottleneck for therapeutic applications, and reports of siRNA immunogenicity and off-targeting also need immediate addressing.

As described in Sect. 1.5, RNAi triggers may also be encoded by RNAi plasmids or viral vectors, which are introduced into the cell nucleus and here utilize the cellular transcription apparatus to ensure a continuous production of the intended RNAi substrates. This potentially allows for long-lasting RNAi and can also ensure efficient delivery of RNAi constructs into cells that are otherwise difficult to target. Especially viral vectors hold therapeutic potential: Whereas retro- and lentiviral vectors may have limited use in therapeutics due to their integration into the host genome (i.e., risk of insertional mutagenesis), the nonintegrating viral vectors such as adeno- and adeno-associated viral vectors may hold greater therapeutic potential [50, 51].

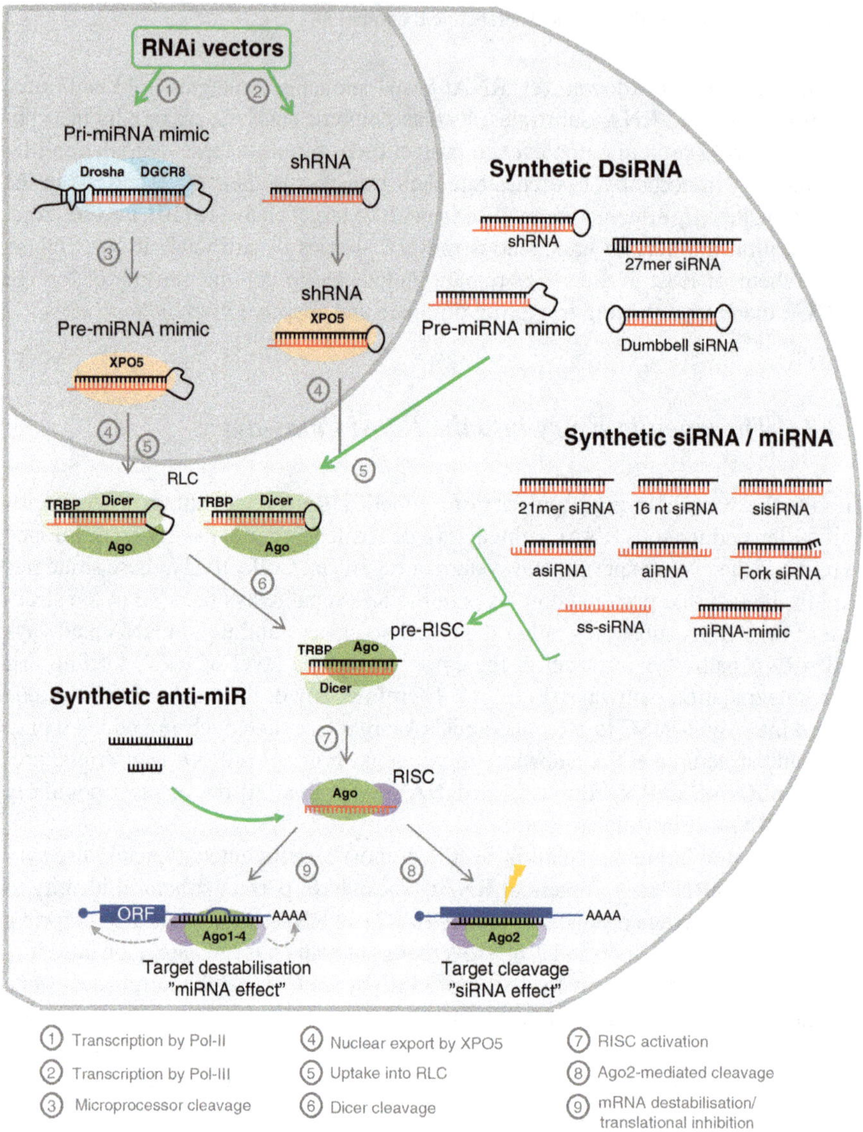

Fig. 1.2 Harnessing endogenous RNAi pathways for gene silencing therapeutics. Artificial triggers of RNAi can be introduced via RNAi vectors that are transcribed in the nucleus; pol III-transcribed cassettes will typically encode shRNAs that are directly transported to the cytoplasm and loaded into RLC (*right side*). Alternatively, RNAi triggers are delivered as pol II-transcribed pri-miRNA mimics, which are cleaved by the microprocessor prior to loading into RLC in the cytoplasm. Synthetic triggers of RNAi are instead introduced directly into the cell cytoplasm; Dicer substrate siRNAs (DsiRNAs), either in the form of shRNAs, pre-miRNA mimics, or 27mer dsRNA, can enter the RLC and are subsequently loaded into pre-RISC upon Dicer cleavage. Various design versions of synthetic 21–23mer siRNAs, and notably also miRNA mimics, are instead loaded into pre-RISC without Dicer cleavage and can subsequently guide Ago2-mediated target cleavage (referred to as on-targeting for siRNAs) or Ago-mediated RNA destabilization/translational inhibition of targets sharing perfect seed matching (the natural miRNA function which is referred to as of-targeting for siRNAs). Endogenous miRNAs can also be inhibited by introducing 12–25-nt chemically engineered antisense oligonucleotides, named anti-miRs or antagomirs, into the cell cytoplasm where they bind and inhibit the single-stranded miRNAs loaded into active RISC. Refer to text for more detailed description

In addition to silencing mRNA by custom siRNAs, recent RNAi strategies have focused on modulating the levels of endogenous miRNAs as these are often found to be up- or downregulated in human diseases such as cancer [52]. The so-called miRNA replacement therapy aims to introduce mimics of natural miRNAs into the RNAi pathway [53]. Here they will exert the functions of their natural counterpart to simultaneously target multiple mRNAs through seed-mediated base pairing to their 3′UTR (rather than cleaving a single target as intended for siRNAs). miRNA mimic's are very similar to siRNAs and can be introduced as synthetic dsRNA molecules or via DNA vectors (described in the following sections). In practice, synthetic miRNA mimics are often designed as perfectly base-pairing siRNA-like molecules where the guiding strand is identical to a given endogenous miRNA. Therefore, delivery and chemical optimization strategies are basically similar to siRNA design; however, modifications that reduce siRNA off-targeting (i.e., miRNA effects) should obviously not be applied in miRNA mimic design.

Finally, endogenous miRNA function can be inhibited by introducing 12–25-nt single-stranded oligonucleotides designed antisense to the given miRNA. Upon introduction into the cell cytoplasm, such molecules, referred to as anti-miRs or antagomirs [54, 55], will bind to RISC-loaded miRNAs and block their natural functions. Being single-stranded, anti-miR typically needs extensive chemical modification both to resist degradation by nucleases and also to enhance their binding affinity to target miRNAs (e.g., by locked nucleic acids (LNA) [56]), to trap the RISC-loaded miRNAs in a nonfunctional state or even promote their degradation [57]. Significant progress has already been made using anti-miR designs in primates [56, 58]. As an alternative, the so-called miRNA sponges transcribed from DNA vectors introduced into the cell nucleus have allowed long-lasting inhibition of endogenous miRNA function [59]. In practice, such sponges are mimicking the structure of natural miRNA targets as RNA transcripts expressed from strong pol II promoters designed to contain multiple binding sites, often heptameric sequences complementary to the target miRNA seed. Notably, emerging evidence suggests that the unknown function of human pseudogenes or long noncoding RNAs may indeed be to function as natural sponges for endogenous miRNAs [60].

1.4 The Application of Synthetic siRNA as RNAi-Based Therapeutics

Synthetic siRNAs are by far the most widely used triggers of RNAi in cell culture in which they are introduced into the cell cytoplasm and directly loaded into RISC. Although efficient transfection can be achieved in most cell culture systems by the use of commercial transfection reagents, in vivo delivery of synthetic siRNAs into the target cells is far more challenging. Synthetic siRNA suffers from a number of drawbacks compared to most small-molecule drugs; namely, they are macromolecules (~14,000 Da) and hydrophilic due to their anionic phosphodiester backbone that restricts entry across the cellular membrane required for target interaction.

Furthermore, they are susceptible to serum nucleases and have poor pharmacokinetic properties as a consequence of e.g., rapid renal clearance upon intravenous injection [61]. Moreover, siRNA silencing specificity and safety has been challenged by the general finding of siRNA off-targeting [62, 63] and potential immunogenicity [64, 65] resulting in the realization that further clinical progress will require greater investments than initially envisaged to overcome these issues. Encouragingly, this process is well underway; synthetic siRNA technology is now dramatically improved by chemical modification that will most likely fulfill its potential if allowed to mature in a similar manner to development and the emerging success of ASO technology [66–68].

1.4.1 Synthetic siRNA Types

The most simple and popular siRNA design today mimics natural Dicer cleavage products and comprises a 21-nt guiding strand antisense to a given RNA target and a complementary passenger strand annealed to form a siRNA duplex with a 19-bp dsRNA stem and 2-nt 3′ overhangs at both ends [4, 5]. Other siRNA designs mimic Dicer substrates to enhance incorporation into RNAi pathways [69] and siRNA potency [referred to as Dicer substrate siRNAs (DsiRNAs)] [70]; especially synthetic 27mer DsiRNAs [71–75] can have very high activity; yet, concerns of unpredictable efficiency and potential immunogenicity [76] need to be addressed [73–75]. A variety of alternative siRNA designs exhibiting different architectures have been developed such as Dicer-independent short shRNAs with RNA stems ≤19 bp [77, 78], blunt 19-bp siRNAs [79, 80], blunt fork-siRNAs [82, 83], single-stranded siRNAs (ss-siRNAs) [84–86], dumbbell-shaped circular siRNAs [87], asymmetric siRNAs (aiRNA) and asymmetric shorter-duplex siRNA (asiRNA) harboring a shortened SS [78, 88, 89], bulge-siRNA [90], and sisiRNAs [91] (Fig. 1.2). Some of these aim to enhance siRNA potency and specificity by ensuring the preferential loading of the guide strand in RISC and/or rendering passenger strand nonfunctional. Examples include the asymmetric siRNAs (asiRNAs) which utilize a 5′ end truncated passenger strand [78, 88] and the sisiRNAs which utilize two short 10–12-nt passenger strands, all of which cannot contribute to siRNA off-targeting [91]. Also, fork-siRNA contains mismatched bases in the 3′ end of the passenger strand which enhance loading of the siRNA guide strands due to a lower thermodynamic stability of its 5′ duplex end. Finally, the use of asymmetric siRNA overhangs also ensures preferential guide strand loading and improved siRNA activity [92]. Other designs aim to enhance siRNA nuclease resistance to prepare unmodified siRNA for usage in vivo: Blunt 19-bp siRNAs and dumbbell-shaped circular siRNAs are both reported to be more resistant to nuclease degradation, even when unmodified, as they contain no free 3′ overhang [80, 87]. Recently, siRNAs has also been incorporated in larger nucleic acid structures with the prospect of enhancing delivery and bioavailability [93, 94].

1.4.2 Maximizing siRNA Activity

Maximizing siRNA silencing activity minimizes the siRNA dose required for efficient gene silencing which is important when considering the difficulties of siRNA delivery in vivo. The siRNA sequence is a primary determinant of siRNA activity [36, 95, 96], and large screens have identified several design rules [97–103] such as a low GC-content of 30–50% [99, 104], especially in the guide strand seed region [99, 102] and preferences for low internal thermostability [99, 101, 102]. Importantly, siRNA sequences should be chosen to render the siRNA duplex thermodynamically asymmetric as the siRNA strand having the 5′ end engaged in the thermodynamically least stable part of the duplex will, as noted above, be preferentially utilized as guiding strand in RISC [36, 37]. Furthermore, siRNA target sites should be accessible and not hidden in stable secondary structures [105–109] nor occupied by RNA-binding proteins [110], and in effect, efficient target sites are often found in AU-rich regions [111]. SiRNA chemical modification screens show that careful chemical engineering of siRNA can enhance their activity beyond unmodified siRNAs [112]; a few examples of dramatic potency improvements have been reported [148]; yet, typically, only modest (less than twofold) improvements in siRNA potency are seen, for example, upon moderate modification of siRNA strand 3′ ends [112–118]. Instead, modifications that favor loading of the guide strand into Ago2-RISC seem to be a reliable, sequence-unspecific strategy to enhance siRNA potency either by introducing optimal siRNA thermodynamic asymmetry using stabilizing modifications (e.g., LNA [119], 2-thiouracil [120], 2′-F) in the passenger strand 5′ end or by introducing destabilizing modifications (such as OXE, ethylamino, UNA, dihydrouracil, or PS [112, 119, 121]) in its 3′ end. Furthermore, chemically modified 3′ overhangs that are favored and unfavored during strand selection by RISC have been identified and can easily be incorporated into the guide and passenger strands of the siRNA, respectively [92, 112]. As a note of caution, the industry standard and popular 2-nt DNA overhang dTdT has been suggested to reduce silencing longevity significantly [122].

1.4.3 Motivation for Chemically Modified siRNA

Unmodified synthetic siRNAs seem most suited for short-term gene silencing experiments in cell culture, where adverse side effects such as siRNA immunogenicity and off-targeting (see below) go unnoticed or are disregarded. Higher standards for siRNA performance and safety are required to establish siRNA-based therapeutics. These concerns can be resolved by chemical modification [81, 91, 112, 123–126]. Unmodified, naked siRNAs are highly labile in biological fluids due to their degradation by ribonucleases, and their poor pharmacokinetic properties in vivo reduce intracellular delivery [46, 61, 127]. Furthermore, siRNAs may be immunogenic [64, 65, 128–130] and inherently trigger non-intended off-target regulation of genes harboring seed matches of either siRNA strand (i.e., each siRNA strand has the potential to function as a miRNA, thereby potentially affecting hundreds of mRNAs) [62]. A wealth of modification types have been tested to circumvent these shortcomings including modification of the phosphodiester backbone (e.g., by phosphorothioate linkages (PS) [131–138]), substitution of the ribose 2′-OH group [e.g., by 2`-*O*-Methyl (OMe), 2′-Flouro (F), 2′-Methoxymethyl (MOE)], or using bridged nucleic acids (such as LNA [139], carbocyclic-LNA [112, 140], ENA [141], carbocyclic-ENA [112, 140], and oxetane (OXE) [112, 142]), substitution of the ribose unit with six carbon sugars (such as ANA, HNA, 2′-F-ANA, and CeNA [112–117]), by disrupting the ribose ring structure (such as in unlocked nucleic acids (UNAs) [124, 143]), or modification of nucleoside bases (5-iodo-, 2-thio- and pseudouracil [120, 131, 144]) (for a more comprehensive review of chemical modification types, see [126]). Modifications, however, must be compatible with siRNA function, and positional tolerances for siRNA modifications have been well established by empirical testing; by rule of thumb, the guide strands 5′ phosphate, 5′ end (seed region), and central positions are particularly sensitive, especially to several or bulky modifications. In contrast, the entire length of the passenger strand, the 3′ end, and overhang of the guide strand are fairly tolerant and can be chemically modified to enhance siRNA performance [112, 126, 131–136, 141].

1.4.4 Improving siRNA Nuclease Resistance

Most unmodified siRNAs are degraded by ribonucleases within minutes in mammalian serum [46, 112, 131, 135, 145, 146], and enhancing siRNA stability has long been considered essential for siRNA function in vivo. Indeed, extensive chemical stabilization was found essential for successful silencing in mouse livers upon low-pressure intravenous injection of naked siRNA, a strategy relevant to siRNA therapeutics [147]. Yet, only few extensively or fully modified siRNAs are reported to be both highly stable and potent [147–149] as extensive chemical modification of siRNAs will typically reduce their activity [112]. Instead, moderate modification levels using phosphorothioates [131–136], thermostabilization of the siRNA duplex stem by LNA [112, 119, 135, 136, 150], OMe or 4′ thioribose [151], or combinations of these modifications have been successful in creating stable and potent siRNA for applications in vivo [46, 152, 153].

Recently, suggestions to modify only nuclease hypersensitive positions in the siRNA were put forward to limit modification levels and preserve silencing [154, 155]. As modification of siRNA 3′ overhangs are very well tolerated by the RNAi machinery, numerous modification types will provide 3′ exonuclease resistance and modestly enhance siRNA stability [112, 131–136]. As most dsRNA-specific endoribonucleases are preferentially recognizing UpA, UpG, and CpA dinucleotide motifs [154–157], further siRNA stabilization by modifying these vulnerable positions, for example, by 2′-OMe, is a straightforward approach to significantly enhance stability [154, 155].

It is worth noting that the benefits of enhancing siRNA nuclease resistance seem primarily to originate from effects during siRNA delivery prior to siRNA internalization or enhanced interaction with the intracellular RNAi machinery [158]. Once inside the cytoplasm, siRNAs are more stable, some likely protected by RISC incorporation [159], and silencing can persist for 30–90 days in slowly or nondividing cells [158, 160]. Therefore, chemical modification of siRNAs to enhance nuclease resistance is primarily needed in applications using "naked" siRNA and less when using shielding delivery agents such as nanoparticles.

1.4.5 Reducing siRNA Immunogenicity

Synthetic siRNAs were initially considered to be non-immunogenic in mammalian cells, due to their structural mimicry of endogenous Dicer cleavage products [5]. However, with increasing investigations, siRNAs are now reported to induce innate immune responses through mechanisms dependent on cell type, delivery route, siRNA structure, and sequence [76]. Specific single-stranded sequence motifs such as GUCCUUCAA [64], UGUGU [129], UGGC [161], and GU [162] have been reported to render siRNA immunogenic. These are recognized by Toll-like receptors 7 and 8 (TLR7/8), transmembrane receptors found in the endosomes of immune cell populations [163]. These responses can be potentiated with the use of transfection agents that facilitate endosomal delivery. Recently, uridine content has been reported to correlate with TLR7/8 activation [164], thereby severely complicating siRNA design. Encouragingly, TLR7/8 activation may be largely avoided by chemically modifying or shielding immune-stimulatory sequences: Several modification types, especially 2′-modified nucleotides (DNA, 2′-OMe, 2′-F, LNA), can abrogate siRNA immunogenicity [130, 147, 165, 166], and modification of uridines only with either 2′-F or 2′-OMe [167] or DNA [168] may be sufficient. In particular, 2′-OMe modification of the passenger strand has been proposed as a universal approach to avoid TLR7 activation [169].

siRNA duplex length also seems to affect siRNA immunogenicity; several studies suggest that 21mer siRNAs are not immunogenic in several cell lines even at high concentrations, whereas 25mer (or longer) siRNAs trigger concentration-dependent immunogenicity in HeLa S3, DU145, and MCF7 cells, but not in HeLa and HEK293 cells. Consequently, it has been suggested to avoid non-modified 27–29mer DsiRNA design for siRNA applications, at least in vivo [76]. siRNA sequence length is sequence-independent monitored by the transmembrane TLR3 (among others) expressed on the surface and in endosomes in the dendritic subpopulation of the leukocytes but also in many primary cell types and popular cell lines [76, 170]. Notably, TLR3 activation by the popular 21mer siRNA has been reported upon intraocular injection in mice regardless of their sequence and 2′-OMe modification, whereas a shorter 19mer siRNA design was safe [171]. Another sensor of siRNA length is protein kinase R (PKR), present in all cells and stimulated by cytoplasmic dsRNA longer than 30 bp [172]. 25–30mer siRNAs did not activate PKR in HEK293 cells [71]; yet, canonical 21mer siRNAs have been shown to bind or trigger modest PKR activation in murine microglial N9 cells [173], T98G cells [174–176], and HeLa cells [175]; therefore, the impact of PKR activation needs to be further established. Usually, however, only few precautions are taken in typical cell culture experiment to avoid activation of PKR. Another sensor of cytoplasmic dsRNA is RIG-I, which recognizes in part the poly-uridine richness of dsRNA and in part the composition of dsRNA termini; 5′ end triphosphates and blunt 21–27mer siRNA trigger RIG-I activation, whereas the standard 21mer siRNA design having two 2-nt 3′ overhangs is tolerated [174].

1.4.6 Reducing siRNA Off-Target Effects

The shared handling of exogenous siRNAs and endogenous miRNAs by RISC in the cytoplasm inherently forces all siRNAs to behave as miRNAs and trigger unintended downregulation (typically less than twofold) of hundreds of endogenous mRNAs sharing sequence complementarity in their 3′UTR to the siRNA seed region [62, 177, 178]. These "off-target" effects can result in toxic phenotypes [161] and compromise the interpretation, outcome, and safety of the particular siRNA application. Although siRNA sequences with low seed match frequency are predictable in silico, off-targeting cannot be fully avoided but significantly reduced by specific chemical modification of the guide strand seed region; particularly, introducing a highly destabilizing UNA modification at position 7 of the guide strand dramatically reduces off-targeting with minimal loss of on-target activity [123]; yet, also 2′-OMe modification of position 2 [179] or substituting position 1–8 with DNA [180] reduces off-target albeit on-target activity may concurrently be slightly decreased.

It is similarly important to minimize the contribution of the passenger strand to off-targeting; most well-designed siRNAs will lead to preferential loading of the guide strand into RISC; however, the passenger strand contribution cannot be fully abrogated; chemically blocking the passenger 5′ phosphate, e.g., by 5′-OMe modification [181] or inclusion of an additional UNA residue [182], abrogates incorporation of the passenger strand into RISC. Another approach to avoid passenger strand selection is the sisiRNA design in which the passenger strand is composed of two shorter 10–12-nt RNA strands incapable of RNAi function [91]. It should also be noted that excess RNAi substrates (such as various siRNA designs) may also unintendedly impact endogenous gene regulation by competing with endogenous miRNA for rate-limiting steps in the RNAi pathway [95], thereby disturbing endogenous miRNA production or function [29, 183].

1.5 Vector-Borne RNAi

Vector-based RNAi is less widely used than synthetic siRNAs yet allows RNAi triggers to be delivered via standard transgenic approaches [184, 185] and especially viral vectors [29, 186–198] to overcome delivery obstacles and allow longer-lasting silencing and temporal-spatial control of gene silencing. In fact, engineered viruses constitute the majority of vectors used in clinical trials due to their high efficiency in vivo, thereby making vector-based RNAi highly relevant for RNAi therapeutics. Compared to chemically synthesized siRNAs, vector-borne RNAi does not allow chemical modification of the siRNA itself to enhance its biogenesis and activity or reduce siRNA off-target effects and immunogenicity (see Sect. 1.4). Therefore, careful considerations have to be made to ensure adequate expression levels without altering endogenous miRNA function.

1.5.1 First-Generation RNAi Vectors (Producing pre-miRNA Mimics)

First-generation RNAi vectors aimed to encode artificial small dsRNA resembling either 21–27mer siRNA or short hairpin RNA (shRNAs) mimicking pre-miRNA structures by using pol III promoter-based transcription units, most often designed using H1 or U6 promoters [199–207]. These promoters are optimal for transcribing short and defined RNAs; transcription is initiated at a single position and terminates after the second T of a templated "TTTTT" stretch, thereby leaving two terminal Us in the resulting RNA transcript. The conceptually most simple setups employ a "tandem design" in which siRNA strands are transcribed from individual units [200, 201] or convergent transcription of a single siRNA sequence flanked by pol III promoters [208, 209]. The transcribed strands anneal to form siRNA, which is exported to the cytoplasm and loaded into RISC. Most researchers prefer a more efficient short hairpin RNA (shRNA) design in which the 19–21mer siRNA strands are tethered by a short 5–12-nt loop sequence in a single transcript, which somewhat mimics nuclear pre-miRNAs [199, 202, 203, 205]. General design rules for the shRNA stem are as for siRNAs (Sect. 1.5); yet, U6 and H1 promoters favor a G or A residue, respectively, at the first position of the transcribed sequence [210, 211]. Also, the shRNA structure and particularly the length of the shRNA duplex stem and choice of loop affect shRNA processing by RNAi proteins and consequently their silencing efficiencies. Early designs encoded 19–21-bp RNA stems connected by 5–9-nt artificial loops; however, slightly longer stems and natural miRNA loops are both shown to improve shRNA processing [49, 212].

Pol III promoters were initially considered ideal for expression of shRNA/pre-miRNA mimics as they are highly and ubiquitously expressed [213, 214] and allow for the use of very small transcriptional units (<200 bps including promoters) well suited for, e.g., retroviral delivery strategies [186–193]. Indeed, successful RNAi using pol III-based transcriptional units has been reported, e.g., in genetically modified mice [184–186, 215, 216]; however, the number of successful applications is still relatively low. Notably, the high transcriptional activity of pol III promoters may flood endogenous RNAi pathways with artificial and poorly processable shRNAs, and pol III-transcribed shRNA can block endogenous miRNA production by saturating limiting amounts of XPO5 in mouse livers [29], an observation subsequently confirmed in other tissues [196, 217, 218]. Also, the use of shRNA vectors has been shown to induce cellular immune responses [176, 219]. Therefore, for strategies utilizing pol III-dependent shRNA expression, great care should be taken to ensure low, nontoxic shRNA expression levels, e.g., by keeping RNA vector copy numbers low.

1.5.2 Second-Generation RNAi Vectors (Entering as Pri-miRNA Mimics)

The observation that natural miRNAs are transcribed as parts of long pri-miRNA transcripts, which are processed in the nucleus by the microprocessor into pre-miRNA [25, 220, 221], inspired researches to deliver siRNA as part of pol II-dependent transcriptional units. Importantly, this allows the use of well-characterized tissue-specific promoters and tightly controlled inducible systems, such as the Tet-On and Tet-Off technologies [222], to obtain temporal-spatial control of siRNA production [223]. As the structural requirements for recognition by the nuclear microprocessor are not fully established, researches have embedded siRNA within a pol II-transcribed artificial pri-miRNA mimic simply by replacing the natural miRNA stem sequence [224], and a number of popular miRNA scaffolds currently exist building on the context of endogenous miRNA [225–227]. A great advantage of these second-generation designs is reduced toxicity and more reliable knockdown phenotypes; expression levels can be fine-tuned to less toxic levels, and pri-miRNA mimics are generally more efficiently processed [212, 228, 229] and exported [230], resulting in more potent gene silencing phenotype, less saturation of RNAi pathways, and cellular toxicity [217, 223, 231]. Finally, pol II-based designs allow the inclusion of several pri-miRNA mimics and protein-coding sequences (CDS) in a single polycistronic transcript [25, 226] also compatible with (lenti) viral delivery [232].

1.6 Future Perspectives

There is little doubt that manipulation of gene expression via RNAi is getting closer to clinical successes, and much anticipated proof of therapeutic gene silencing by siRNAs was recently established in humans [7]. So far, important progress has been made in terms of engineering the synthetic RNAi triggers, such as siRNA, miRNAs, and anti-miRs, to be more nuclease-resistant, potent, specific, and safe. Still, the lack of efficient means of targeted delivery represents a bottleneck for broader clinical application and the concurrent evaluation and refinement of RNAi triggers in vivo. Consequently, researches may very well focus on implementing the pallet of promising local delivery systems to achieve immediate successes in vivo, while universal systemic delivery strategies are a more distant goal.

While awaiting the clinical success of siRNA therapeutics, our knowledge of natural gene silencing pathways, RNAi-mediated or not, is expanding. In particular, the mapping of regulation and deregulation of miRNA expression in human diseases holds both diagnostic/prognostic and therapeutic value. Strategies for replenishing miRNAs downregulated or lost in human disease by introducing miRNA mimics or conversely inhibiting aberrantly expressed miRNAs via synthetic anti-miRs are most likely to gain even greater popularity in the near future. Also, the combination of siRNAs targeting particular detrimental mRNA and miRNA mimics/anti-miRs restoring natural miRNA pathway functions is likely to prove a powerful therapeutic cocktail.

Although the reported transcriptional silencing by promoter-targeting siRNAs [233] seem to require more investigations to be relevant for RNA therapeutics [234], the investigations of epigenetic remodeling by natural short and long noncoding RNA (ncRNA) may very well establish novel siRNA targets or even bedrock novel silencing molecules [235, 236].

Finally, non-RNAi-mediated gene silencing is currently also flourishing; important progress in vivo is already being made with RNAi-independent chemically engineered antisense oligonucleotides (ASO) which are short ~20-nt single-stranded antisense oligonucleotides designed to trigger degradation of a given target mRNA upon binding [237, 238]. Alternatively, several types of so-called steric-blocking oligonucleotides, most typically 20–25-nt single-stranded chemically modified oligonucleotides or phosphorodiamidate morpholino oligos (PMOs), have proven capable of blocking RNA-binding protein, reducing mRNA translation and modulate mRNA splicing and folding, with some success in clinical trials [66, 239–243].

The arsenal of promising nucleic acid-based therapeutics seems quite powerful ~10 years after the first successful application of siRNA in human cells [244]. We believe nucleic acid-based therapeutics have now properly matured and are on the verge of clinical breakthroughs, a natural progression not dissimilar to development of monoclonal antibody therapeutics pioneered in the 1970–1980s.

References

1. Fire A, Xu S, Montgomery MK, Kostas SA, Driver SE, Mello CC (1998) Potent and specific genetic interference by double-stranded RNA in Caenorhabditis elegans. Nature 391:806–811
2. Hamilton AJ, Baulcombe DC (1999) A species of small antisense RNA in posttranscriptional gene silencing in plants. Science 286:950–952
3. Zamore PD, Tuschl T, Sharp PA, Bartel DP (2000) RNAi: double-stranded RNA directs the ATP-dependent cleavage of mRNA at 21 to 23 nucleotide intervals. Cell 101:25–33
4. Elbashir SM, Lendeckel W, Tuschl T (2001) RNA interference is mediated by 21- and 22-nucleotide RNAs. Genes Dev 15:188–200
5. Elbashir SM, Harborth J, Lendeckel W, Yalcin A, Weber K, Tuschl T (2001) Duplexes of 21-nucleotide RNAs mediate RNA interference in cultured mammalian cells. Nature 411:494–498
6. Davidson BL, McCray PB Jr (2011) Current prospects for RNA interference-based therapies. Nat Rev Genet 12:329–340

7. Davis ME, Zuckerman JE, Choi CH, Seligson D, Tolcher A, Alabi CA, Yen Y, Heidel JD, Ribas A (2010) Evidence of RNAi in humans from systemically administered siRNA via targeted nanoparticles. Nature 464:1067–1070
8. Cerutti H, Casas-Mollano JA (2006) On the origin and functions of RNA-mediated silencing: from protists to man. Curr Genet 50:81–99
9. Ruiz MT, Voinnet O, Baulcombe DC (1998) Initiation and maintenance of virus-induced gene silencing. Plant Cell 10:937–946
10. Dernburg AF, Zalevsky J, Colaiacovo MP, Villeneuve AM (2000) Transgene-mediated cosuppression in the C. elegans germ line. Genes Dev 14:1578–1583
11. Robalino J, Browdy CL, Prior S, Metz A, Parnell P, Gross P, Warr G (2004) Induction of antiviral immunity by double-stranded RNA in a marine invertebrate. J Virol 78:10442–10448
12. Napoli C, Lemieux C, Jorgensen R (1990) Introduction of a chimeric chalcone synthase gene into petunia results in reversible co-suppression of homologous genes in trans. Plant Cell 2:279–289
13. Metzlaff M, O'Dell M, Cluster PD, Flavell RB (1997) RNA-mediated RNA degradation and chalcone synthase A silencing in petunia. Cell 88:845–854
14. Pal-Bhadra M, Bhadra U, Birchler JA (1997) Cosuppression in Drosophila: gene silencing of Alcohol dehydrogenase by white-Adh transgenes is Polycomb dependent. Cell 90:479–490
15. Ketting RF, Haverkamp TH, van Luenen HG, Plasterk RH (1999) Mut-7 of C. elegans, required for transposon silencing and RNA interference, is a homolog of Werner syndrome helicase and RNaseD. Cell 99:133–141
16. Tabara H, Sarkissian M, Kelly WG, Fleenor J, Grishok A, Timmons L, Fire A, Mello CC (1999) The rde-1 gene, RNA interference, and transposon silencing in C. elegans. Cell 99:123–132
17. Zamore PD (2002) Ancient pathways programmed by small RNAs. Science 296:1265–1269
18. Lee RC, Feinbaum RL, Ambros V (1993) The C. elegans heterochronic gene lin-4 encodes small RNAs with antisense complementarity to lin-14. Cell 75:843–854
19. Lagos-Quintana M, Rauhut R, Lendeckel W, Tuschl T (2001) Identification of novel genes coding for small expressed RNAs. Science 294:853–858
20. Lau NC, Lim LP, Weinstein EG, Bartel DP (2001) An abundant class of tiny RNAs with probable regulatory roles in Caenorhabditis elegans. Science 294:858–862
21. Lee RC, Ambros V (2001) An extensive class of small RNAs in Caenorhabditis elegans. Science 294:862–864
22. Hutvagner G, Zamore PD (2002) A microRNA in a multiple-turnover RNAi enzyme complex. Science 297:2056–2060
23. Doench JG, Petersen CP, Sharp PA (2003) siRNAs can function as miRNAs. Genes Dev 17:438–442
24. Rodriguez A, Griffiths-Jones S, Ashurst JL, Bradley A (2004) Identification of mammalian microRNA host genes and transcription units. Genome Res 14:1902–1910
25. Cai X, Hagedorn CH, Cullen BR (2004) Human microRNAs are processed from capped, polyadenylated transcripts that can also function as mRNAs. RNA 10:1957–1966
26. Gregory RI, Yan KP, Amuthan G, Chendrimada T, Doratotaj B, Cooch N, Shiekhattar R (2004) The Microprocessor complex mediates the genesis of microRNAs. Nature 432:235–240
27. Denli AM, Tops BB, Plasterk RH, Ketting RF, Hannon GJ (2004) Processing of primary microRNAs by the Microprocessor complex. Nature 432:231–235
28. Lund E, Guttinger S, Calado A, Dahlberg JE, Kutay U (2004) Nuclear export of microRNA precursors. Science 303:95–98
29. Grimm D, Streetz KL, Jopling CL, Storm TA, Pandey K, Davis CR, Marion P, Salazar F, Kay MA (2006) Fatality in mice due to oversaturation of cellular microRNA/short hairpin RNA pathways. Nature 441:537–541
30. Zeng Y, Cullen BR (2004) Structural requirements for pre-microRNA binding and nuclear export by Exportin 5. Nucleic Acids Res 32:4776–4785

31. Wang HW, Noland C, Siridechadilok B, Taylor DW, Ma E, Felderer K, Doudna JA, Nogales E (2009) Structural insights into RNA processing by the human RISC-loading complex. Nat Struct Mol Biol 16:1148–1153
32. MacRae IJ, Ma E, Zhou M, Robinson CV, Doudna JA (2008) In vitro reconstitution of the human RISC-loading complex. Proc Natl Acad Sci U S A 105:512–517
33. Maniataki E, Mourelatos Z (2005) A human, ATP-independent, RISC assembly machine fueled by pre-miRNA. Genes Dev 19:2979–2990
34. Zhang H, Kolb FA, Brondani V, Billy E, Filipowicz W (2002) Human Dicer preferentially cleaves dsRNAs at their termini without a requirement for ATP. EMBO J 21:5875–5885
35. Hammond SM, Bernstein E, Beach D, Hannon GJ (2000) An RNA-directed nuclease mediates post-transcriptional gene silencing in Drosophila cells. Nature 404:293–296
36. Khvorova A, Reynolds A, Jayasena SD (2003) Functional siRNAs and miRNAs exhibit strand bias. Cell 115:209–216
37. Schwarz DS, Hutvagner G, Du T, Xu Z, Aronin N, Zamore PD (2003) Asymmetry in the assembly of the RNAi enzyme complex. Cell 115:199–208
38. Matranga C, Tomari Y, Shin C, Bartel DP, Zamore PD (2005) Passenger-strand cleavage facilitates assembly of siRNA into Ago2-containing RNAi enzyme complexes. Cell 123: 607–620
39. Rand TA, Petersen S, Du F, Wang X (2005) Argonaute2 cleaves the anti-guide strand of siRNA during RISC activation. Cell 123:621–629
40. Leuschner PJ, Ameres SL, Kueng S, Martinez J (2006) Cleavage of the siRNA passenger strand during RISC assembly in human cells. EMBO Rep 7:314–320
41. Liu J, Carmell MA, Rivas FV, Marsden CG, Thomson JM, Song JJ, Hammond SM, Joshua-Tor L, Hannon GJ (2004) Argonaute2 is the catalytic engine of mammalian RNAi. Science 305:1437–1441
42. Elbashir SM, Martinez J, Patkaniowska A, Lendeckel W, Tuschl T (2001) Functional anatomy of siRNAs for mediating efficient RNAi in Drosophila melanogaster embryo lysate. EMBO J 20:6877–6888
43. Orban TI, Izaurralde E (2005) Decay of mRNAs targeted by RISC requires XRN1, the Ski complex, and the exosome. RNA 11:459–469
44. Fabian MR, Mathonnet G, Sundermeier T, Mathys H, Zipprich JT, Svitkin YV, Rivas F, Jinek M, Wohlschlegel J, Doudna JA et al (2009) Mammalian miRNA RISC recruits CAF1 and PABP to affect PABP-dependent deadenylation. Mol cell 35:868–880
45. Pillai RS, Bhattacharyya SN, Artus CG, Zoller T, Cougot N, Basyuk E, Bertrand E, Filipowicz W (2005) Inhibition of translational initiation by Let-7 MicroRNA in human cells. Science 309:1573–1576
46. Soutschek J, Akinc A, Bramlage B, Charisse K, Constien R, Donoghue M, Elbashir S, Geick A, Hadwiger P, Harborth J et al (2004) Therapeutic silencing of an endogenous gene by systemic administration of modified siRNAs. Nature 432:173–178
47. Zamecnik PC, Stephenson ML (1978) Inhibition of Rous sarcoma virus replication and cell transformation by a specific oligodeoxynucleotide. Proc Natl Acad Sci U S A 75:280–284
48. Yi R, Qin Y, Macara IG, Cullen BR (2003) Exportin-5 mediates the nuclear export of pre-microRNAs and short hairpin RNAs. Genes Dev 17:3011–3016
49. Zeng Y, Cullen BR (2003) Sequence requirements for micro RNA processing and function in human cells. RNA 9:112–123
50. Franich NR, Fitzsimons HL, Fong DM, Klugmann M, During MJ, Young D (2008) AAV vector-mediated RNAi of mutant huntingtin expression is neuroprotective in a novel genetic rat model of Huntington's disease. Mol Ther 16:947–956
51. Koornneef A, van Logtenstein R, Timmermans E, Pisas L, Blits B, Abad X, Fortes P, Petry H, Konstantinova P, Ritsema T (2011) AAV-mediated in vivo knockdown of luciferase using combinatorial RNAi and U1i. Gene Ther 18:929–935
52. Ventura A, Jacks T (2009) MicroRNAs and cancer: short RNAs go a long way. Cell 136:586–591

53. Bader AG, Brown D, Winkler M (2010) The promise of microRNA replacement therapy. Cancer Res 70:7027–7030
54. Krutzfeldt J, Rajewsky N, Braich R, Rajeev KG, Tuschl T, Manoharan M, Stoffel M (2005) Silencing of microRNAs in vivo with 'antagomirs'. Nature 438:685–689
55. Stenvang J, Petri A, Lindow M, Obad S, Kauppinen S (2012) Inhibition of microRNA function by antimiR oligonucleotides. Silence 3:1
56. Elmen J, Lindow M, Schutz S, Lawrence M, Petri A, Obad S, Lindholm M, Hedtjarn M, Hansen HF, Berger U et al (2008) LNA-mediated microRNA silencing in non-human primates. Nature 452:896–899
57. Ameres SL, Horwich MD, Hung JH, Xu J, Ghildiyal M, Weng Z, Zamore PD (2010) Target RNA-directed trimming and tailing of small silencing RNAs. Science 328:1534–1539
58. Rayner KJ, Esau CC, Hussain FN, McDaniel AL, Marshall SM, van Gils JM, Ray TD, Sheedy FJ, Goedeke L, Liu X et al (2011) Inhibition of miR-33a/b in non-human primates raises plasma HDL and lowers VLDL triglycerides. Nature 478:404–407
59. Ebert MS, Neilson JR, Sharp PA (2007) MicroRNA sponges: competitive inhibitors of small RNAs in mammalian cells. Nat Methods 4:721–726
60. Ebert MS, Sharp PA (2010) Emerging roles for natural microRNA sponges. Curr Biol 20:R858–R861
61. Merkel OM, Librizzi D, Pfestroff A, Schurrat T, Behe M, Kissel T (2009) In vivo SPECT and real-time gamma camera imaging of biodistribution and pharmacokinetics of siRNA delivery using an optimized radiolabeling and purification procedure. Bioconjug Chem 20:174–182
62. Jackson AL, Bartz SR, Schelter J, Kobayashi SV, Burchard J, Mao M, Li B, Cavet G, Linsley PS (2003) Expression profiling reveals off-target gene regulation by RNAi. Nat Biotechnol 21:635–637
63. Jackson AL, Burchard J, Schelter J, Chau BN, Cleary M, Lim L, Linsley PS (2006) Widespread siRNA "off-target" transcript silencing mediated by seed region sequence complementarity. RNA 12:1179–1187
64. Judge AD, Sood V, Shaw JR, Fang D, McClintock K, MacLachlan I (2005) Sequence-dependent stimulation of the mammalian innate immune response by synthetic siRNA. Nat Biotechnol 23:457–462
65. Kariko K, Bhuyan P, Capodici J, Weissman D (2004) Small interfering RNAs mediate sequence-independent gene suppression and induce immune activation by signaling through toll-like receptor 3. J Immunol 172:6545–6549
66. Cirak S, Arechavala-Gomeza V, Guglieri M, Feng L, Torelli S, Anthony K, Abbs S, Garralda ME, Bourke J, Wells DJ et al (2011) Exon skipping and dystrophin restoration in patients with Duchenne muscular dystrophy after systemic phosphorodiamidate morpholino oligomer treatment: an open-label, phase 2, dose-escalation study. Lancet 378:595–605
67. Monia BP, Lesnik EA, Gonzalez C, Lima WF, McGee D, Guinosso CJ, Kawasaki AM, Cook PD, Freier SM (1993) Evaluation of 2′-modified oligonucleotides containing 2′-deoxy gaps as antisense inhibitors of gene expression. J Biol Chem 268:14514–14522
68. Gupta N, Fisker N, Asselin MC, Lindholm M, Rosenbohm C, Orum H, Elmen J, Seidah NG, Straarup EM (2010) A locked nucleic acid antisense oligonucleotide (LNA) silences PCSK9 and enhances LDLR expression in vitro and in vivo. PLoS One 5:e10682
69. Tanudji M, Machalek D, Arndt GM, Rivory L (2009) Competition between siRNA duplexes: impact of RISC loading efficiency and comparison between conventional-21 bp and Dicer-substrate siRNAs. Oligonucleotides 20:27–32
70. Siolas D, Lerner C, Burchard J, Ge W, Linsley PS, Paddison PJ, Hannon GJ, Cleary MA (2005) Synthetic shRNAs as potent RNAi triggers. Nat Biotechnol 23:227–231
71. Kim DH, Behlke MA, Rose SD, Chang MS, Choi S, Rossi JJ (2005) Synthetic dsRNA Dicer substrates enhance RNAi potency and efficacy. Nat Biotechnol 23:222–226
72. Amarzguioui M, Lundberg P, Cantin E, Hagstrom J, Behlke MA, Rossi JJ (2006) Rational design and in vitro and in vivo delivery of Dicer substrate siRNA. Nat Protoc 1:508–517
73. Hefner E, Clark K, Whitman C, Behlke MA, Rose SD, Peek AS, Rubio T (2008) Increased potency and longevity of gene silencing using validated Dicer substrates. J Biomol Tech 19:231–237

74. Rose SD, Kim DH, Amarzguioui M, Heidel JD, Collingwood MA, Davis ME, Rossi JJ, Behlke MA (2005) Functional polarity is introduced by Dicer processing of short substrate RNAs. Nucleic Acids Res 33:4140–4156
75. Collingwood MA, Rose SD, Huang L, Hillier C, Amarzguioui M, Wiiger MT, Soifer HS, Rossi JJ, Behlke MA (2008) Chemical modification patterns compatible with high potency dicer-substrate small interfering RNAs. Oligonucleotides 18:187–200
76. Reynolds A, Anderson EM, Vermeulen A, Fedorov Y, Robinson K, Leake D, Karpilow J, Marshall WS, Khvorova A (2006) Induction of the interferon response by siRNA is cell type- and duplex length-dependent. RNA 12:988–993
77. Ge Q, Dallas A, Ilves H, Shorenstein J, Behlke MA, Johnston BH (2009) Effects of chemical modification on the potency, serum stability, and immunostimulatory properties of short shRNAs. RNA 16:118–130
78. Chu CY, Rana TM (2008) Potent RNAi by short RNA triggers. RNA 14:1714–1719
79. Prakash TP, Allerson CR, Dande P, Vickers TA, Sioufi N, Jarres R, Baker BF, Swayze EE, Griffey RH, Bhat B (2005) Positional effect of chemical modifications on short interference RNA activity in mammalian cells. J Med Chem 48:4247–4253
80. Czauderna F, Fechtner M, Dames S, Aygun H, Klippel A, Pronk GJ, Giese K, Kaufmann J (2003) Structural variations and stabilising modifications of synthetic siRNAs in mammalian cells. Nucleic Acids Res 31:2705–2716
81. Glud SZ, Bramsen JB, Dagnaes-Hansen F, Wengel J, Howard KA, Nyengaard JR, Kjems J (2009) Naked siLNA-mediated gene silencing of lung bronchoepithelium EGFP expression after intravenous administration. Oligonucleotides 19:163–168
82. Hohjoh H (2002) RNA interference (RNA(i)) induction with various types of synthetic oligonucleotide duplexes in cultured human cells. FEBS Lett 521:195–199
83. Hohjoh H (2004) Enhancement of RNAi activity by improved siRNA duplexes. FEBS Lett 557:193–198
84. Holen T, Amarzguioui M, Babaie E, Prydz H (2003) Similar behaviour of single-strand and double-strand siRNAs suggests they act through a common RNAi pathway. Nucleic Acids Res 31:2401–2407
85. Martinez J, Patkaniowska A, Urlaub H, Luhrmann R, Tuschl T (2002) Single-stranded antisense siRNAs guide target RNA cleavage in RNAi. Cell 110:563–574
86. Hall AH, Wan J, Spesock A, Sergueeva Z, Shaw BR, Alexander KA (2006) High potency silencing by single-stranded boranophosphate siRNA. Nucleic Acids Res 34: 2773–2781
87. Abe N, Abe H, Ito Y (2007) Dumbbell-shaped nanocircular RNAs for RNA interference. J Am Chem Soc 129:15108–15109
88. Sun X, Rogoff HA, Li CJ (2008) Asymmetric RNA duplexes mediate RNA interference in mammalian cells. Nat Biotechnol 26:1379–1382
89. Chang CI, Yoo JW, Hong SW, Lee SE, Kang HS, Sun X, Rogoff HA, Ban C, Kim S, Li CJ et al (2009) Asymmetric shorter-duplex siRNA structures trigger efficient gene silencing with reduced nonspecific effects. Mol Ther 17:725–732
90. Dua P, Yoo JW, Kim S, Lee DK (2011) Modified siRNA structure with a single nucleotide bulge overcomes conventional siRNA-mediated off-target silencing. Mol Ther 19:1676–1687
91. Bramsen JB, Laursen MB, Damgaard CK, Lena SW, Babu BR, Wengel J, Kjems J (2007) Improved silencing properties using small internally segmented interfering RNAs. Nucleic Acids Res 35:5886–5897
92. Sano M, Sierant M, Miyagishi M, Nakanishi M, Takagi Y, Sutou S (2008) Effect of asymmetric terminal structures of short RNA duplexes on the RNA interference activity and strand selection. Nucleic Acids Res 36:5812–5821
93. Afonin KA, Grabow WW, Walker FM, Bindewald E, Dobrovolskaia MA, Shapiro BA, Jaeger L (2011) Design and self-assembly of siRNA-functionalized RNA nanoparticles for use in automated nanomedicine. Nat Protoc 6:2022–2034
94. Grabow WW, Zakrevsky P, Afonin KA, Chworos A, Shapiro BA, Jaeger L (2011) Self-assembling RNA nanorings based on RNAI/II inverse kissing complexes. Nano Lett 11:878–887

95. Yoo JW, Kim S, Lee DK (2008) Competition potency of siRNA is specified by the 5′-half sequence of the guide strand. Biochem Biophys Res Commun 367:78–83
96. Koller E, Propp S, Murray H, Lima W, Bhat B, Prakash TP, Allerson CR, Swayze EE, Marcusson EG, Dean NM (2006) Competition for RISC binding predicts in vitro potency of siRNA. Nucleic Acids Res 34:4467–4476
97. Ladunga I (2007) More complete gene silencing by fewer siRNAs: transparent optimized design and biophysical signature. Nucleic Acids Res 35:433–440
98. Huesken D, Lange J, Mickanin C, Weiler J, Asselbergs F, Warner J, Meloon B, Engel S, Rosenberg A, Cohen D et al (2005) Design of a genome-wide siRNA library using an artificial neural network. Nat Biotechnol 23:995–1001
99. Reynolds A, Leake D, Boese Q, Scaringe S, Marshall WS, Khvorova A (2004) Rational siRNA design for RNA interference. Nat Biotechnol 22:326–330
100. Shabalina SA, Spiridonov AN, Ogurtsov AY (2006) Computational models with thermodynamic and composition features improve siRNA design. BMC Bioinforma 7:65
101. Jagla B, Aulner N, Kelly PD, Song D, Volchuk A, Zatorski A, Shum D, Mayer T, De Angelis DA, Ouerfelli O et al (2005) Sequence characteristics of functional siRNAs. RNA 11:864–872
102. Ui-Tei K, Naito Y, Takahashi F, Haraguchi T, Ohki-Hamazaki H, Juni A, Ueda R, Saigo K (2004) Guidelines for the selection of highly effective siRNA sequences for mammalian and chick RNA interference. Nucleic Acids Res 32:936–948
103. Li W, Cha L (2007) Predicting siRNA efficiency. Cell Mol Life Sci 64:1785–1792
104. Holen T (2005) Mechanisms of RNAi: mRNA cleavage fragments may indicate stalled RISC. J RNAi Gene Silencing 1:21–25
105. Brown KM, Chu CY, Rana TM (2005) Target accessibility dictates the potency of human RISC. Nat Struct Mol Biol 12:469–470
106. Schubert S, Grunweller A, Erdmann VA, Kurreck J (2005) Local RNA target structure influences siRNA efficacy: systematic analysis of intentionally designed binding regions. J Mol Biol 348:883–893
107. Overhoff M, Alken M, Far RK, Lemaitre M, Lebleu B, Sczakiel G, Robbins I (2005) Local RNA target structure influences siRNA efficacy: a systematic global analysis. J Mol Biol 348:871–881
108. Tafer H, Ameres SL, Obernosterer G, Gebeshuber CA, Schroeder R, Martinez J, Hofacker IL (2008) The impact of target site accessibility on the design of effective siRNAs. Nat Biotechnol 26:578–583
109. Shao Y, Chan CY, Maliyekkel A, Lawrence CE, Roninson IB, Ding Y (2007) Effect of target secondary structure on RNAi efficiency. RNA 13:1631–1640
110. Kretschmer-Kazemi Far R, Sczakiel G (2003) The activity of siRNA in mammalian cells is related to structural target accessibility: a comparison with antisense oligonucleotides. Nucleic Acids Res 31:4417–4424
111. Yuan B, Latek R, Hossbach M, Tuschl T, Lewitter F (2004) siRNA Selection Server: an automated siRNA oligonucleotide prediction server. Nucleic Acids Res 32:W130–W134
112. Bramsen JB, Laursen MB, Nielsen AF, Hansen TB, Bus C, Langkjær N, Babu BR, Højland T, Abramov M, Van Aerschot A et al (2009) A large-scale chemical modification screen identifies design rules to generate siRNAs with high activity, high stability and low toxicity. Nucleic Acids Res 37:2867–2881
113. Dowler T, Bergeron D, Tedeschi AL, Paquet L, Ferrari N, Damha MJ (2006) Improvements in siRNA properties mediated by 2′-deoxy-2′-fluoro-beta-D-arabinonucleic acid (FANA). Nucleic Acids Res 34:1669–1675
114. Fisher M, Abramov M, Van Aerschot A, Rozenski J, Dixit V, Juliano RL, Herdewijn P (2009) Biological effects of hexitol and altritol-modified siRNAs targeting B-Raf. Eur J Pharmacol 606:38–44
115. Watts JK, Choubdar N, Sadalapure K, Robert F, Wahba AS, Pelletier J, Pinto BM, Damha MJ (2007) 2′-fluoro-4′-thioarabino-modified oligonucleotides: conformational switches linked to siRNA activity. Nucleic Acids Res 35:1441–1451

116. Fisher M, Abramov M, Van Aerschot A, Xu D, Juliano RL, Herdewijn P (2007) Inhibition of MDR1 expression with altritol-modified siRNAs. Nucleic Acids Res 35:1064–1074
117. Nauwelaerts K, Fisher M, Froeyen M, Lescrinier E, Aerschot AV, Xu D, DeLong R, Kang H, Juliano RL, Herdewijn P (2007) Structural characterization and biological evaluation of small interfering RNAs containing cyclohexenyl nucleosides. J Am Chem Soc 129:9340–9348
118. Deleavey GF, Watts JK, Alain T, Robert F, Kalota A, Aishwarya V, Pelletier J, Gewirtz AM, Sonenberg N, Damha MJ (2010) Synergistic effects between analogs of DNA and RNA improve the potency of siRNA-mediated gene silencing. Nucleic Acids Res 38(13): 4547–4557
119. Elmén J, Thonberg H, Ljungberg K, Frieden M, Westergaard M, Xu Y, Wahren B, Liang Z, Ørum H, Koch T et al (2005) Locked nucleic acid (LNA) mediated improvements in siRNA stability and functionality. Nucleic Acids Res 33:439–447
120. Sipa K, Sochacka E, Kazmierczak-Baranska J, Maszewska M, Janicka M, Nowak G, Nawrot B (2007) Effect of base modifications on structure, thermodynamic stability, and gene silencing activity of short interfering RNA. RNA 13:1301–1316
121. Li ZY, Mao H, Kallick DA, Gorenstein DG (2005) The effects of thiophosphate substitutions on native siRNA gene silencing. Biochem Biophys Res Commun 329:1026–1030
122. Strapps WR, Pickering V, Muiru GT, Rice J, Orsborn S, Polisky BA, Sachs A, Bartz SR (2010) The siRNA sequence and guide strand overhangs are determinants of in vivo duration of silencing. Nucleic Acids Res 38:4788–4797
123. Bramsen JB, Pakula MM, Hansen TB, Bus C, Langkjaer N, Odadzic D, Smicius R, Wengel SL, Chattopadhyaya J, Engels JW et al (2010) A screen of chemical modifications identifies position-specific modification by UNA to most potently reduce siRNA off-target effects. Nucleic Acids Res 38:5761–5773
124. Laursen MB, Pakula MM, Gao S, Fluiter K, Mook OR, Baas F, Langklaer N, Wengel SL, Wengel J, Kjems J et al (2010) Utilization of unlocked nucleic acid (UNA) to enhance siRNA performance in vitro and in vivo. Mol Biosyst 6:862–870
125. Odadzic D, Bramsen JB, Smicius R, Bus C, Kjems J, Engels JW (2008) Synthesis of 2′-O-modified adenosine building blocks and application for RNA interference. Bioorg Med Chem 16:518–529
126. Bramsen JB, Kjems J (2011) Chemical modification of small interfering RNA. Methods Mol Biol 721:77–103
127. Gao S, Dagnaes-Hansen F, Nielsen EJ, Wengel J, Besenbacher F, Howard KA, Kjems J (2009) The effect of chemical modification and nanoparticle formulation on stability and biodistribution of siRNA in mice. Mol Ther 17:1225–1233
128. Goodchild A, Nopper N, King A, Doan T, Tanudji M, Arndt GM, Poidinger M, Rivory LP, Passioura T (2009) Sequence determinants of innate immune activation by short interfering RNAs. BMC Immunol 10:40
129. Sioud M (2006) Single-stranded small interfering RNA are more immunostimulatory than their double-stranded counterparts: a central role for 2′-hydroxyl uridines in immune responses. Eur J Immunol 36:1222–1230
130. Hornung V, Guenthner-Biller M, Bourquin C, Ablasser A, Schlee M, Uematsu S, Noronha A, Manoharan M, Akira S, de Fougerolles A et al (2005) Sequence-specific potent induction of IFN-alpha by short interfering RNA in plasmacytoid dendritic cells through TLR7. Nat Med 11:263–270
131. Chiu YL, Rana TM (2003) siRNA function in RNAi: a chemical modification analysis. RNA 9:1034–1048
132. Amarzguioui M, Holen T, Babaie E, Prydz H (2003) Tolerance for mutations and chemical modifications in a siRNA. Nucleic Acids Res 31:589–595
133. Choung S, Kim YJ, Kim S, Park HO, Choi YC (2006) Chemical modification of siRNAs to improve serum stability without loss of efficacy. Biochem Biophys Res Commun 342:919–927
134. Harborth J, Elbashir SM, Vandenburgh K, Manninga H, Scaringe SA, Weber K, Tuschl T (2003) Sequence, chemical, and structural variation of small interfering RNAs and short hairpin RNAs and the effect on mammalian gene silencing. Antisense Nucleic Acid Drug Dev 13:83–105

135. Braasch DA, Jensen S, Liu Y, Kaur K, Arar K, White MA, Corey DR (2003) RNA interference in mammalian cells by chemically-modified RNA. Biochemistry 42:7967–7975
136. Grunweller A, Wyszko E, Bieber B, Jahnel R, Erdmann VA, Kurreck J (2003) Comparison of different antisense strategies in mammalian cells using locked nucleic acids, 2′-O-methyl RNA, phosphorothioates and small interfering RNA. Nucleic Acids Res 31:3185–3193
137. Hall AH, Wan J, Shaughnessy EE, Ramsay Shaw B, Alexander KA (2004) RNA interference using boranophosphate siRNAs: structure-activity relationships. Nucleic Acids Res 32:5991–6000
138. Braasch DA, Paroo Z, Constantinescu A, Ren G, Oz OK, Mason RP, Corey DR (2004) Biodistribution of phosphodiester and phosphorothioate siRNA. Bioorg Med Chem Lett 14:1139–1143
139. Wengel J, Petersen M, Nielsen KE, Jensen GA, Hakansson AE, Kumar R, Sorensen MD, Rajwanshi VK, Bryld T, Jacobsen JP (2001) LNA (locked nucleic acid) and the diastereoisomeric alpha-L-LNA: conformational tuning and high-affinity recognition of DNA/RNA targets. Nucleosides Nucleotides Nucleic Acids 20:389–396
140. Srivastava P, Barman J, Pathmasiri W, Plashkevych O, Wenska M, Chattopadhyaya J (2007) Five- and six-membered conformationally locked 2′,4′-carbocyclic ribo-thymidines: synthesis, structure, and biochemical studies. J Am Chem Soc 129:8362–8379
141. Hamada M, Ohtsuka T, Kawaida R, Koizumi M, Morita K, Furukawa H, Imanishi T, Miyagishi M, Taira K (2002) Effects on RNA interference in gene expression (RNAi) in cultured mammalian cells of mismatches and the introduction of chemical modifications at the 3′-ends of siRNAs. Antisense Nucleic Acid Drug Dev 12:301–309
142. Pradeepkumar PI, Amirkhanov NV, Chattopadhyaya J (2003) Antisense oligonuclotides with oxetane-constrained cytidine enhance heteroduplex stability, and elicit satisfactory RNase H response as well as showing improved resistance to both exo and endonucleases. Org Biomol Chem 1:81–92
143. Langkjær N, Pasternak A, Wengel J (2009) UNA (unlocked nucleic acid): a flexible RNA mimic that allows engineering of nucleic acid duplex stability. Bioorg Med Chem 17:5420–5425
144. Hornung V, Ellegast J, Kim S, Brzozka K, Jung A, Kato H, Poeck H, Akira S, Conzelmann KK, Schlee M et al (2006) 5′-Triphosphate RNA is the ligand for RIG-I. Science 314:994–997
145. Chiu YL, Rana TM (2002) RNAi in human cells: basic structural and functional features of small interfering RNA. Mol Cell 10:549–561
146. Layzer JM, McCaffrey AP, Tanner AK, Huang Z, Kay MA, Sullenger BA (2004) In vivo activity of nuclease-resistant siRNAs. RNA 10:766–771
147. Morrissey DV, Blanchard K, Shaw L, Jensen K, Lockridge JA, Dickinson B, McSwiggen JA, Vargeese C, Bowman K, Shaffer CS, Polisky BA, Zinnen S (2005) Activity of stabilized short interfering RNA in a mouse model of hepatitis B virus replication. Hepatology 41:1349–1356
148. Allerson CR, Sioufi N, Jarres R, Prakash TP, Naik N, Berdeja A, Wanders L, Griffey RH, Swayze EE, Bhat B (2005) Fully 2′-modified oligonucleotide duplexes with improved in vitro potency and stability compared to unmodified small interfering RNA. J Med Chem 48:901–904
149. Blidner RA, Hammer RP, Lopez MJ, Robinson SO, Monroe WT (2007) Fully 2′-deoxy-2′-fluoro substituted nucleic acids induce RNA interference in mammalian cell culture. Chem Biol Drug Des 70:113–122
150. Mook OR, Baas F, de Wissel MB, Fluiter K (2007) Evaluation of locked nucleic acid-modified small interfering RNA in vitro and in vivo. Mol Cancer Ther 6:833–843
151. Dande P, Prakash TP, Sioufi N, Gaus H, Jarres R, Berdeja A, Swayze EE, Griffey RH, Bhat B (2006) Improving RNA interference in mammalian cells by 4′-thio-modified small interfering RNA (siRNA): effect on siRNA activity and nuclease stability when used in combination with 2′-O-alkyl modifications. J Med Chem 49:1624–1634
152. Zimmermann TS, Lee AC, Akinc A, Bramlage B, Bumcrot D, Fedoruk MN, Harborth J, Heyes JA, Jeffs LB, John M et al (2006) RNAi-mediated gene silencing in non-human primates. Nature 441:111–114

153. Nishina K, Unno T, Uno Y, Kubodera T, Kanouchi T, Mizusawa H, Yokota T (2008) Efficient in vivo delivery of siRNA to the liver by conjugation of alpha-tocopherol. Mol Ther 16:734–740
154. Volkov AA, Kruglova NS, Meschaninova MI, Venyaminova AG, Zenkova MA, Vlassov VV, Chernolovskaya EL (2009) Selective protection of nuclease-sensitive sites in siRNA prolongs silencing effect. Oligonucleotides 19:191–202
155. Turner JJ, Jones SW, Moschos SA, Lindsay MA, Gait MJ (2007) MALDI-TOF mass spectral analysis of siRNA degradation in serum confirms an RNAse A-like activity. Mol Biosyst 3:43–50
156. Sorrentino S (1998) Human extracellular ribonucleases: multiplicity, molecular diversity and catalytic properties of the major RNase types. Cell Mol Life Sci 54:785–794
157. Qiu L, Moreira A, Kaplan G, Levitz R, Wang JY, Xu C, Drlica K (1998) Degradation of hammerhead ribozymes by human ribonucleases. Mol Gen Genet 258:352–362
158. Bartlett DW, Davis ME (2007) Effect of siRNA nuclease stability on the in vitro and in vivo kinetics of siRNA-mediated gene silencing. Biotechnol Bioeng 97:909–921
159. Hoerter JA, Krishnan V, Lionberger TA, Walter NG (2011) siRNA-like double-stranded RNAs are specifically protected against degradation in human cell extract. PLoS One 6:e20359
160. Song E, Lee SK, Dykxhoorn DM, Novina C, Zhang D, Crawford K, Cerny J, Sharp PA, Lieberman J, Manjunath N et al (2003) Sustained small interfering RNA-mediated human immunodeficiency virus type 1 inhibition in primary macrophages. J Virol 77:7174–7181
161. Fedorov Y, Anderson EM, Birmingham A, Reynolds A, Karpilow J, Robinson K, Leake D, Marshall WS, Khvorova A (2006) Off-target effects by siRNA can induce toxic phenotype. Rna 12(7):1188–1196
162. Diebold SS, Kaisho T, Hemmi H, Akira S, Reis e Sousa C (2004) Innate antiviral responses by means of TLR7-mediated recognition of single-stranded RNA. Science 303: 1529–1531
163. Heil F, Hemmi H, Hochrein H, Ampenberger F, Kirschning C, Akira S, Lipford G, Wagner H, Bauer S (2004) Species-specific recognition of single-stranded RNA via toll-like receptor 7 and 8. Science 303:1526–1529
164. Diebold SS, Massacrier C, Akira S, Paturel C, Morel Y, Reis e Sousa C (2006) Nucleic acid agonists for Toll-like receptor 7 are defined by the presence of uridine ribonucleotides. Eur J Immunol 36:3256–3267
165. Judge AD, Bola G, Lee AC, MacLachlan I (2006) Design of noninflammatory synthetic siRNA mediating potent gene silencing in vivo. Mol Ther 13:494–505
166. Kariko K, Buckstein M, Ni H, Weissman D (2005) Suppression of RNA recognition by Toll-like receptors: the impact of nucleoside modification and the evolutionary origin of RNA. Immunity 23:165–175
167. Cekaite L, Furset G, Hovig E, Sioud M (2007) Gene expression analysis in blood cells in response to unmodified and 2′-modified siRNAs reveals TLR-dependent and independent effects. J Mol Biol 365:90–108
168. Flatekval GF, Sioud M (2009) Modulation of dendritic cell maturation and function with mono- and bifunctional small interfering RNAs targeting indoleamine 2,3-dioxygenase. Immunology 128:e837–e848
169. Hamm S, Latz E, Hangel D, Muller T, Yu P, Golenbock D, Sparwasser T, Wagner H, Bauer S (2009) Alternating 2′-O-ribose methylation is a universal approach for generating non-stimulatory siRNA by acting as TLR7 antagonist. Immunobiology 215:559–569
170. Kariko K, Bhuyan P, Capodici J, Ni H, Lubinski J, Friedman H, Weissman D (2004) Exogenous siRNA mediates sequence-independent gene suppression by signaling through toll-like receptor 3. Cells Tissues Organs 177:132–138
171. Kleinman ME, Yamada K, Takeda A, Chandrasekaran V, Nozaki M, Baffi JZ, Albuquerque RJC, Yamasaki S, Itaya M, Pan YZ et al (2008) Sequence- and target-independent angiogenesis suppression by siRNA via TLR3. Nature 452:591–597
172. Manche L, Green SR, Schmedt C, Mathews MB (1992) Interactions between double-stranded RNA regulators and the protein kinase DAI. Mol Cell Biol 12:5238–5248

173. Zhang Z, Weinschenk T, Guo K, Schluesener HJ (2006) siRNA binding proteins of microglial cells: PKR is an unanticipated ligand. J Cell Biochem 97:1217–1229
174. Marques JT, Devosse T, Wang D, Zamanian-Daryoush M, Serbinowski P, Hartmann R, Fujita T, Behlke MA, Williams BR (2006) A structural basis for discriminating between self and nonself double-stranded RNAs in mammalian cells. Nat Biotechnol 24:559–565
175. Puthenveetil S, Whitby L, Ren J, Kelnar K, Krebs JF, Beal PA (2006) Controlling activation of the RNA-dependent protein kinase by siRNAs using site-specific chemical modification. Nucleic Acids Res 34:4900–4911
176. Sledz CA, Holko M, de Veer MJ, Silverman RH, Williams BR (2003) Activation of the interferon system by short-interfering RNAs. Nat Cell Biol 5:834–839
177. Birmingham A, Anderson EM, Reynolds A, Ilsley-Tyree D, Leake D, Fedorov Y, Baskerville S, Maksimova E, Robinson K, Karpilow J et al (2006) 3′ UTR seed matches, but not overall identity, are associated with RNAi off-targets. Nat Methods 3:199–204
178. Lim LP, Lau NC, Garrett-Engele P, Grimson A, Schelter JM, Castle J, Bartel DP, Linsley PS, Johnson JM (2005) Microarray analysis shows that some microRNAs downregulate large numbers of target mRNAs. Nature 433:769–773
179. Jackson AL, Burchard J, Leake D, Reynolds A, Schelter J, Guo J, Johnson JM, Lim L, Karpilow J, Nichols K et al (2006) Position-specific chemical modification of siRNAs reduces "off-target" transcript silencing. RNA 12:1197–1205
180. Ui-Tei K, Naito Y, Zenno S, Nishi K, Yamato K, Takahashi F, Juni A, Saigo K (2008) Functional dissection of siRNA sequence by systematic DNA substitution: modified siRNA with a DNA seed arm is a powerful tool for mammalian gene silencing with significantly reduced off-target effect. Nucleic Acids Res 36:2136–2151
181. Chen PY, Weinmann L, Gaidatzis D, Pei Y, Zavolan M, Tuschl T, Meister G (2008) Strand-specific 5′-O-methylation of siRNA duplexes controls guide strand selection and targeting specificity. RNA 14:263–274
182. Vaish N, Chen F, Seth S, Fosnaugh K, Liu Y, Adami R, Brown T, Chen Y, Harvie P, Johns R et al (2011) Improved specificity of gene silencing by siRNAs containing unlocked nucleobase analogs. Nucleic Acids Res 39:1823–1832
183. Khan AA, Betel D, Miller ML, Sander C, Leslie CS, Marks DS (2009) Transfection of small RNAs globally perturbs gene regulation by endogenous microRNAs. Nat Biotechnol 27:549–555
184. Hasuwa H, Kaseda K, Einarsdottir T, Okabe M (2002) Small interfering RNA and gene silencing in transgenic mice and rats. FEBS Lett 532:227–230
185. Seibler J, Kuter-Luks B, Kern H, Streu S, Plum L, Mauer J, Kuhn R, Bruning JC, Schwenk F (2005) Single copy shRNA configuration for ubiquitous gene knockdown in mice. Nucleic Acids Res 33:e67
186. Kunath T, Gish G, Lickert H, Jones N, Pawson T, Rossant J (2003) Transgenic RNA interference in ES cell-derived embryos recapitulates a genetic null phenotype. Nat Biotechnol 21:559–561
187. Rubinson DA, Dillon CP, Kwiatkowski AV, Sievers C, Yang L, Kopinja J, Rooney DL, Ihrig MM, McManus MT, Gertler FB et al (2003) A lentivirus-based system to functionally silence genes in primary mammalian cells, stem cells and transgenic mice by RNA interference. Nat Genet 33:401–406
188. Tiscornia G, Singer O, Ikawa M, Verma IM (2003) A general method for gene knockdown in mice by using lentiviral vectors expressing small interfering RNA. Proc Natl Acad Sci U S A 100:1844–1848
189. Klinghoffer RA, Magnus J, Schelter J, Mehaffey M, Coleman C, Cleary MA (2010) Reduced seed region-based off-target activity with lentivirus-mediated RNAi. RNA 16:879–884
190. Yuan MK, Tao Y, Yu WZ, Kai W, Jiang YR (2010) Lentivirus-mediated RNA interference of vascular endothelial growth factor in monkey eyes with iris neovascularization. Mol Vis 16:1743–1753
191. Liu YP, Vink MA, Westerink JT, Ramirez de Arellano E, Konstantinova P, Ter Brake O, Berkhout B (2010) Titers of lentiviral vectors encoding shRNAs and miRNAs are reduced by different mechanisms that require distinct repair strategies. RNA 16:1328–1339

192. Harper SQ, Staber PD, Beck CR, Fineberg SK, Stein C, Ochoa D, Davidson BL (2006) Optimization of feline immunodeficiency virus vectors for RNA interference. J Virol 80:9371–9380
193. Li M, Rossi JJ (2008) Lentiviral vector delivery of siRNA and shRNA encoding genes into cultured and primary hematopoietic cells. Methods Mol Biol 433:287–299
194. Heilbronn R, Weger S (2010) Viral vectors for gene transfer: current status of gene therapeutics. Handb Exp Pharmacol 197:143–170
195. Borel F, van Logtenstein R, Koornneef A, Maczuga P, Ritsema T, Petry H, van Deventer SJ, Jansen PL, Konstantinova P (2011) In vivo knock-down of multidrug resistance transporters ABCC1 and ABCC2 by AAV-delivered shRNAs and by artificial miRNAs. J RNAi Gene Silencing 7:434–442
196. Ehlert EM, Eggers R, Niclou SP, Verhaagen J (2010) Cellular toxicity following application of adeno-associated viral vector-mediated RNA interference in the nervous system. BMC Neurosci 11:20
197. Yang X, Haurigot V, Zhou S, Luo G, Couto LB (2010) Inhibition of hepatitis C virus replication using adeno-associated virus vector delivery of an exogenous anti-hepatitis C virus microRNA cluster. Hepatology 52:1877–1887
198. Xia H, Mao Q, Eliason SL, Harper SQ, Martins IH, Orr HT, Paulson HL, Yang L, Kotin RM, Davidson BL (2004) RNAi suppresses polyglutamine-induced neurodegeneration in a model of spinocerebellar ataxia. Nat Med 10:816–820
199. Brummelkamp TR, Bernards R, Agami R (2002) A system for stable expression of short interfering RNAs in mammalian cells. Science 296:550–553
200. Lee NS, Dohjima T, Bauer G, Li H, Li MJ, Ehsani A, Salvaterra P, Rossi J (2002) Expression of small interfering RNAs targeted against HIV-1 rev transcripts in human cells. Nat Biotechnol 20:500–505
201. Miyagishi M, Taira K (2002) U6 promoter-driven siRNAs with four uridine 3′ overhangs efficiently suppress targeted gene expression in mammalian cells. Nat Biotechnol 20:497–500
202. Paddison PJ, Caudy AA, Bernstein E, Hannon GJ, Conklin DS (2002) Short hairpin RNAs (shRNAs) induce sequence-specific silencing in mammalian cells. Genes Dev 16:948–958
203. Paul CP, Good PD, Winer I, Engelke DR (2002) Effective expression of small interfering RNA in human cells. Nat Biotechnol 20:505–508
204. Sui G, Soohoo C, Affar el B, Gay F, Shi Y, Forrester WC, Shi Y (2002) A DNA vector-based RNAi technology to suppress gene expression in mammalian cells. Proc Natl Acad Sci U S A 99:5515–5520
205. Yu JY, DeRuiter SL, Turner DL (2002) RNA interference by expression of short-interfering RNAs and hairpin RNAs in mammalian cells. Proc Natl Acad Sci U S A 99:6047–6052
206. Bogenhagen DF, Brown DD (1981) Nucleotide sequences in Xenopus 5 S DNA required for transcription termination. Cell 24:261–270
207. Tazi J, Forne T, Jeanteur P, Cathala G, Brunel C (1993) Mammalian U6 small nuclear RNA undergoes 3′ end modifications within the spliceosome. Mol Cell Biol 13:1641–1650
208. Tran N, Cairns MJ, Dawes IW, Arndt GM (2003) Expressing functional siRNAs in mammalian cells using convergent transcription. BMC Biotechnol 3:21
209. Jian R, Peng T, Deng S, Jiang J, Hu F, An J, Cheng X (2006) A simple strategy for generation of gene knockdown constructs with convergent H1 and U6 promoters. Eur J Cell Biol 85:433–440
210. Lobo SM, Ifill S, Hernandez N (1990) cis-acting elements required for RNA polymerase II and III transcription in the human U2 and U6 snRNA promoters. Nucleic Acids Res 18:2891–2899
211. Baer M, Nilsen TW, Costigan C, Altman S (1990) Structure and transcription of a human gene for H1 RNA, the RNA component of human RNase P. Nucleic Acids Res 18:97–103
212. Boden D, Pusch O, Silbermann R, Lee F, Tucker L, Ramratnam B (2004) Enhanced gene silencing of HIV-1 specific siRNA using microRNA designed hairpins. Nucleic Acids Res 32:1154–1158
213. Weinberg RA, Penman S (1968) Small molecular weight monodisperse nuclear RNA. J Mol Biol 38:289–304

214. Noonberg SB, Scott GK, Garovoy MR, Benz CC, Hunt CA (1994) In vivo generation of highly abundant sequence-specific oligonucleotides for antisense and triplex gene regulation. Nucleic Acids Res 22:2830–2836
215. Yu J, McMahon AP (2006) Reproducible and inducible knockdown of gene expression in mice. Genesis 44:252–261
216. Seibler J, Zevnik B, Kuter-Luks B, Andreas S, Kern H, Hennek T, Rode A, Heimann C, Faust N, Kauselmann G et al (2003) Rapid generation of inducible mouse mutants. Nucleic Acids Res 31:e12
217. McBride JL, Boudreau RL, Harper SQ, Staber PD, Monteys AM, Martins I, Gilmore BL, Burstein H, Peluso RW, Polisky B et al (2008) Artificial miRNAs mitigate shRNA-mediated toxicity in the brain: implications for the therapeutic development of RNAi. Proc Natl Acad Sci U S A 105:5868–5873
218. Fish RJ, Kruithof EK (2004) Short-term cytotoxic effects and long-term instability of RNAi delivered using lentiviral vectors. BMC Mol Biol 5:9
219. Bridge AJ, Pebernard S, Ducraux A, Nicoulaz AL, Iggo R (2003) Induction of an interferon response by RNAi vectors in mammalian cells. Nat Genet 34:263–264
220. Lee Y, Kim M, Han J, Yeom KH, Lee S, Baek SH, Kim VN (2004) MicroRNA genes are transcribed by RNA polymerase II. EMBO J 23:4051–4060
221. Lee Y, Jeon K, Lee JT, Kim S, Kim VN (2002) MicroRNA maturation: stepwise processing and subcellular localization. EMBO J 21:4663–4670
222. Gossen M, Freundlieb S, Bender G, Muller G, Hillen W, Bujard H (1995) Transcriptional activation by tetracyclines in mammalian cells. Science 268:1766–1769
223. Dickins RA, McJunkin K, Hernando E, Premsrirut PK, Krizhanovsky V, Burgess DJ, Kim SY, Cordon-Cardo C, Zender L, Hannon GJ et al (2007) Tissue-specific and reversible RNA interference in transgenic mice. Nat Genet 39:914–921
224. Zeng Y, Wagner EJ, Cullen BR (2002) Both natural and designed micro RNAs can inhibit the expression of cognate mRNAs when expressed in human cells. Mol Cell 9:1327–1333
225. Stegmeier F, Hu G, Rickles RJ, Hannon GJ, Elledge SJ (2005) A lentiviral microRNA-based system for single-copy polymerase II-regulated RNA interference in mammalian cells. Proc Natl Acad Sci U S A 102:13212–13217
226. Chung KH, Hart CC, Al-Bassam S, Avery A, Taylor J, Patel PD, Vojtek AB, Turner DL (2006) Polycistronic RNA polymerase II expression vectors for RNA interference based on BIC/miR-155. Nucleic Acids Res 34:e53
227. McManus MT, Petersen CP, Haines BB, Chen J, Sharp PA (2002) Gene silencing using micro-RNA designed hairpins. RNA 8:842–850
228. Silva JM, Li MZ, Chang K, Ge W, Golding MC, Rickles RJ, Siolas D, Hu G, Paddison PJ, Schlabach MR et al (2005) Second-generation shRNA libraries covering the mouse and human genomes. Nat Genet 37:1281–1288
229. Chang K, Elledge SJ, Hannon GJ (2006) Lessons from nature: microRNA-based shRNA libraries. Nat Methods 3:707–714
230. Kawasaki H, Taira K (2003) Short hairpin type of dsRNAs that are controlled by tRNA(Val) promoter significantly induce RNAi-mediated gene silencing in the cytoplasm of human cells. Nucleic Acids Res 31:700–707
231. Boudreau RL, Martins I, Davidson BL (2009) Artificial microRNAs as siRNA shuttles: improved safety as compared to shRNAs in vitro and in vivo. Mol Ther 17:169–175
232. Amendola M, Passerini L, Pucci F, Gentner B, Bacchetta R, Naldini L (2009) Regulated and multiple miRNA and siRNA delivery into primary cells by a lentiviral platform. Mol Ther 17:1039–1052
233. Morris KV, Chan SW, Jacobsen SE, Looney DJ (2004) Small interfering RNA-induced transcriptional gene silencing in human cells. Science 305:1289–1292
234. Moses J, Goodchild A, Rivory LP (2010) Intended transcriptional silencing with siRNA results in gene repression through sequence-specific off-targeting. RNA 16:430–441
235. Morris KV (2009) RNA-directed transcriptional gene silencing and activation in human cells. Oligonucleotides 19:299–306

236. Beisel C, Paro R (2011) Silencing chromatin: comparing modes and mechanisms. Nat Rev Genet 12:123–135
237. Goodchild J (2011) Therapeutic oligonucleotides. Methods Mol Biol 764:1–15
238. Crooke ST (1998) Vitravene–another piece in the mosaic. Antisense Nucleic Acid Drug Dev 8:vii–viii
239. van Deutekom JC, Janson AA, Ginjaar IB, Frankhuizen WS, Aartsma-Rus A, Bremmer-Bout M, den Dunnen JT, Koop K, van der Kooi AJ, Goemans NM et al (2007) Local dystrophin restoration with antisense oligonucleotide PRO051. N Engl J Med 357:2677–2686
240. Goemans NM, Tulinius M, van den Akker JT, Burm BE, Ekhart PF, Heuvelmans N, Holling T, Janson AA, Platenburg GJ, Sipkens JA et al (2011) Systemic administration of PRO051 in Duchenne's muscular dystrophy. N Engl J Med 364:1513–1522
241. Wheeler TM, Sobczak K, Lueck JD, Osborne RJ, Lin X, Dirksen RT, Thornton CA (2009) Reversal of RNA dominance by displacement of protein sequestered on triplet repeat RNA. Science 325:336–339
242. Nakamori M, Gourdon G, Thornton CA (2011) Stabilization of expanded (CTG)*(CAG) repeats by antisense oligonucleotides. Mol Ther 19:2222–2227
243. Stephenson ML, Zamecnik PC (1978) Inhibition of Rous sarcoma viral RNA translation by a specific oligodeoxyribonucleotide. Proc Natl Acad Sci U S A 75:285–288
244. Elbashir SM, Harborth J, Weber K, Tuschl T (2002) Analysis of gene function in somatic mammalian cells using small interfering RNAs. Methods 26:199–213

Chapter 2
Synthetic Dicer-Substrate siRNAs as Triggers of RNA Interference

Scott D. Rose and Mark A. Behlke

Abstract The first synthetic oligonucleotides used to suppress gene expression in mammalian cells via RNA interference were 21-nucleotide (nt) RNA duplexes having symmetric 2-nt 3′-overhangs and were designed to mimic the natural products of Dicer processing of long RNA substrates. Synthetic RNA duplexes which are longer than 23-nt length are substrates for processing by Dicer and can show increased potency as artificial triggers of RNA interference, particularly at a low concentration. Longer duplexes, however, can have variable cleavage patterns following Dicer processing which can adversely affect potency. Optimized synthetic Dicer substrates are asymmetric duplexes having a 25-nt passenger strand and a 27-nt guide strand with a single 2-nt 3′-overhang on the guide strand and modified bases at the 3′-end of the passenger strand. This modified design results in predictable patterns of Dicer processing and shows improved activity. The development of this design strategy and use of Dicer-substrate RNAs to trigger gene suppression in a variety of systems will be reviewed in this chapter.

2.1 Introduction

RNA interference (RNAi) is a highly conserved mechanism of gene regulation that extends broadly across phyla [1]. RNAi encompasses two general mechanisms of gene suppression, one where the target mRNA is degraded and a second where protein translation is inhibited [2, 3]. Both routes reduce levels of the protein product made from the targeted gene. Translational suppression is typically mediated by microRNAs (miRNAs), which form imperfect hybrids with the target mRNA [4].

S.D. Rose, PhD • M.A. Behlke, MD, PhD (✉)
Integrated DNA Technologies, Inc., 1710 Commercial Park, Coralville, IA, USA
e-mail: mbehlke@idtdna.com

K.A. Howard (ed.), *RNA Interference from Biology to Therapeutics*,
Advances in Delivery Science and Technology, DOI 10.1007/978-1-4614-4744-3_2,

Degradative RNAi, on the other hand, is typically mediated by small interfering RNAs (siRNAs), which form perfect or near-perfect hybrids with the target mRNA [5]. There is substantial "cross talk" between the pathways, and miRNAs can lead to mRNA degradation, and siRNAs can lead to translational suppression [6].

Endogenous siRNAs are symmetric 21-nt RNA duplexes having a 19-nt duplex domain and 2-nt 3′-overhangs that are processed from longer double-stranded (ds) RNAs by the endoribonuclease Dicer [7–9]. Dicer functions as a heterodimer with a second RNA-binding protein, R2D2 in Drosophila or TRBP (the human immunodeficiency virus transactivating response RNA-binding protein) in humans [10–12]. Dicer is a large protein with multiple functional domains, including two RNase H family nuclease domains, which perform substrate cleavage, and a PAZ domain, which binds short single-stranded RNA overhangs [13, 14]. Binding of an RNA overhang by the PAZ domain orients the substrate within Dicer. The first nuclease domain is separated from the PAZ domain by a "connector helix," which determines the distance between the PAZ binding site and the cleavage site, which is 21–22 bases for mammalian Dicer enzymes. Following cleavage, the nascent siRNA remains associated with Dicer and TRBP. An Argonaute (Ago) family protein then associates with the complex, forming a functional RNA-induced silencing complex (RISC) [15, 16]. The siRNA is transferred from Dicer/TRBP to the Ago protein, where the siRNA is converted to single-stranded (ss) form by either cleavage/degradation of one strand or unwinding by a helicase activity [17–19]. The ejected or degraded strand is called "the passenger strand," and the retained strand is called "the guide strand." The guide strand directs the sequence specificity of all subsequent gene suppression activity of the complex. Like Dicer, the Ago proteins possess a PAZ domain which binds the 3′-single-stranded overhang of the siRNA and orientates it within the complex [20]. There are four Argonaute proteins in humans that perform different effector functions in RISC [21–24]. In particular, Ago2 is an endoribonuclease which cleaves the target mRNA as directed by sequence complementarity to the siRNA guide strand bound in RISC [25–27] and is the key protein responsible for degradative RNAi.

The first generation of chemically synthesized siRNAs was designed to mimic the natural products of Dicer, i.e., 21-nt RNA duplexes with 2-nt 3′-overhangs [28]. This design remains the dominant form of synthetic siRNAs in use today. During the 10 years since this initial discovery, a variety of artificial designs have been proposed to improve upon some aspect of the RNAi process, which were discussed in a review by Chang and colleagues [29]. This chapter will review development and use of Dicer-substrate siRNAs (DsiRNAs) as a trigger of RNAi.

2.2 Development of Dicer-Substrate siRNA Technology

Dicer is involved in RISC loading, so it is possible that using a synthetic RNA duplex that is a substrate for Dicer and thus engages Dicer prior to RISC assembly may show different properties as a trigger for RNAi rather than an RNA duplex that mimics a Dicer product. This hypothesis was tested in a series of collaborative experiments performed in the laboratories of John Rossi at the Beckman Research Institute of the City of Hope National Medical Center and at Integrated DNA Technologies [30, 31].

2.2.1 Characterization of Synthetic Dicer-Substrate siRNAs

2.2.1.1 Duplex Length and Structure

A series of blunt-ended RNA oligonucleotide duplexes were tested for cleavage in an in vitro dicing assay [30]. The synthetic dsRNAs were incubated with recombinant human Dicer, desalted, and subjected to electrospray ionization (ESI) mass spectrometry. Duplexes as short as 23-nt length were cleaved to 21-nt length, and duplexes from 23- to 30-nt length all showed efficient cleavage. Cleavage efficiency decreased as length was extended to 35, 40, and 45 nts. Thus, synthetic RNA duplexes can function as Dicer substrates, and the optimal length for in vitro dicing was estimated to be in the 25–30-nt range.

The effect of duplex end structure was investigated, and Dicer cleavage occurred whether the duplex had blunt ends, 5′-overhangs, or 3′-overhangs; however, functional potency varied significantly with structure (see below). The effect of end modification (i.e., the addition of non-nucleotide moieties) was investigated by placing a bulky fluorescein group at the 5′-end, 3′-end, or both ends of each strand of the duplex. 5′-modification was well tolerated, but dicing efficiency was markedly reduced by the introduction of a single 3′-modification on either end of the duplex. Cleavage was entirely blocked if the duplex was modified at both 3′-ends. Functional potency in gene knockdown correlated with dicing efficiency. These observations are consistent with a mechanism where the Dicer PAZ domain first binds the substrate RNA at the 3′-end and then cleavage follows; any modification of structure that interferes with 3′-end binding in these very short substrate RNAs disrupts processing.

2.2.1.2 Optimized Design of Dicer-Substrate siRNAs

Functional potency of a series of anti-*EGFP* RNA duplexes was tested by transfection into EGFP-expressing HEK297 cells [30]. The length of the RNA duplexes was varied from 21 to 30 nts, having either 5′-overhangs, 3′-overhangs, or blunt ends [30]. All of the duplexes tested showed effective suppression of EGFP fluorescence when used at a high concentration (50 nM), but only the longer duplexes showed efficacy when the concentration used was reduced to subnanomolar levels (50–200 pM). At this site, a 27-nt blunt duplex was the most potent compound tested, and the EC_{50} shifted from 20 nM for the 21-nt siRNA to 200 pM for the 27-nt blunt duplex. A prolonged duration of silencing was also observed for the 27-nt duplex, with detectable EGFP suppression increasing from 4 to 10 days; the increased duration of silencing may simply reflect the higher potency of the compound. Significant

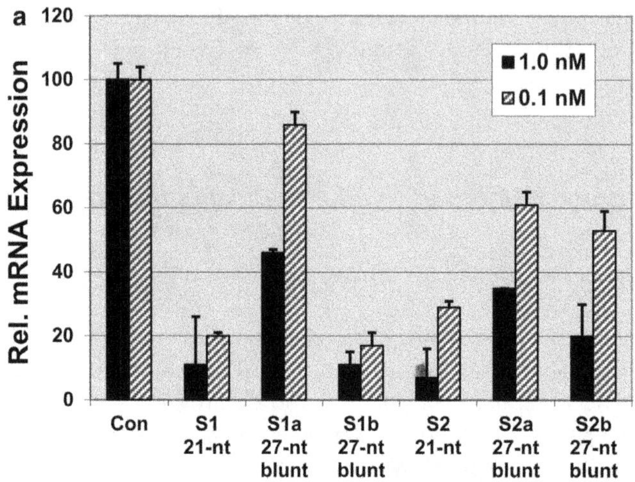

b

S1	21-nt	5′	GGAGGAAUUGGAACAGAAAtt	3′
		3′	ttCCUCCUUAACCUUGUCUUU	5′
S1a	27-nt	5′	AAAGUUGGAGGAAUUGGAACAGAAAUA	3′
		3′	UUUCAACCUCCUUAACCUUGUCUUUAU	5′
S1b	27-nt	5′	UUGGAGGAAUUGGAACAGAAAUACACC	3′
		3′	AACCUCCUUAACCUUGUCUUUAUGUGG	5′
S2	21-nt	5′	GCACCAGAGCCAAUGGAACtt	3′
		3′	ttCGUGGUCUCGGUUACCUUG	5′
S2a	27-nt	5′	AAGGUUGCACCAGAGCCAAUGGAACUU	3′
		3′	UUCCAACGUGGUCUCGGUUACCUUGAA	5′
S2b	27-nt	5′	UUGCACCAGAGCCAAUGGAACUUGAUG	3′
		3′	AACGUGGUCUCGGUUACCUUGAACUAC	5′

Fig. 2.1 Suppression of *STAT1* expression. The ability of 21-nt siRNAs and 27-nt blunt dsRNA to suppress *STAT1* expression was studied at two sites. (**a**) *STAT1* (NM_007315)-specific dsRNAs were transfected into HeLa cells at 1 nM or 0.1 nM concentration, and total RNA was isolated 24 h post-transfection. RT-qPCR was performed on the STAT1 mRNA, and results were normalized to an internal *HPRT1* control. (**b**) Sequences of the STAT1-specific dsRNAs employed are shown with the 27-nt duplexes aligned under the 21-nt siRNA. RNA bases are *uppercase* and DNA bases are *lowercase*

but less dramatic increases in potency were observed at other sites within the *EGFP* gene and also within a set of dsRNAs of varying length that targeted the Sjogren's syndrome antigen B (*SSB*) gene and the heterogeneous nuclear ribonucleoprotein H1 (*HNRNPH1*) gene [30]. Thus, increased potency was observed using Dicer-substrate siRNAs at multiple sites in three different genes in this study.

As the number of sites studied using this "first generation" blunt 27-nt design was expanded, the situation become more complex. Inconsistent performance was sometimes observed between sites such that some duplexes showed higher potency in 27-nt blunt dsRNA form than in 21-nt siRNA form, others showed similar potency between forms, while yet other sites showed higher potency in 21-nt siRNA form. An example of this behavior is shown in Fig. 2.1, where functional potency of knockdown at two different sites in the *STAT1* gene are compared between a 21-nt siRNA and two blunt 27-nt dsRNAs that overlap the 21-nt sequence. At site 1, one of the two 27-nt blunt dsRNAs showed low potency, while the other showed high potency, similar to a 21-nt siRNA at this site. At site 2, both of the 27-nt dsRNAs showed lower potency than the 21-nt siRNA.

To investigate these seemingly inconsistent results, an assay was developed at Integrated DNA Technologies to study the products of Dicer processing of the synthetic RNA duplexes, hoping to find some correlation between Dicer processing patterns and functional potency. The synthetic dsRNAs were incubated with recombinant human Dicer, desalted, and subjected to electrospray ionization (ESI) mass spectrometry. Using these methods, single Dalton accuracy in mass measurement was achieved, which, in most cases, permitted precise identification of the species produced by Dicer cleavage [31]. It was found that blunt 27-nt substrate dsRNAs were usually cleaved by Dicer into several products. Sometimes as many as five or more different species were observed, which varied in size from 20 to 23 nts. Furthermore, the precise pattern of cleavage could not be readily predicted from sequence. Significant differences in potency are sometimes seen between siRNAs that are shifted by as little as a single base along a target sequence; thus, small differences in dicing patterns could account for large differences in observed potency. These observations also imply that a direct comparison of potency between a blunt 27-nt dsRNA and the "cognate" 21-nt siRNA is really only possible if the dominant 21-nt siRNA produced by Dicer cleavage is identified by a mass spectrometry assay (or other means). Without this information, there is no way to know which of the possible 21-nt siRNAs that could potentially be cleaved from a 27-nt dsRNA substrate is the correct species to employ in a comparative study.

In summary, the mass spectrometry and functional studies discussed above indicated that it was not possible to predict the potency of blunt 27-nt synthetic dsRNAs; sometimes Dicer cleavage resulted in highly potent 21-nt siRNAs from a substrate, and other times low potency species were produced. To improve on these results, a thorough structure–activity relationship (SAR) study was performed where duplexes of different length and design were studied for cleavage patterns and functional potency with the goal of finding a dsRNA substrate that gave predictable results with Dicer processing.

The Dicer-substrate SAR studies identified a design that showed particular promise [31]. The lead compound was an asymmetric dsRNA with a 25-nt sense (passenger) strand and 27-nt antisense (guide) strand. This design has a single 3′-overhang which resides on the guide strand and is blunt at the other end; in addition, two DNA bases were placed at the 3′-end of the passenger strand. Dicer processing of this substrate usually resulted in production of a single 21-nt siRNA species which spanned from the original 3′-overhang in the substrate to a cleavage point 21 bases away. Henceforth, this asymmetric 25/27-nt RNA duplex will be called a "DsiRNA" (optimized Dicer-substrate siRNA). While it is not certain why this design was so effective at encouraging uniform processing, the authors speculated that the single 3′-overhang was bound by the Dicer PAZ domain which positions the substrate at a specific distance from the active endonuclease site, thereby resulting in more uniform processing than was seen using blunt substrates where there is no 3′-overhang for PAZ binding. Further, the PAZ domain preferentially binds RNA 3′-overhangs, and the inclusion of DNA bases at the 3′-blunt end may further discourage PAZ binding at that end. These ideas are consistent with some of the structural features and biochemistry of Dicer which were subsequently elucidated [13, 14].

Note that an asymmetric 25/27-nt RNA duplex could be designed in two ways, both of which overlap the desired target site and both of which are predicted to produce the same final 21-nt siRNA after Dicer cleavage: one where the passenger strand is 27 nt and one where the guide strand is 27-nt length. Both of these substrate designs were tested at different sites in several genes using the mass spectrometry dicing assay, and it was verified that the same 21-nt siRNA was indeed produced by Dicer processing from these two related but different substrates. For convenience, the asymmetric 25/27-nt RNA duplex with the 27-nt sequence on the guide strand was called the right ("R") form (keeping with the convention that the sense strand is "top," the Dicer PAZ domain binds the single 3′-overhang on the left side of the duplex, and cleavage proceeds to the right "R"). Conversely, the asymmetric 25/27-nt RNA duplex with the 27-nt sequence on the passenger strand was called the left ("L") form (the Dicer PAZ domain binds the single 3′-overhang on the right side of the duplex, and cleavage proceeds to the left "L"). In spite of the fact that both forms result in the same 21-nt daughter siRNA species, very different functional potencies were observed for actual knockdown of a target gene. The "R" form was almost always more potent than the "L" form. An example of this interesting observation is shown in Fig. 2.2, which is reprinted from Rose et al. [31]. A site in EGFP was selected for study, and the potency of a 21-nt siRNA was compared with "L" and "R" form asymmetric 25/27-nt RNA duplexes at the same site. The duplexes were transfected into HEK293 cells that expressed EGFP, and fluorescence was measured 24 h after transfection. The 21-nt siRNA and the "L" form 25/27-nt duplex showed similar potency with around a 70% reduction in EGFP signal seen using 2 nM duplex. In contrast, the "R" form 25/27-nt duplex showed much higher potency, with almost 80% suppression of EGFP fluorescence signal seen at 200 pM; EGFP fluorescence signal was nearly undetectable using a 2 nM concentration of the "R" form duplex.

Using a luciferase-based assay where either a sense target or an antisense target was cloned into the 3′-UTR of firefly luciferase, Rose and colleagues demonstrated that the "L" and "R" forms of the 25/27-nt duplex exhibited differential loading of the guide vs. passenger strands [31]. The 27-nt strand with the 3′-overhang generally shows a relative increase in RISC loading compared with the 25-nt strand. Thus, the "R" form, where the 3′-overhang is on the guide (antisense) strand, generally shows higher functional potency for suppressing expression of the target gene than does the "L" form, which has a loading bias in favor of the passenger strand. These design features are schematically illustrated in Fig. 2.3. Interestingly, 3 years later, a similar effect was reported for 21-nt siRNAs: use of an asymmetric design with a single 3′-overhang on the guide strand improved loading of that strand and increased functional potency [32]. A number of factors influence which strand of an RNA duplex loads into RISC and functions as the guide strand and which strand is

a

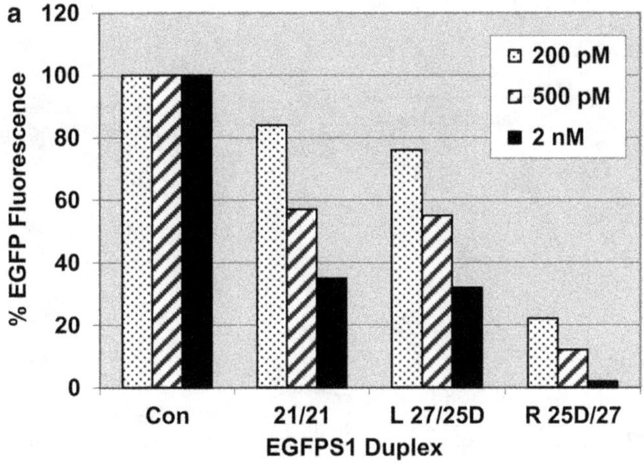

b

```
5′ CAAGCUGACCCUGAAGUUCAUCUGCACCACCGGCAA     EGFP Site-1

5′          ACCCUGAAGUUCAUCUGCACC            EGFPS1 21/21
3′        ACUGGGACUUCAAGUAGACGU

5′         pACCCUGAAGUUCAUCUGCACCACcg        EGFPS1 R 25D/27
3′        ACUGGGACUUCAAGUAGACGUGGUGGC

5′  AAGCUGACCCUGAAGUUCAUCUGCACC             EGFPS1 L 27/25D
3′  ttCGACUGGGACUUCAAGUAGACGUp
```

Fig. 2.2 "R" form duplexes are more potent than "L" form duplexes. (**a**) An EGFP expression plasmid was transfected into HEK293 cells. Anti-EGFP or control dsRNAs were transfected 24 h later at the indicated concentrations. At 24 h post-transfection, relative fluorescence was measured setting the control cultures at 100%. (**b**) The EGFP site 1 mRNA target sequence is shown at the top with 21-nt and asymmetric 25/27-nt DsiRNAs aligned below. RNA bases are *uppercase* and DNA bases are *lowercase*. "p" indicates a 5′-phosphate. Reprinted from Rose et al. [31] with permission from Oxford University Press

ejected as the passenger strand. The primary feature contributing to loading bias relates to thermodynamic end asymmetry [33–35]; it now appears that end structure (3′-overhang vs. blunt) also plays a role [31, 32]. Interestingly, a recent report from the Doudna group demonstrated that Dicer is integrally involved in strand selection during RISC formation, lending additional credibility to the functional polarity observed using Dicer-substrate siRNAs in the studies discussed above [36].

2.2.1.3 Functional Potency of DsiRNAs

A study was performed at the Bio-Rad laboratories by Eli Hefner, Teresa Rubio, and colleagues to validate the performance of the newly optimized DsiRNA design [37]. A set of five genes were selected as targets, including *ACTB*, *AKT1*, *RAF1*, *CDK2*, and *TP53*. For each gene, an asymmetric 25/27-nt DsiRNA targeting that gene and the cognate 21-nt siRNA were compared for potency and duration of silencing.

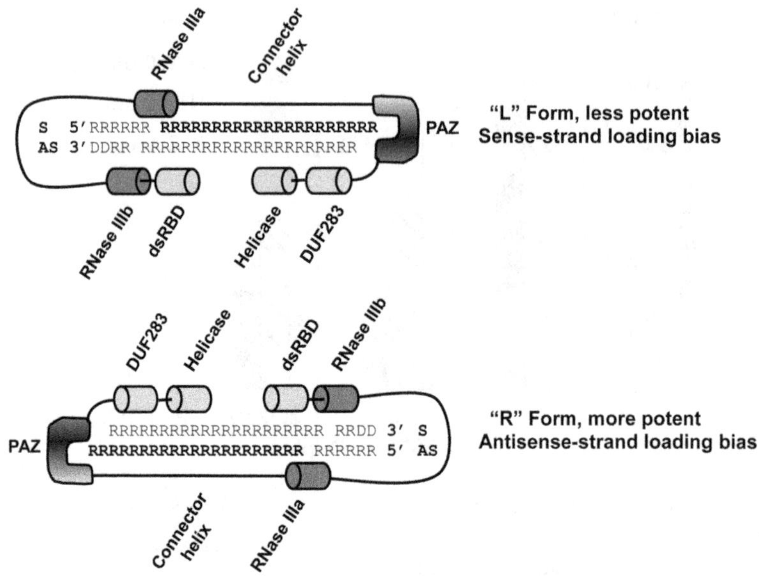

Fig. 2.3 Functional polarity is introduced by Dicer processing. Two forms of DsiRNAs are shown: the "L" form duplex with the single 3′-overhang on the sense strand (S) and the "R" form duplex with the single 3′-overhang on the antisense strand (AS). Sites of Dicer cleavage are indicated by a gap in the sequence; R=RNA, D=DNA. A schematic image of Dicer showing the different functional domains is overlaid on the dsRNA substrate, positioning the RNase III domains at the staggered cut sites and the PAZ domain at the single 3′-overhang. The strand of the cleavage product which is favored for loading into RISC as the "guide strand" is highlighted in *bold*, demonstrating how functional polarity could be introduced by differential positioning of the antisense strand in Dicer between the "L" and "R" forms

Each RNA duplex was transfected into HeLa cells using the siLentFect™ cationic lipid reagent at concentrations of 50 nM, 5 nM, 1 nM, and 100 pM. RNA was extracted 24 h post-transfection, and RT-qPCR was performed to assess relative knockdown of the target mRNA. In four of the five genes studied (*ACTB*, *RAF1*, *CDK2*, and *TP53*), the DsiRNA showed higher potency than the cognate 21-nt siRNA, especially at the lower doses. For the other target (*AKT1*), the DsiRNA and siRNA showed identical potency. The dose response results for the anti-*TP53* DsiRNA and siRNA are shown in Fig. 2.4a, and the sequences employed are shown in Fig. 2.4c.

A comparison of the duration of silencing was also performed. The DsiRNA and siRNA pairs were individually transfected into HeLa cells at 5 nM concentration, and cultures were sampled at days 1, 2, 4, and 6 post-transfection. RNA was extracted, and RT-qPCR was performed to assess relative knockdown of the target mRNA. The results paralleled the dose–response data discussed previously, and, for four of the five genes studied (*ACTB*, *RAF1*, *CDK2*, and *TP53*), the DsiRNAs showed longer duration of silencing than their cognate siRNAs. For the other target

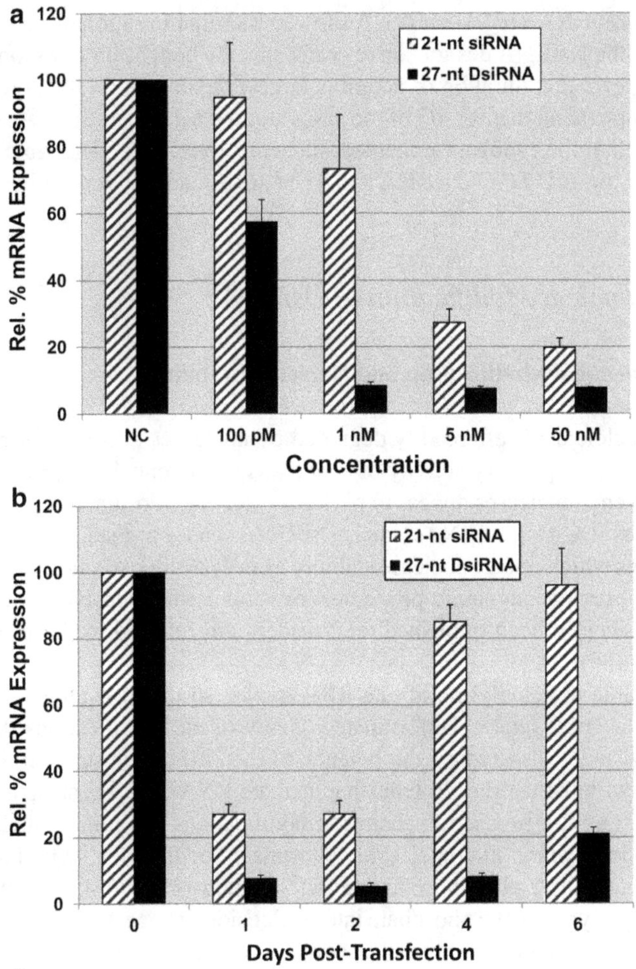

c

```
21-nt    5'     ACCAGGGCAGCUACGGUUUtt        3'
siRNA    3'   ttUGGUCCCGUCGAUGCCAAA          5'

27-nt    5'    pACCAGGGCAGCUACGGUUUCCGUct    3'
DsiRNA   3'   GAUGGUCCCGUCGAUGCCAAAGGCAGA    5'
```

Fig. 2.4 Comparison of potency of an anti-*TP53* DsiRNA and the cognate siRNA. (**a**) An asymmetric 25/27-nt DsiRNA and its cognate 21-nt siRNA targeting TP53 (NM_000546) were transfected into HeLa cells at the indicated concentrations. RNA was isolated at 24 h post-transfection, and RT-qPCR was performed to measure relative *TP53* expression levels normalized to an internal control gene, *GAPDH*. NC = negative control. (**b**) The same DsiRNA and siRNA were transfected into HeLa cells, and cultures were maintained for 1–6 days as indicated. RNA was isolated, and RT-qPCR was performed to measure TP53 expression levels as before. (**c**) Sequences of the dsRNAs are shown; RNA bases are *uppercase* and DNA bases are *lowercase*. "p" indicates a 5′-phosphate. Adapted from Hefner et al. [37] with permission from the Association of Biomolecular Resource Facilities (ABRF)

(*AKT1*), the DsiRNA and 21-nt siRNA showed a similar duration of action. It is not thought that the DsiRNA design conveys any specific benefit for duration of action; rather, it appears that duration of action is largely dependent on the potency of the silencing reagent and that for 4/5 of the cases studied here, the DsiRNAs were more potent and therefore showed extended silencing over time. The time course of silencing for the anti-*TP53* DsiRNA and 21-nt siRNA are shown in Fig. 2.4b.

2.2.2 Chemical Modification of DsiRNAs

2.2.2.1 Chemical Modification and Nuclease Stability

Synthetic nucleic acids are readily degraded by nucleases present in serum and in cells. In serum, the primary activity of concern is a 3′-exonuclease [38], whereas in cell extracts endonucleases appear to play a greater role. Fortunately, antisense oligonucleotides (ASOs), siRNAs, and DsiRNAs can be made using chemical modifications which impart nuclease stability as well as improve their safety profiles and general pharmacodynamic properties. Several comprehensive reviews of this topic have recently been published, and readers are referred to these sources for more details [39–41].

It is possible to heavily modify an RNA duplex so that it is almost completely resistant to nuclease attack. Unfortunately, many of the modifications that convey nuclease resistance also reduce the potency of the siRNA, presumably by altering interactions with the various proteins that mediate RNAi in cells, such as Dicer and Argonaute 2 (Ago2). In general, chemical modifications that impart nuclease resistance either involve the internucleoside phosphate bonds or the sugar backbone of the nucleic acid. The phosphorothioate modification (PS) substitutes sulfur for a non-bridging oxygen in the phosphate internucleoside linkage. This conveys relative resistance to many nucleases but can also make the oligonucleotide more "sticky" to proteins and can significantly alter function in undesired ways [42]. While it is common practice to completely modify ASOs with PS bonds, this modification is usually used sparingly in siRNAs. SiRNAs with high PS content can perform poorly, and altered interactions between the RNA and the protein machinery in RISC may be implicated [43]. Selective incorporation of PS linkages near the ends of the duplex and, particularly in the 3′-overhangs, helps protect these vulnerable sites and is a modification strategy commonly used today [40]. Single-stranded RNA domains, such as the siRNA 3′-overhangs, are highly susceptible to degradation. An inverted dT base, PS bonds, or other modifications are often placed at this site.

Modification at the 2′-position of the ribose usually decreases the susceptibility of the neighboring phosphate bonds to nuclease attack, and a wide variety of 2′-modified residues are routinely employed to modify siRNAs. In particular, 2′-*O*-methyl (2′-OMe) RNA is a naturally occurring chemical modification that is found in mammalian tRNAs and rRNAs. This modification is relatively inexpensive to incorporate into synthetic nucleic acids and has no known toxicity. Other

2′-modifications in common use include 2′-fluoro (2′-F) or locked nucleic acids (LNAs), which are a bicyclic nucleic acid with a methylene bridge linking the 2′- and 4′-positions in the ribose ring.

The 2′-OMe modification can be placed in the sense strand, antisense strand, or both strands of a siRNA [44–48]. Complete modification usually results in an inactive siRNA, and use of alternating (or less) 2′-OMe groups is commonly employed. Although not necessary, the 2′-OMe modification is often employed in conjunction with other modifications, such as 2′-F residues. The 2′-F modification is not natural; however, it appears to be generally safe to administer to cells or live animals and can help stabilize siRNAs and improve function [49–52]. In particular, use of 2′-OMe purines with 2′-pyrimidine residues can result in a highly stabilized siRNA with improved performance in vivo [53, 54]. The relative potency of siRNAs having this kind of extensive modification pattern shows sequence dependence and thus may not work effectively at all sites. LNA modifications have an even greater impact on structure and potency of siRNAs and thus are generally used sparingly as modifications. Synthetic oligonucleotides that are heavily LNA modified can show some hepatic toxicity in mice [55], although this effect appears to be sequence dependent and some LNA-modified oligonucleotides are well tolerated [56–58].

The modification strategies discussed above were developed and validated using 21-nt siRNAs. Longer dsRNAs, such as DsiRNAs, appear to naturally show greater resistance to nuclease degradation than short siRNAs [59, 60]. This may in part be due to the higher thermodynamic stability seen for longer duplexes, which may limit the amount of transient single-stranded character in AU-rich regions that are more susceptible to attack by endogenous endoribonucleases, such as RNase A [61]. The same modification strategies employed in 21-nt siRNAs can generally be directly applied to DsiRNAs, except that a small internal domain needs to remain unmodified for Dicer cleavage to occur (the only kind of nuclease attack that is actually desired). It is possible to synthesize DsiRNA duplexes that show high levels of serum stability while retaining the ability to be processed by Dicer. A structural map of the different functional DsiRNA domains is schematically shown in Fig. 2.5. In this figure, the top strand is the passenger strand, and sequence to the left of the Dicer cleavage site comprises the final 21-nt siRNA. This region can generally be modified in ways similar to other synthetic 21-nt siRNAs, as described above. Sequence to the right of the Dicer cleavage site can also be modified; note that this short sequence is discarded and is not part of the final product that enters RISC. As shown in Fig. 2.5, the favored site to add bulky end modifications to a DsiRNA is at the 3′-end of the passenger strand (labeled "ligand"); modifications can also be added to the 5′-end of the guide strand. Using this approach, a bulky modifying group such as a fluorescent dye or ligand that may aid delivery (such as cholesterol, cell-penetrating peptides, etc.) can be attached to the RNA duplex in a way that it is "disposable"; any group connected to this end is cleaved off the DsiRNA and discarded and so does not remain on the mature siRNA and thus does not enter RISC and cannot affect RISC loading.

Collingwood and colleagues reported a systematic survey of various modification patterns in DsiRNAs, focusing on the use of the 2′-OMe and 2′-F modifications [62]. The initial survey was performed at a site in the human *STAT1* gene (corre-

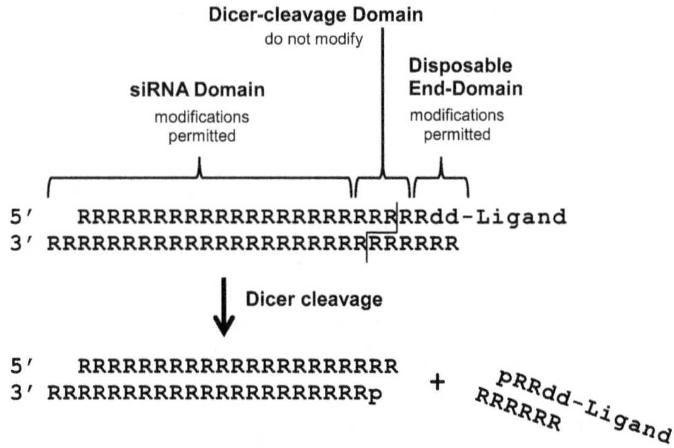

Fig. 2.5 Schematic of DsiRNA domains for chemical modification. An asymmetric 25/27-nt DsiRNA is shown on the *top line* with the domains suitable for chemical modification indicated by *brackets*. A small interior domain offset to the *right of center* is *not bracketed*, which is the site of Dicer cleavage; this region should remain unmodified. The preferred attachment site of ligands is shown on the *right*, connecting to the 3′-end of the passenger strand. The 21-nt siRNA that results from Dicer processing is shown along with the discarded cleavage fragment. RNA bases are *uppercase* and DNA bases are *lowercase*; "p" indicates a 5′-phosphate

sponding to the STAT1 site 2 shown in Fig. 2.1). Like 21-nt siRNAs, some of the more heavily modified duplexes showed significant impairment of functional potency, while less highly modified patterns remained potent. 2′-OMe modification of the sense and/or antisense strands in the "siRNA Domain" (Fig. 2.5) in an alternating pattern was particularly effective, and this approach to modification was also shown to work well at additional sites in the human *HPRT1*, mouse *F3*, and *EGFP* genes. The mass spectrometry dicing assay was used to examine processing of a set of anti-*HPRT1*-modified DsiRNAs, and as long as the modifications did not extend into the Dicer-cleavage domain, the expected siRNA products were made following in vitro dicing. Further, DsiRNAs modified with only 11 2′-OMe residues on the antisense strand showed a significant improvement in stability when incubated in serum compared to unmodified 21-nt siRNAs or 25/27-nt DsiRNAs. While this simple modification pattern is often effective, it can impair potency in a sequence-specific fashion at some sites, so additional optimization of the precise placement of modified bases can be beneficial.

Nishina and colleagues described use of a modified asymmetric 27/29-nt DsiRNA to suppress *Apob* expression in mouse liver [63]. 2′-OMe RNA residues and PS bonds were placed at optimized locations in the sense and antisense strands, avoiding modification of the Dicer cleavage domain. Vitamin E (α-tocopherol) was attached to the 5′-end of the antisense strand via a phosphate linkage. The sequence and modification pattern of this compound are shown in Fig. 2.6. The modification pattern employed six 2′-OMe residues and a single PS bond in the sense strand and nine 2′-OMe residues with five PS bonds in the antisense strand and achieved sufficient stability to be used via direct naked intravenous injection in mice. The

```
5'      GUCAUCACACUGAAUACCAAU| GC U GG*A        3'
3'  C*A*CAGUAGUGUGACUUAUGGU|UA*CG*A*CC U-VitE  5'
```

Fig. 2.6 Chemical modification pattern of an anti-*Apob* DsiRNA. Sequence of a modified DsiRNA targeting the *Apob* gene (NM_009693) is shown [63]. RNA bases are *uppercase* and 2′-OMe RNA bases are *underlined*; "*asterisk*" indicates a PS (phosphorothioate) internucleotide linkage. Vitamin E (α-tocopherol) was attached to the 5′-end of the antisense strand via a phosphate linkage. The sites of Dicer cleavage are indicated

vitamin E group is a lipid-soluble antioxidant vitamin that can gain entry to all mammalian cells. Similar to cholesterol, vitamin E associates with serum lipoproteins; the SCARB1 scavenger receptor and LDL receptor are involved in cellular uptake [64]. By attaching vitamin E to the 5′-end of the antisense strand of a DsiRNA, the ligand became part of the disposable "end domain" of the DsiRNA, which is cleaved off of the RNA duplex and does not remain attached to the mature siRNA and, therefore, poses no risk to interfere with RISC entry. *Apob* mRNA levels were reduced by 80% at a dose of 32 mg/kg using this approach.

Kubo and colleagues studied modification of the ends of DsiRNAs with an aliphatic amino modifier and found that addition of this simple group could improve stability and potency of an otherwise unmodified duplex [65]. In particular, modification at the 3′-end of the sense strand had favorable effects. Consistent with the earlier finding reported by Kim [30], placement of a single modifying group at the 3′-end of the antisense strand impaired Dicer processing, while modification of both of the 3′-ends prevented Dicer cleavage. This group also conjugated palmitic acid (C16) at the 3′-end of the sense strand, a location that permits removal of the ligand so the modifier is not present in the mature 21-nt siRNA following Dicer processing (see Fig. 2.5). The 3′-palmitic acid modification led to additional stabilization of the duplex, and an otherwise unmodified DsiRNA survived 48-h incubation in 10% fetal calf serum. Addition of the C16 aliphatic chain also promoted naked delivery of the modified DsiRNA to HeLa cells in tissue culture when used at a relatively high dose (200 nM).

2.2.2.2 Chemical Modification and Immune Stimulation

The mammalian innate immune system employs a fixed repertoire of receptors which recognize structures that are usually associated with pathogens (pathogen-associated molecular patterns or PAMPs), such as bacterial flagella or bacterial lipopolysaccharide. The innate immune system is also capable of recognizing nucleic acids, including single-stranded and double-stranded RNA (ssRNA, dsRNA); this system probably evolved as a fast-response pathway to viral infection. Several members of the Toll-like receptor family (TLRs) recognize RNA, such as TLR3 which binds dsRNA and TLR7 and TLR8 which bind ssRNA. TLR3, 7, and 8 primarily reside in endosomal compartments which limits their contact with endogenous RNAs, helping to limit the risk of an autoimmune response. Further, these receptors preferentially recognize unmodified RNA. RNAs bearing several modifications that are common in mammalian cells, such as 2′-OMe RNA, pseudouridine, and others, typically help the RNA molecule avoid detection [66]. In fact, 2′-OMe RNA is a competitive inhibitor of TLR7 and can block recognition of unmodified RNAs by this receptor, even in *trans* [67]. Thus, employing the 2′-OMe RNA modification, even in only one strand, can block recognition of a siRNA by TLR7. Other 2′-modifications, such as 2′-F and LNA, also help avoid immune detection. Additional sensors exist in the cytoplasm that can also trigger a response to foreign RNAs, such as PKR, OAS, RIG-I, and MDA5. Many mammalian cells that have adapted to growth in cell culture have lost the ability to respond to foreign nucleic acids; however, the risk of triggering an immune response in vivo, where all cell types are present, is quite high. Strategies to evade immune detection are essential to all methods that use synthetic nucleic acids in mammals, including siRNAs. The reader is referred to some excellent recent reviews on this topic for more information [68, 69].

Like all synthetic RNAs, DsiRNAs can trigger an unwanted immune response when introduced into mammalian cells. In fact, long RNAs are generally more potent at stimulating an immune response than are short RNAs. Thus, there was initially some concern that dsRNAs in the 27-nt length range would present a greater risk for triggering immune responses than traditional short siRNAs. In support of this idea, Reynolds and colleagues studied the ability of different dsRNAs to trigger immune responses in a variety of cell types and found that cell type, the length of the synthetic RNA, and method of delivery were all contributing factors to immune stimulation and other off-target effects [70, 71]. Reynolds studied blunt 27-nt dsRNAs and found that these structures showed a significantly higher propensity to trigger immune responses in several cell types than 21-nt siRNAs (with a 2-base 3′-overhang on both ends). However, it later became evident that end structure is an additional important feature to consider in triggering immune responses. Marques and colleagues reported that transfection of 27-nt dsRNAs with 3′-overhangs on both ends into T98G glioblastoma cells did not trigger an immune response while use of blunt 27-nt dsRNAs resulted in a brisk immune response [72]. The optimized DsiRNA design (asymmetric 25/27 nt with a single 3′-overhang on one side and a blunt end with two 3′-DNA bases on the other end) was a relatively weak immune trigger. Collingwood and colleagues further demonstrated that addition of 2′-OMe residues to the optimized asymmetric 25/27-nt DsiRNA design prevented stimulation of interferon-α (IFN-α) secretion from human peripheral blood mononuclear cells (PBMCs) [62]. PBMCs are considered a sensitive cell population to employ when testing the potential for a synthetic nucleic acid to trigger an immune response as PBMCs are a mixed population of cells which comprise a wide variety of normal immune-competent white blood cells. Using cationic lipid-mediated transfection, unmodified 21-nt siRNAs and 25/27-nt DsiRNAs both stimulated significant levels of IFN-α secretion in human PBMCs, whereas 2′-OMe-modified versions of the same sequences did not. Thus, the combined use of chemical modification and optimized end structure enabled safe use of 25/27-nt DsiRNAs.

2.3 Studies Using DsiRNAs In Vivo

Many studies have been published that employ siRNAs in animals. The scientific issues involved in conducting in vivo RNAi studies in mammals and overviews of significant articles in this rapidly growing field have been well covered in recent reviews [73, 74]. Some recent successes using DsiRNAs in vivo will be discussed below. The in vivo studies discussed herein are summarized in Table 2.1.

Like all classes of synthetic dsRNAs, delivery is the most significant hurdle to widespread use of DsiRNAs in vivo. While the systemic effects associated with intravenous injection are often desirable, local administration is usually easier to perform and suffers from fewer toxicity issues. Several studies have been performed with DsiRNAs using local delivery, including intraperitoneal (IP) and intrathecal (IT) routes of administration. Amarzguioui and colleagues described methods to deliver DsiRNAs using an inexpensive commercially available cationic lipid in mice via IP injection [85]. It was observed that the lipid TransIT-TKO™ was particularly effective in delivering DsiRNAs into macrophages, both in cell culture and in vivo. The peritoneal cavity is a reservoir for monocyte lineage cells, and IP injection offers an opportunity to easily introduce DsiRNA reagents into this cell population via local administration. Immune cells like macrophages are mobile: the transfected macrophages can be recruited to sites of inflammation outside of the peritoneum, thereby achieving systemic effects using a local administration strategy.

Lundberg, Cantin, and colleagues employed this strategy to suppress production of tumor necrosis factor alpha (TNF-α) by macrophages during acute herpes simplex virus (HSV) infection in the mouse central nervous system (CNS) [75]. Studies were performed in C57BL/6, a strain of mice that normally survives herpes encephalitis (HSE). TNF-α is an important factor in the resistance of this strain to HSE, and C57BL/6 mice that are double knockouts for the two known TNF receptors ($p55^{-/-}$, $p75^{-/-}$) show increased fatality following HSE. Mice were administered the anti-*Tnf* DsiRNA DsiRNA complexed with the cationic lipid TransIT-TKO™ as a single 2 μg dose (0.1 mg/kg) in 200 μL volume by IP injection immediately prior to infection with HSV, then received five additional doses of 4 μg (0.2 mg/kg) on days 1, 2, 4, 6, and 8 [for a total of six doses with a total cumulative dose of 22 μg (or 1.1 mg/kg) of the DsiRNA]. Surprisingly, suppression of TNF-α production using an anti-*Tnf* DsiRNA resulted in significantly increased mortality rates, suggesting that the presence of TNF-α somehow alters the pathophysiology of HSE even in the absence of the two known TNF receptors. The mechanism of this unexpected observation remains under investigation. In this case, it is not believed the IP administered DsiRNA directly entered the mouse CNS. Rather, the DsiRNA were taken up by macrophages, which were subsequently recruited by the inflammatory process ongoing in the brain due to the HSV infection, thereby achieving CNS effects from IP administration of a compound that normally cannot cross the blood brain barrier (BBB).

Lundberg, Cantin, and colleagues later used this same system (IP injection of anti-*Tnf* DsiRNA complexed with TransIT-TKO™ to transfect macrophages) to alter the course of fatal hepatic necrosis during endotoxin-induced "toxic shock"

Table 2.1 Studies employing DsiRNAs in vivo

Citation	Delivery route	Carrier	Target	System
Lundberg [75]	IP	Lipid	*Tnf*	Mouse, CNS herpes simplex encephalitis
Lundberg [76]	IP	Lipid	*Tnf*	Mouse liver, endotoxin shock
Howard [77]	IP	Chitosan	*Tnf*	Mouse, collagen-induced arthritis
Nawroth [78]	IP	Chitosan	*Tnf*	Mouse, radiation-induced fibrosis
Dore-Savard [79]	IT	Lipid	*Ntsr2*	Rat CNS, nociception
LaCroix-Fralish [80]	IT	Lipid	*Atp1b3*	Mouse CNS, nociception
Sato [81]	IV	Vitamin A targeted liposome	*Serpinh1*	Rat, acute hepatic cirrhosis
Kortylewski [82]	IV	CpG oligo conjugate	*Stat3*	Mouse, metastatic melanoma
Nishina [63]	IV	Vitamin E conjugate	*Apob*	Mouse liver, cholesterol metabolism
Neff [83]	IV	Aptamer conjugate	HIV *tat/rev*	Humanized mouse, HIV
Zhou [84]	IV	PAMAM dendrimer	HIV *tat/rev, CD4, TMPO3*	Humanized mouse, HIV

Studies are shown in the order presented in the text. Routes of administration, *IP* intraperitoneal, *IT* intrathecal, *IV* intravenous

[76]. Bacterial lipopolysaccharide (LPS) binds the receptor TLR4 and triggers release of a number of pro-inflammatory cytokines, including large amounts of TNF-α by macrophages. The "cytokine storm" induced by LPS can lead to hypotension, shock, and death. One early sign of the severity of this shock state is hepatic necrosis. Mice were administered the anti-*Tnf* DsiRNA complexed with TransIT-TKO™ as a single 5 μg dose (0.25 mg/kg) in 200 μL volume by IP injection immediately prior to injection of the low dose LPS/D-GalN cocktail. Downregulation of TNF-α production by peritoneal mononuclear cells following this treatment reduced the magnitude of liver damage and delayed mortality. This study again demonstrated the ability to achieve systemic effects in an inflammatory disease state using IP administration of a DsiRNA targeting a cytokine expressed in macrophages.

Howard, Kjems, and colleagues employed this same strategy to treat collagen-induced arthritis (CIA) in mice [77]. In this case, a novel chitosan-based nanoparticle delivery system was employed [86]. The chitosan nanoparticles employed in this study exhibit a net positive charge that may lead to serum-induced aggregation after IV injection but are ideal for use with IP injection and are efficiently taken up by peritoneal macrophages. TNF-α secreted by macrophages is an important factor in the development and pathophysiology of inflammatory arthritis, and anti-TNF-α antibodies are presently employed to treat rheumatoid arthritis. Mice were immunized with an anthrogen-CIA collagen emulsion and displayed the onset of arthritis within 28–35 days postinjection, at which point DsiRNA treatments were initiated. Mice received either 5 μg of chitosan-complexed unmodified anti-*Tnf* DsiRNAs or control DsiRNAs (0.25 mg/kg) or 2.5 μg of chitosan-complexed 2′-OMe-modified anti-*Tnf* DsiRNA (0.125 mg/kg) in 200 μL volume by IP injection on days 1, 3, 5, 7, and 9. Arthritis scores were measured, and joints were examined by histopathology. Suppression of TNF-α by IP administration of an anti-*Tnf* DsiRNA reduced joint swelling, and mice receiving this treatment showed minimal erosion of the articular surfaces of affected joints, while control mice showed severe cartilage destruction. Overall, use of 2′-OMe-modified DsiRNAs showed the most effective biological responses.

The same group used the chitosan-based nanoparticle delivery system to treat radiation-induced fibrosis (RIF) in mice using the IP administration route [78]. RIF arises from damage and scarring within tissue exposed to high levels of ionizing radiation and is characterized by reduced mobility, strictures, pain, and even tissue necrosis. Morbidity from RIF can limit the dose of radiation that can be administered in the treatment of various cancers, and methods to reduce or prevent RIF would have significant value. There is some evidence to suggest TNF-α plays a potentially significant role in this disease. The hind limbs of mice were given a single 45-Gy dose of gamma radiation. Chitosan-complexed unmodified control or anti-*Tnf* DsiRNAs were administered at a dose of 5 μg (0.25 mg/kg) in 200 μL volume by IP injection for a varying periods of time, given 2 days prior to irradiation, 1 day post-irradiation, and biweekly thereafter. Mice receiving anti-*Tnf* DsiRNA therapy for <3 weeks all developed RIF, whereas none of the mice that received continuous therapy for 3 weeks or longer developed RIF. All mice that received the scrambled control DsiRNA developed RIF. This work suggests that intervention of TNF-α production is a valid therapeutic approach to prevent the development of RIF during antitumor therapy.

Two groups have employed the cationic lipid iFECT™ to deliver DsiRNA into the CNS via IT injection. The BBB presents an obstacle for delivery of large molecule compounds to the CNS using IV injection; direct injection into the CNS is the most direct solution to this problem. Dore-Savard and colleagues used anti-*Ntsr2* DsiRNAs to study the function of neurotensin receptor 2 (a GPCR) in thermal nociception in rats [79]. Ntsr2 mediates analgesia, and administration of the synthetic neurotensin-2 agonist JMV-431 reduces perception of pain. Downregulation of the Ntsr2 receptor by anti-*Ntsr2* DsiRNAs should block the expected analgesic effects following JMV-431 administration. The anti-Ntsr2 DsiRNAs were given as two IT injections at spinal levels L5/6 1 day apart using a dose of 1 μg in iFECT™ (0.005 mg/kg). The neurotensin agonist JMV-431 was administered daily at days 1–4 following the DsiRNA injections, and the rats were studied for nociceptive behavior using the thermal tail-flick test. Analgesic effects from the JMV-431 were absent on days 1 and 2 and slowly returned to baseline over the next several days. Reductions in both Ntsr2 mRNA and protein were observed, consistent with an RNAi-mediated suppression of *Ntsr2* gene expression.

LaCroix-Fralish and colleagues studied function of the β3 subunit of the Na^+–K^+-ATPase pump (*Atp1b3*) in the pain response to formalin footpad burns in mice [80]. The mouse strain C57BL/6 shows a higher pain response in this test compared to the A/J strain. Traditional genetic approaches and QTL analysis had previously implicated the *Atp1b3* gene as possibly being involved in this interstrain variability; however, no role for this gene in nociception had ever been demonstrated. Anti-*Atp1b3* DsiRNAs formulated in iFECT™ were given by IT injection at a dose of 0.5 μg (0.025 mg/kg) once a day for 3 days, and the mice were, thereafter, studied for pain responses to formalin footpad injection. Mice given the anti-*Atp1b3* DsiRNA showed similar pain responses between strains. Within the control DsiRNA cohort, the C57BL/6 mice showed higher sensitivity than the A/J mice (i.e., a wild-type response was seen). These results confirmed a role for *Atp1b3* in nociception in the mouse spinal cord.

Sato and colleagues used symmetric 27/27-nt Dicer-substrate siRNAs (with a 2-nt 3′-overhang on both ends) to treat hepatic cirrhosis in rats using a targeted liposomal delivery system [81]. It is thought that collagen production by the hepatic stellate cells is crucial to development of hepatic cirrhosis following acute or chronic injury. Targeted liposomes were made using the Lipotrust lipid reagent system conjugated to vitamin A as a means to facilitate delivery to the stellate cells. Symmetric Dicer-substrate siRNAs targeting the collagen chaperone heat shock protein 47 gene (*Serpinh1*, gp46) were administered IV at doses up to 0.75 mg/kg three times weekly. This treatment, in contrast to control reagents, reduced collagen production by the stellate cells and prevented fibrosis in several different models of acute cirrhosis, including bile duct ligation and chemical injury with dimethylnitrosamine or carbon tetrachloride.

Kortylewski and colleagues described use of a novel method to improve delivery of DsiRNAs to cells expressing TLR9 known to bind "CpG-motif" DNA (an unmethylated cytosine–guanine dinucleotide) [82]. TLR9 is expressed by some immune cells, such as plasmacytoid dendritic cells and B cells. Mice also express TLR9 on myeloid dendritic cells and monocytes. Like the RNA-binding receptors TLR3, 7, and 8, TLR9 primarily resides in the endosomal compartment. A DsiRNA targeting Stat3 was conjugated to a phosphorothioate-modified single-stranded DNA sequence termed CpG 1668, a known potent class B TLR9 agonist [87]. Stat3 expression is increased in some tumors where it promotes cell division and tumor growth. It also reduces local immune responses to the tumor, thereby further aiding tumor survival. Thus, suppressing Stat3 could be beneficial and reduce tumor cell survival by several mechanisms. The anti-*Stat3*+CpG DsiRNAs were taken up by cells into the endosomal compartment via a TLR9-independent mechanism. Cells lacking TLR9 did not show any gene suppression effects. For cells expressing TLR9, some fraction of the material escaped into the cytoplasm and led to suppression of *Stat3*. This reagent was used to treat mice bearing subcutaneous implants of syngeneic B16 melanoma cells. Direct peritumoral local injections of the anti-*Stat3*+CpG DsiRNAs were performed in the tumor nodules at a dose of 0.78 nmol per injection (20 μg, or ~1 mg/kg). After three daily injections, significant regression of the B16 tumor implants was observed. Next, B16 tumor cells were injected IV, and the growth of pulmonary nodules was measured over time (an established method to mimic the behavior of metastatic melanoma). The anti-*Stat3*+CpG DsiRNAs were administered IV at a dose of 0.78 nmol per injection (20 μg, or ~1 mg/kg). The intravenous injections were given every other day for 2 weeks and led to a significant reduction of both the size and number of B16 implants detectable in the lungs of mice receiving the anti-*Stat3* therapy but not in the control mice. Another example for ligand-mediated delivery, Nishina and colleagues used direct IV injection of a highly modified anti-*Apob* DsiRNA conjugated to vitamin E (α-tocopherol) to reduce expression of apolipoprotein B in liver and reduce serum cholesterol levels [63]. Refer to discussion in Sect. 2.2.2.1 for additional details.

Aptamers are highly structured nucleic acid molecules which form conformations that can bind a target ligand with high affinity and specificity, much like antibodies. Aptamers can be conjugated to other compounds to facilitate targeted delivery. For more information, see recent review articles by Thiel [88], Zhou [89], and Syed [90]. Aptamers have been employed to facilitate delivery of DsiRNAs both in vitro and in vivo.

Zhou and colleagues developed an aptamer specific for the HIV-1 gp120 protein expressed on the surface of HIV-infected cells [91, 92]. This anti-gp120 aptamer was fused to a DsiRNA specific for the HIV *tat/rev* gene. The chimeric aptamer–DsiRNA molecule was synthesized by in vitro transcription (IVT) using 2′-F pyrimidine bases to improve nuclease stability, thereby permitting use of the fusion construct in serum without the necessity for additional protection. The aptamer component was shown to target HIV-infected cells and facilitate DsiRNA uptake. Following uptake, the aptamer and DsiRNA are separated by Dicer processing to liberate an active 21-nt siRNA that directs suppression of the HIV *tat/rev* gene. Neff and colleagues studied use of the gp120-aptamer anti-*tat/rev* DsiRNA fusion molecule to suppress HIV infection in a humanized mouse immune system [83]. The humanized Rag2$^{-/-}\gamma c^{-/-}$ (RAG-hu) mouse is a rodent model system that allows the study of sustained chronic HIV infection. Over time, infected mice show declining

$CD4^+$ T cell numbers similar to what is seen in human HIV infection. The gp120-aptamer anti-*tat/rev* DsiRNA fusion molecule was administered intravenously to the RAG-hu mice at a dose of 0.25 nmol (0.38 mg/kg) given as daily injections for 2 days followed by weekly injections for 4 weeks. Mice were monitored for HIV titers and $CD4^+$ T cell counts during therapy and for 3–9 weeks after treatment ended. The anti-gp120 aptamer alone lowered viral titers; however, the DsiRNA fusion construct showed the highest efficacy, reducing HIV titers by several logs and preventing the decline of $CD4^+$ T cells. The beneficial effects persisted for several weeks following the last dose. Zhou and colleagues employed the same anti-*tat/rev* DsiRNA in RAG-hu mice with a poly(amidoamine) (PAMAM) dendrimer-based nanoparticle to facilitate delivery [84]. DsiRNAs specific for HIV host factors *CD4* and *TNPO3* were also employed. The DsiRNA/dendrimer particles were administered IV at a dose of 0.25 nmol (DsiRNA) given as daily injections for 2 days followed by weekly injections for 4 weeks. Similar to earlier results using aptamer-based delivery, IV administration of the anti-*tat/rev* DsiRNA in HIV-infected humanized mice resulted in a significant reduction of viral titers. The most promising results were obtained using a cocktail of all three DsiRNAs: an anti-*tat/rev* plus anti-*CD4* plus anti-*TNPO3* DsiRNA mix. The PAMAM dendrimer nanoparticles were shown to accumulate in PBMCs and the liver without evidence for toxicity.

2.4 Future Perspectives

Development of nucleic acid-based drugs to treat human disease is an area receiving considerable attention today. RNAi-based therapeutics are currently under development by a number of biotechnology and pharmaceutical companies. Recent reviews by Davidson [93] and Burnett [94] provide a comprehensive overview of human clinical trials that have been completed or are in progress as of mid-2011. Dicerna Pharmaceuticals (Watertown, MA, USA) is specifically focusing on use of DsiRNA as a platform technology for drug development. Working in partnership with the Japanese pharmaceutical company Kyowa Hakko Kirin, oncology and immunological/inflammatory diseases are current areas of focus. DsiRNAs have not yet been used in humans; however, DsiRNA drug development is proceeding forward, and it is hoped that some of the lead compounds will enter clinical trials in the not too distant future.

The DsiRNAs that move forward as therapeutic candidates will likely be more highly modified than the compounds described in this chapter. In particular, more extensive use of nuclease stabilizing groups such as 2′-OMe RNA will be employed with modification of both the sense and antisense strands while still leaving an unmodified Dicer cleavage site available (Fig. 2.5). Increasing the degree of modification will improve half-life of the compounds and reduce risk of immune stimulation. Given the known effects that sequence context has on which precise modification patterns are tolerated, it is likely that individual modification patterns will be optimized for each clinical candidate rather than using a simple "one size fits all" universal modification approach.

It is possible to suppress the expression of any desired gene using existing RNAi methods in vitro. The single greatest impediment to widespread adoption of in vivo use of these reagents for both research and medical applications is the availability of effective delivery tools capable of carrying a highly charged anionic RNA duplex across the cell membrane and deliver it intact in an active form to the cytoplasm with low toxicity. A wide variety of delivery tools are under development which may enable use of dsRNAs as drugs. These methods and chemical compositions have been the subject of numerous excellent reviews in recent years, and the reader is referred to these sources for additional details [88, 95–105]. There is no reason to believe that 21-nt siRNAs and 27-nt DsiRNAs will show any difference in their relative efficiency of delivery using cationic lipid or polyplex nanoparticles, so the same tools should be readily applied across both platforms. The DsiRNA platform, however, may offer some advantages when used with some other delivery technologies. For example, it is possible to covalently conjugate carrier molecules to DsiRNAs so that the modifier is removed by Dicer processing and does not remain attached to the final mature 21-nt siRNA that actually enters RISC (e.g., at the 3′-end of the sense strand, see Fig. 2.5). This may offer an advantage for using DsiRNAs when employing delivery tools such as cell-penetrating peptides, aptamers, or other high molecular weight ligands which might interfere with RISC entry. Indeed, pilot in vivo studies discussed above in Sect. 2.3 demonstrated the successful use of DsiRNAs covalently conjugated with both large aptamer and CpG-motif oligonucleotides to facilitate delivery.

Synthesis of siRNA and DsiRNA duplexes is available from a variety of commercial sources from very small scale to multi-gram scales to support all research and preclinical needs. A smaller number of suppliers are certified to produce the cGMP quality synthetic oligonucleotides needed for pharmaceutical use in humans. Currently, most cGMP manufacturers employ commercially available synthesis platforms, such as the GE Healthcare OligoPilot™ and OligoProcess™ synthesizers, or the Asahi Kasei TechniKrom® platform, which are capable of doing syntheses in millimole to mole scales, resulting in greater than kilogram yields of final product. Methods to produce cGMP quality synthetic oligonucleotides are well established, largely thanks to the many years of experience gained from clinical trials done using single-stranded antisense oligonucleotides [106, 107].

Acknowledgments The authors thank Kim Lennox for critical reading of the manuscript and Dr. Joe Dobosy for assistance with the figures. We further thank Dr. John Rossi, Dr. Dongho Kim, and all members of the research laboratories at Integrated DNA Technologies for their contributions towards the development of the Dicer-substrate siRNA technology.

Conflict of Interest Statement MAB and SDR are employed by Integrated DNA Technologies (IDT), which sells compounds similar to those described herein. IDT is, however, not a publicly traded company, and the authors do not hold any stock or equity in IDT. MAB is a scientific cofounder of Dicerna Pharmaceuticals and is a member of their Scientific Advisory Board.

References

1. Mello CC, Conte D Jr (2004) Revealing the world of RNA interference. Nature 431(7006):338–342
2. Siomi H, Siomi MC (2009) On the road to reading the RNA-interference code. Nature 457(7228):396–404
3. Carthew RW, Sontheimer EJ (2009) Origins and mechanisms of miRNAs and siRNAs. Cell 136(4):642–655
4. Ambros V (2004) The functions of animal microRNAs. Nature 431(7006):350–355
5. Meister G, Tuschl T (2004) Mechanisms of gene silencing by double-stranded RNA. Nature 431(7006):343–349
6. Vickers TA, Lima WF, Wu H, Nichols JG, Linsley PS, Crooke ST (2009) Off-target and a portion of target-specific siRNA mediated mRNA degradation is Ago2 'Slicer' independent and can be mediated by Ago1. Nucleic Acids Res 37(20):6927–6941
7. Bernstein E, Caudy AA, Hammond SM, Hannon GJ (2001) Role for a bidentate ribonuclease in the initiation step of RNA interference. Nature 409(6818):363–366
8. Zhang H, Kolb FA, Brondani V, Billy E, Filipowicz W (2002) Human Dicer preferentially cleaves dsRNAs at their termini without a requirement for ATP. EMBO J 21(21):5875–5885
9. Okamura K, Lai EC (2008) Endogenous small interfering RNAs in animals. Nat Rev Mol Cell Biol 9(9):673–678
10. Liu Q, Rand TA, Kalidas S, Du F, Kim HE, Smith DP et al (2003) R2D2, a bridge between the initiation and effector steps of the Drosophila RNAi pathway. Science 301(5641): 1921–1925
11. Haase AD, Jaskiewicz L, Zhang H, Laine S, Sack R, Gatignol A et al (2005) TRBP, a regulator of cellular PKR and HIV-1 virus expression, interacts with Dicer and functions in RNA silencing. EMBO Rep 6(10):961–967
12. Chendrimada TP, Gregory RI, Kumaraswamy E, Norman J, Cooch N, Nishikura K et al (2005) TRBP recruits the Dicer complex to Ago2 for microRNA processing and gene silencing. Nature 436(7051):740–744
13. MacRae IJ, Zhou K, Li F, Repic A, Brooks AN, Cande WZ et al (2006) Structural basis for double-stranded RNA processing by Dicer. Science 311(5758):195–198
14. MacRae IJ, Zhou K, Doudna JA (2007) Structural determinants of RNA recognition and cleavage by Dicer. Nat Struct Mol Biol 14(10):934–940
15. Maniataki E, Mourelatos Z (2005) A human, ATP-independent, RISC assembly machine fueled by pre-miRNA. Genes Dev 19(24):2979–2990
16. Sontheimer EJ (2005) Assembly and function of RNA silencing complexes. Nat Rev Mol Cell Biol 6(2):127–138
17. Matranga C, Tomari Y, Shin C, Bartel DP, Zamore PD (2005) Passenger-strand cleavage facilitates assembly of siRNA into Ago2-containing RNAi enzyme complexes. Cell 123(4):607–620
18. Rand TA, Petersen S, Du F, Wang X (2005) Argonaute2 cleaves the anti-guide strand of siRNA during RISC activation. Cell 123(4):621–629
19. Gregory RI, Chendrimada TP, Cooch N, Shiekhattar R (2005) Human RISC couples MicroRNA biogenesis and posttranscriptional gene silencing. Cell 123(4):631–640
20. Lingel A, Simon B, Izaurralde E, Sattler M (2004) Nucleic acid 3'-end recognition by the Argonaute2 PAZ domain. Nat Struct Mol Biol 11(6):576–577
21. Okamura K, Ishizuka A, Siomi H, Siomi MC (2004) Distinct roles for Argonaute proteins in small RNA-directed RNA cleavage pathways. Genes Dev 18(14):1655–1666
22. Peters L, Meister G (2007) Argonaute proteins: mediators of RNA silencing. Mol Cell 26(5):611–623
23. Hutvagner G, Simard MJ (2008) Argonaute proteins: key players in RNA silencing. Nat Rev Mol Cell Biol 9(1):22–32

24. Wang B, Li S, Qi HH, Chowdhury D, Shi Y, Novina CD (2009) Distinct passenger strand and mRNA cleavage activities of human Argonaute proteins. Nat Struct Mol Biol 16(12): 1259–1266
25. Meister G, Landthaler M, Patkaniowska A, Dorsett Y, Teng G, Tuschl T (2004) Human Argonaute2 mediates RNA cleavage targeted by miRNAs and siRNAs. Mol Cell 15(2):185–197
26. Ameres SL, Martinez J, Schroeder R (2007) Molecular basis for target RNA recognition and cleavage by human RISC. Cell 130(1):101–112
27. Wang HW, Noland C, Siridechadilok B, Taylor DW, Ma E, Felderer K et al (2009) Structural insights into RNA processing by the human RISC-loading complex. Nat Struct Mol Biol 16(11):1148–1153
28. Elbashir SM, Harborth J, Lendeckel W, Yalcin A, Weber K, Tuschl T (2001) Duplexes of 21-nucleotide RNAs mediate RNA interference in cultured mammalian cells. Nature 411(6836):494–498
29. Chang CI, Kim HA, Dua P, Kim S, Li CJ, Lee DK (2011) Structural diversity repertoire of gene silencing small interfering RNAs. Nucleic Acid Ther 21(3):125–131
30. Kim DH, Behlke MA, Rose SD, Chang MS, Choi S, Rossi JJ (2005) Synthetic dsRNA Dicer substrates enhance RNAi potency and efficacy. Nat Biotechnol 23(2):222–226
31. Rose SD, Kim DH, Amarzguioui M, Heidel JD, Collingwood MA, Davis ME et al (2005) Functional polarity is introduced by Dicer processing of short substrate RNAs. Nucleic Acids Res 33(13):4140–4156
32. Sano M, Sierant M, Miyagishi M, Nakanishi M, Takagi Y, Sutou S (2008) Effect of asymmetric terminal structures of short RNA duplexes on the RNA interference activity and strand selection. Nucleic Acids Res 36(18):5812–5821
33. Aza-Blanc P, Cooper CL, Wagner K, Batalov S, Deveraux QL, Cooke MP (2003) Identification of modulators of TRAIL-induced apoptosis via RNAi-based phenotypic screening. Mol Cell 12(3):627–637
34. Khvorova A, Reynolds A, Jayasena SD (2003) Functional siRNAs and miRNAs exhibit strand bias. Cell 115(2):209–216
35. Schwarz DS, Hutvagner G, Du T, Xu Z, Aronin N, Zamore PD (2003) Asymmetry in the assembly of the RNAi enzyme complex. Cell 115(2):199–208
36. Noland CL, Ma E, Doudna JA (2011) siRNA repositioning for guide strand selection by human Dicer complexes. Mol Cell 43(1):110–121
37. Hefner E, Clark K, Whitman C, Behlke MA, Rose SD, Peek AS et al (2008) Increased potency and longevity of gene silencing using validated Dicer substrates. J Biomol Tech 19(4):231–237
38. Eder PS, DeVine RJ, Dagle JM, Walder JA (1991) Substrate specificity and kinetics of degradation of antisense oligonucleotides by a 3' exonuclease in plasma. Antisense Res Dev 1(2):141–151
39. Kurreck J (2003) Antisense technologies Improvement through novel chemical modifications. Eur J Biochem 270(8):1628–1644
40. Behlke MA (2008) Chemical modification of siRNAs for in vivo use. Oligonucleotides 18(4):305–320
41. Gaglione M, Messere A (2010) Recent progress in chemically modified siRNAs. Mini Rev Med Chem 10(7):578–595
42. Krieg AM, Stein CA (1995) Phosphorothioate oligodeoxynucleotides: antisense or anti-protein? Antisense Res Dev 5(4):241
43. Braasch DA, Jensen S, Liu Y, Kaur K, Arar K, White MA et al (2003) RNA interference in mammalian cells by chemically-modified RNA. Biochemistry 42(26):7967–7975
44. Amarzguioui M, Holen T, Babaie E, Prydz H (2003) Tolerance for mutations and chemical modifications in a siRNA. Nucleic Acids Res 31(2):589–595
45. Chiu YL, Rana TM (2003) siRNA function in RNAi: a chemical modification analysis. RNA 9(9):1034–1048

46. Czauderna F, Fechtner M, Dames S, Aygun H, Klippel A, Pronk GJ et al (2003) Structural variations and stabilising modifications of synthetic siRNAs in mammalian cells. Nucleic Acids Res 31(11):2705–2716
47. Harborth J, Elbashir SM, Vandenburgh K, Manninga H, Scaringe SA, Weber K et al (2003) Sequence, chemical, and structural variation of small interfering RNAs and short hairpin RNAs and the effect on mammalian gene silencing. Antisense Nucleic Acid Drug Dev 13(2):83–105
48. Choung S, Kim YJ, Kim S, Park HO, Choi YC (2006) Chemical modification of siRNAs to improve serum stability without loss of efficacy. Biochem Biophys Res Commun 342(3):919–927
49. Allerson CR, Sioufi N, Jarres R, Prakash TP, Naik N, Berdeja A et al (2005) Fully 2'-modified oligonucleotide duplexes with improved in vitro potency and stability compared to unmodified small interfering RNA. J Med Chem 48(4):901–904
50. Prakash TP, Allerson CR, Dande P, Vickers TA, Sioufi N, Jarres R et al (2005) Positional effect of chemical modifications on short interference RNA activity in mammalian cells. J Med Chem 48(13):4247–4253
51. Kraynack BA, Baker BF (2006) Small interfering RNAs containing full 2'-O-methylribonucleotide-modified sense strands display Argonaute2/eIF2C2-dependent activity. RNA 12(1):163–176
52. Layzer JM, McCaffrey AP, Tanner AK, Huang Z, Kay MA, Sullenger BA (2004) In vivo activity of nuclease-resistant siRNAs. RNA 10(5):766–771
53. Morrissey DV, Blanchard K, Shaw L, Jensen K, Lockridge JA, Dickinson B et al (2005) Activity of stabilized short interfering RNA in a mouse model of hepatitis B virus replication. Hepatology 41(6):1349–1356
54. Morrissey DV, Lockridge JA, Shaw L, Blanchard K, Jensen K, Breen W et al (2005) Potent and persistent in vivo anti-HBV activity of chemically modified siRNAs. Nat Biotechnol 23(8):1002–1007
55. Swayze EE, Siwkowski AM, Wancewicz EV, Migawa MT, Wyrzykiewicz TK, Hung G et al (2007) Antisense oligonucleotides containing locked nucleic acid improve potency but cause significant hepatotoxicity in animals. Nucleic Acids Res 35(2):687–700
56. Elmen J, Thonberg H, Ljungberg K, Frieden M, Westergaard M, Xu Y et al (2005) Locked nucleic acid (LNA) mediated improvements in siRNA stability and functionality. Nucleic Acids Res 33(1):439–447
57. Elmen J, Lindow M, Silahtaroglu A, Bak M, Christensen M, Lind-Thomsen A et al (2008) Antagonism of microRNA-122 in mice by systemically administered LNA-antimiR leads to up-regulation of a large set of predicted target mRNAs in the liver. Nucleic Acids Res 36(4):1153–1162
58. Elmen J, Lindow M, Schutz S, Lawrence M, Petri A, Obad S et al (2008) LNA-mediated microRNA silencing in non-human primates. Nature 452(7189):896–899
59. Kubo T, Zhelev Z, Ohba H, Bakalova R (2007) Modified 27-nt dsRNAs with dramatically enhanced stability in serum and long-term RNAi activity. Oligonucleotides 17(4): 445–464
60. Kubo T, Zhelev Z, Ohba H, Bakalova R (2008) Chemically modified symmetric and asymmetric duplex RNAs: an enhanced stability to nuclease degradation and gene silencing effect. Biochem Biophys Res Commun 365(1):54–61
61. Turner JJ, Jones SW, Moschos SA, Lindsay MA, Gait MJ (2007) MALDI-TOF mass spectral analysis of siRNA degradation in serum confirms an RNAse A-like activity. Mol Biosyst 3(1):43–50
62. Collingwood MA, Rose SD, Huang L, Hillier C, Amarzguioui M, Wiiger MT et al (2008) Chemical modification patterns compatible with high potency dicer-substrate small interfering RNAs. Oligonucleotides 18(2):187–200
63. Nishina K, Unno T, Uno Y, Kubodera T, Kanouchi T, Mizusawa H et al (2008) Efficient in vivo delivery of siRNA to the liver by conjugation of alpha-tocopherol. Mol Ther 16(4):734–740

64. Rigotti A (2007) Absorption, transport, and tissue delivery of vitamin E. Mol Aspects Med 28(5–6):423–436
65. Kubo T, Takei Y, Mihara K, Yanagihara K, Seyama T (2012) Amino-modified and lipid-conjugated dicer-substrate siRNA enhances RNAi efficacy. Bioconjug Chem 23(2):164–173
66. Kariko K, Buckstein M, Ni H, Weissman D (2005) Suppression of RNA recognition by Toll-like receptors: the impact of nucleoside modification and the evolutionary origin of RNA. Immunity 23(2):165–175
67. Robbins M, Judge A, Liang L, McClintock K, Yaworski E, Maclachlan I (2007) 2'-O-methyl-modified RNAs Act as TLR7 Antagonists. Mol Ther 15(9):1663–1669
68. Gantier MP, Williams BR (2007) The response of mammalian cells to double-stranded RNA. Cytokine Growth Factor Rev 18(5–6):363–371
69. Robbins M, Judge A, MacLachlan I (2009) siRNA and innate immunity. Oligonucleotides 19(2):89–102
70. Reynolds A, Anderson EM, Vermeulen A, Fedorov Y, Robinson K, Leake D et al (2006) Induction of the interferon response by siRNA is cell type- and duplex length-dependent. RNA 12(6):988–993
71. Fedorov Y, King A, Anderson E, Karpilow J, Ilsley D, Marshall W et al (2005) Different delivery methods-different expression profiles. Nat Methods 2(4):241
72. Marques JT, Devosse T, Wang D, Zamanian-Daryoush M, Serbinowski P, Hartmann R et al (2006) A structural basis for discriminating between self and nonself double-stranded RNAs in mammalian cells. Nat Biotechnol 24(5):559–565
73. Behlke MA (2006) Progress towards in vivo use of siRNAs. Mol Ther 13(4):644–670
74. Rettig GR, Behlke MA (2012) Progress towards in vivo use of siRNAs-II. Mol Ther 20:483–512
75. Lundberg P, Welander PV, Edwards CK 3rd, van Rooijen N, Cantin E (2007) Tumor necrosis factor (TNF) protects resistant C57BL/6 mice against herpes simplex virus-induced encephalitis independently of signaling via TNF receptor 1 or 2. J Virol 81(3):1451–1460
76. Lundberg P, Yang H-J, Jung S-J, Behlke MA, Rose SD, Cantin EM (2012) Protection against TNFα-dependent liver toxicity by intraperitoneal liposome delivered DsiRNA targeting TNFα in vivo. J Control Release 160:194–199
77. Howard KA, Paludan SR, Behlke MA, Besenbacher F, Deleuran B, Kjems J (2008) Chitosan/siRNA nanoparticle-mediated TNF-alpha knockdown in peritoneal macrophages for anti-inflammatory treatment in a murine arthritis model. Mol Ther 17(1):162–168
78. Nawroth I, Alsner J, Behlke MA, Besenbacher F, Overgaard J, Howard KA et al (2010) Intraperitoneal administration of chitosan/DsiRNA nanoparticles targeting TNFalpha prevents radiation-induced fibrosis. Radiother Oncol 97(1):143–148
79. Dore-Savard L, Roussy G, Dansereau MA, Collingwood MA, Lennox KA, Rose SD et al (2008) Central delivery of Dicer-substrate siRNA: a direct application for pain research. Mol Ther 16(7):1331–1339
80. LaCroix-Fralish ML, Mo G, Smith SB, Sotocinal SG, Ritchie J, Austin JS et al (2009) The beta3 subunit of the Na+, K+-ATPase mediates variable nociceptive sensitivity in the formalin test. Pain 144(3):294–302
81. Sato Y, Murase K, Kato J, Kobune M, Sato T, Kawano Y et al (2008) Resolution of liver cirrhosis using vitamin A-coupled liposomes to deliver siRNA against a collagen-specific chaperone. Nat Biotechnol 26(4):431–442
82. Kortylewski M, Swiderski P, Herrmann A, Wang L, Kowolik C, Kujawski M et al (2009) In vivo delivery of siRNA to immune cells by conjugation to a TLR9 agonist enhances antitumor immune responses. Nat Biotechnol 27(10):925–932
83. Neff CP, Zhou J, Remling L, Kuruvilla J, Zhang J, Li H et al (2011) An aptamer-siRNA chimera suppresses HIV-1 viral loads and protects from helper CD4(+) T cell decline in humanized mice. Sci Transl Med 3(66):66ra6
84. Zhou J, Neff CP, Liu X, Zhang J, Li H, Smith DD et al (2011) Systemic administration of combinatorial dsiRNAs via nanoparticles efficiently suppresses HIV-1 infection in humanized mice. Mol Ther 19(12):2228–2238

85. Amarzguioui M, Lundberg P, Cantin E, Hagstrom JE, Behlke MA, Rossi JJ (2006) Rational design and in vitro and in vivo delivery of Dicer substrate siRNA. Nat Protoc 1(2):508–517
86. Howard KA, Rahbek UL, Liu X, Damgaard CK, Glud SZ, Andersen MO et al (2006) RNA interference in vitro and in vivo using a novel chitosan/siRNA nanoparticle system. Mol Ther 14(4):476–484
87. Krieg AM, Yi AK, Matson S, Waldschmidt TJ, Bishop GA, Teasdale R et al (1995) CpG motifs in bacterial DNA trigger direct B-cell activation. Nature 374(6522):546–549
88. Thiel KW, Giangrande PH (2009) Therapeutic applications of DNA and RNA aptamers. Oligonucleotides 19(3):209–222
89. Zhou J, Rossi JJ (2010) Aptamer-targeted cell-specific RNA interference. Silence 1(1):4
90. Syed MA, Pervaiz S (2010) Advances in aptamers. Oligonucleotides 20(5):215–224
91. Zhou J, Li H, Li S, Zaia J, Rossi JJ (2008) Novel dual inhibitory function aptamer-siRNA delivery system for HIV-1 therapy. Mol Ther 16(8):1481–1489
92. Zhou J, Swiderski P, Li H, Zhang J, Neff CP, Akkina R et al (2009) Selection, characterization and application of new RNA HIV gp 120 aptamers for facile delivery of Dicer substrate siRNAs into HIV infected cells. Nucleic Acids Res 37(9):3094–3109
93. Davidson BL, McCray PB Jr (2011) Current prospects for RNA interference-based therapies. Nat Rev Genet 12(5):329–340
94. Burnett JC, Rossi JJ, Tiemann K (2011) Current progress of siRNA/shRNA therapeutics in clinical trials. Biotechnol J 6(9):1130–1146
95. De Paula D, Bentley MV, Mahato RI (2007) Hydrophobization and bioconjugation for enhanced siRNA delivery and targeting. RNA 13(4):431–456
96. de Fougerolles AR (2008) Delivery vehicles for small interfering RNA in vivo. Hum Gene Ther 19(2):125–132
97. Howard KA, Kjems J (2007) Polycation-based nanoparticle delivery for improved RNA interference therapeutics. Expert Opin Biol Ther 7(12):1811–1822
98. Howard KA (2009) Delivery of RNA interference therapeutics using polycation-based nanoparticles. Adv Drug Deliv Rev 61(9):710–720
99. Tseng YC, Mozumdar S, Huang L (2009) Lipid-based systemic delivery of siRNA. Adv Drug Deliv Rev 61(9):721–731
100. Davis ME (2009) The first targeted delivery of siRNA in humans via a self-assembling, cyclodextrin polymer-based nanoparticle: from concept to clinic. Mol Pharm 6(3):659–668
101. Eguchi A, Dowdy SF (2009) siRNA delivery using peptide transduction domains. Trends Pharmacol Sci 30(7):341–345
102. Jarver P, Mager I, Langel U (2010) In vivo biodistribution and efficacy of peptide mediated delivery. Trends Pharmacol Sci 31(11):528–535
103. Giljohann DA, Seferos DS, Daniel WL, Massich MD, Patel PC, Mirkin CA (2010) Gold nanoparticles for biology and medicine. Angew Chem Int Ed Engl 49(19):3280–3294
104. Serda RE, Godin B, Blanco E, Chiappini C, Ferrari M (2011) Multi-stage delivery nanoparticle systems for therapeutic applications. Biochim Biophys Acta 1810(3):317–329, Epub 2010/05/25
105. Peer D, Lieberman J (2011) Special delivery: targeted therapy with small RNAs. Gene Ther 18(12):1127–1133
106. Sanghvi YS, Schulte M (2004) Therapeutic oligonucleotides: the state-of-the-art in purification technologies. Curr Opin Drug Discov Devel 7(6):765–776
107. Tedebark U, Scozzari A, Werbitzky O, Capaldi D, Holmberg L (2011) Industrial-scale manufacturing of a possible oligonucleotide cargo CPP-based drug. Methods Mol Biol 683:505–524, Epub 2010/11/06

Chapter 3
The Role of MicroRNAs in Natural Tissue Development and Application in Regenerative Medicine

Morten Østergaard Andersen, Philipp Dillschneider, and Jørgen Kjems

Abstract Many cellular functions rely on the coordinated expression and repression of a large number of messenger RNAs; these are tightly controlled in part by microRNAs (miRNAs) at the posttranscriptional level. The number of characterised miRNAs that are involved in tissue development and repair is steadily increasing, and our understanding of their functions is starting to merge. Modulating miRNA levels through externally applied stimuli enables us to control the translation of numerous mRNAs giving us unprecedented control over cellular events; therefore, we predict that such techniques will revolutionise regenerative medicine. This chapter will introduce miRNA biology and their role in controlling pluripotency, stem cell differentiation, proliferation, senescence, survival, inflammation and angiogenesis. There are several strategies by which miRNA-modulating technologies can be used to specifically target tissue engineering and repair, either in culture or in association with implanted cells and/or implants. We will here summarise these methods providing examples from present literature. Based on previous results, we will also predict more advanced technologies that may deliver miRNA in a spatial/temporally regulated manner that may imitate natural miRNA expression. Furthermore, exogenous miRNAs, carried between cells in secreted vesicles, have recently been characterised and may further increase the role and potential of miRNA in relation to regenerative medicine.

M.Ø. Andersen (✉) • P. Dillschneider • J. Kjems
Department of Molecular Biology and Genetics, Interdisciplinary Nanoscience Center, Aarhus University, C.F. Møllers Allé 3, Building 1130, 8000 Aarhus, Denmark
e-mail: moa@mb.au.dk

K.A. Howard (ed.), *RNA Interference from Biology to Therapeutics*, Advances in Delivery Science and Technology, DOI 10.1007/978-1-4614-4744-3_3,

3.1 Introduction

Regenerative medicine aims to guide and promote the repair of damaged tissues or, if needed, replace them with living tissues created through tissue engineering. MicroRNAs (miRNAs) are a continually expanding class of endogenously expressed non-coding RNAs that, through a phenomenon known as RNA interference (RNAi), regulate a growing number of cellular functions, including many relevant to regenerative medicine. Since exogenously introduced, miRNAs or anti-miRNAs (here commonly referred to as miRNA modulators) can be used to control these events; their role in regenerative medicine continues to rise [1]. MicroRNA modulators can be used to promote tissue regeneration and engineering in a number of ways. If injected systemically or locally, miRNA modulators can affect the cells present at the diseased target site, alleviating the condition either by promoting regeneration by stimulating the tissue-forming cells [2] or by inhibiting further tissue destruction by modulating the immune system [3]. Alternatively, injections of miRNA modulators together with cells can protect and guide the implanted cells to regenerate the damaged tissue [4]. Finally, one can combine cells and miRNA modulators with implant materials to try and completely create the desired tissue de novo [5], known as tissue engineering. The number of studies combining miRNA modulators with cells and implantable materials is still limited; however, information detailing possible applications and methods for future use can be gained from the prior use of closely related drugs, especially small interfering RNA (siRNA) [6]. An overview of these strategies can be seen in Fig. 3.1.

Tissue engineering is showing recent promise. Living blood vessels [7], heart valves [8], bladders [9], tracheas [10] and urethras [11] are just some examples of tissues that have been grown from patient cells and which have subsequently been implanted successfully, curing diseases. Stem cells are often used when the aim is to generate new tissue [12]. Stem cells are undifferentiated cells with the potential to generate specialised tissue cells while retaining a pool of undifferentiated cells through asymmetric cell division. Stem cells can be harvested from embryos (embryonic stem cells) in which case they are pluripotent and can be differentiated into all cell types found in the body. Similar pluripotent stem cells can also be created from somatic adult cells through a process of dedifferentiation, known as induced pluripotent stem cells (IPSC). Different types of stem cells also reside in the adult body, normally in a non-dividing quiescent state, wherefrom they can be activated to undergo asymmetric cell division when there is a need for new cells. Cells isolated from adults are multipotent, meaning that they can be differentiated into a limited number of cell types. Stem cells can be injected directly as stem cell therapy in which case they will home to sites of damage and promote repair either by specialising to replacement cells or by secreting beneficial factors. Stem cells are also commonly used for tissue engineering in which case they are usually seeded on a temporary three-dimensional scaffold; the cells are then induced to differentiate and create a specific tissue, for example, by modulating their miRNA levels. When the desired tissue has been generated, preferably replacing the biodegradable scaffold,

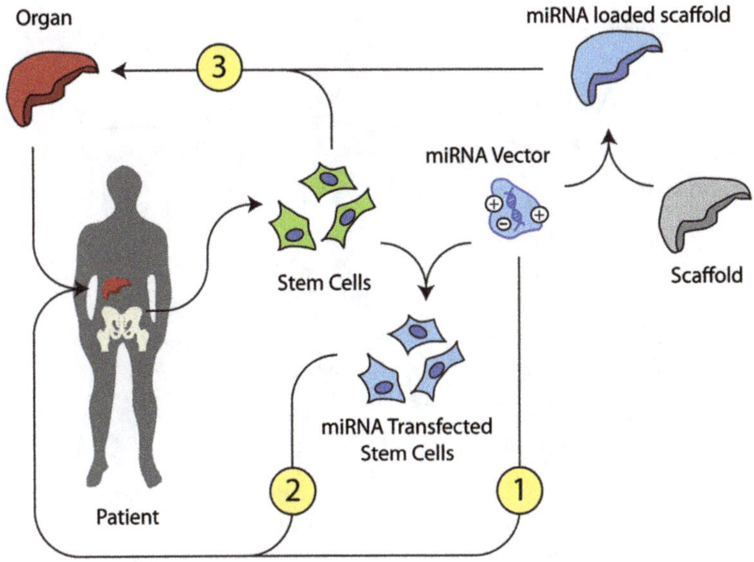

Fig. 3.1 Illustrates three routes for miRNA-mediated tissue regeneration. MicroRNA modulators can be injected directly where they will migrate and affect cells within the patient (route 1). Alternatively, cells can be explanted from a patient and be transfected with miRNA ex vivo. This can be done in standard cell culture (route 2) or on miRNA functionalised scaffolds (route 3); afterwards, the cells are then reinjected (route 2) or transplanted along with the scaffold (route 3)

it is implanted into the patient. Alternatively, the scaffolds loaded with cells, miRNA modulators and/or small molecular cues can be transplanted immediately and grown in vivo.

3.2 RNA Interference

RNA interference (RNAi) is a cellular process, which utilises small RNAs for gene expression regulation and innate defence [13]. RNAi is operational in most eukaryotic species, including fission yeast, plants, flies, nematodes, mammals and humans [14]. Figure 3.2 gives an overview of RNAi in mammals focusing on the miRNA pathway.

The miRNA pathway is initiated in the nucleus by polymerase II transcription of primary precursor miRNAs (pri-miRNAs) either as part of an ordinary protein-coding mRNA or as a unique non-coding RNA. These transcripts are processed by an RNase III enzyme named Drosha, resulting in a ~70-nt-long precursor hairpin (pre-miRNA), exported by Exportin 5 into the cytoplasm where it is cleaved by another RNase III enzyme, named Dicer. The result is a ~19–23 base pair long double-stranded RNA composed of the miRNA and a passenger strand [15], these RNAs can also be introduced into the cytoplasm as synthetic molecules in which case they

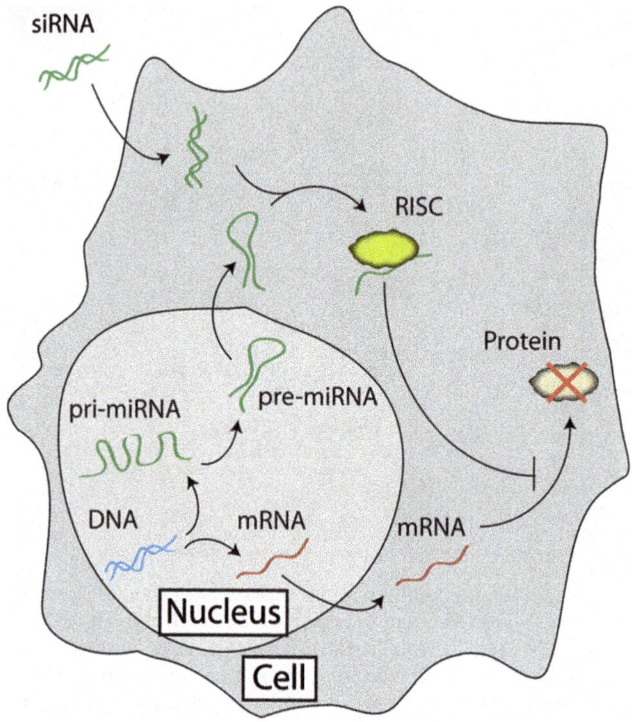

Fig. 3.2 Overview of RNAi cellular pathways with focus on miRNA processing. In the nucleus, pri-miRNA are transcribed and subsequently processed by Drosha into hairpin-structured pre-miRNA. These are exported to the cytoplasm where they are trimmed into double-stranded RNAs of ~19–21 base pairs by Dicer. One of the strands is incorporated into the RISC complex as the mature miRNA. The 5′ 2–8 nucleotides of the mature miRNA constitute the seed sequence which guides the RISC complex to mRNAs by complementary base pairing. The translation of these mRNAs is then abrogated, and the mRNAs are usually destabilised, leading to a decline in protein expression

are called small interfering RNAs (siRNA) or miRNA mimics. Proteins, including the argonaute protein, are recruited to form the precursor RNA-induced silencing complex (pre-RISC). The strand with less stable base pairing at the 5′-end, called the guide strand, remains in the RISC complex as the mature miRNA while the passenger strand is degraded.

This miRNA guides the RISC to the mRNA, usually to a site within the 3′ untranslated region. The specificity is mainly directed by a perfect match to the so-called seed sequence, which spans position 2–8 in the miRNA. This leads to translational repression and usually also increased mRNA degradation. The remaining part of the sequence can modulate the effect, and more extensive matches can lead to Ago-2-mediated cleavage of the target, which is often the case with artificially constructed siRNA [16].

Endogenous miRNA levels can be transiently upregulated by delivering exogenous pre-miRNAs [17] or pri-miRNAs [18]; alternatively, the function of a miRNA can be blocked by anti-miRNAs (also known as antagomirs) [19] or miRNA decoys (also known as sponges) [20]. Synthetic pre-miRNAs are usually hairpin structures that are processed like the endogenous pre-miRNAs in the cytoplasm, and their delivery leads to elevated levels of RISC-associated mature miRNA. Anti-miRNAs are typically modified oligonucleotides with a sequence that is complementary to the guide strand of a mature miRNA, which they bind tightly thereby blocking their subsequent RISC guidance. MicroRNA decoys are transcribed in the nucleus from an exogenously introduced DNA vector and are regular RNA transcripts that contain multiple targets sites for a particular miRNA species, effectively depleting the miRNA from its real target. Modulation of miRNA levels can be prolonged and even persistent if the miRNA or decoy-miRNA is expressed in the nucleus from an exogenously introduced vector [21]. Prolonged expression is typically driven by plasmids or non-integrating viruses such as adenoviral vectors, whereas persistent miRNA modulation can be accomplished using integrating retroviral vectors.

3.3 Applications of miRNA in Tissue Regeneration

Regenerative medicine is currently employing a large variety of drugs to promote tissue repair and stem cell differentiation. Examples include vitamins, hormones, synthetic low molecular weight drugs as well as larger biologically derived peptides, proteins, viruses, plasmids and various RNA species. If we take osteogenic (or bone) regeneration as an example, beta glycerol phosphate, the synthetic steroid dexamethasone, vitamins D and C, growth factors such as the bone morphogenic proteins (BMPs) in both native and plasmid-expressed forms as well as siRNAs have been applied to promote osteogenesis [22–25]. Such drugs are applied by direct injection or ex vivo to guide bone cells to produce more bone or to promote stem cell differentiation into bone. What additional benefits can miRNA bring to regenerative medicine, and are there advantages compared to traditional drugs?

MicroRNAs often exert natural developmental functions by modulating the levels of several mRNAs working in parallel, thereby causing more dramatic effects than changing the level of a single gene or protein. But perhaps the most unique feature of an RNAi-based medicine is the unparalleled flexibility it offers. Once a delivery system for miRNA modulators has been devised that works with the implant and cells of interest, that system can be used to deliver a modulator of any miRNA and, thereby, control numerous cellular functions such as cell survival or differentiation into different cell types. One could, for example, use different miRNAs with the same delivery system to stimulate the formation of different cell types from stem cells. Alternatively, one could promote cell survival with one miRNA and differentiation with another. Tissue development and regeneration is a highly complex process involving numerous miRNAs, and a complete understanding is still lacking. However, in the following sections, we will provide examples of various cellular processes that are regulated by specific miRNAs and exemplify how they can be used for regenerative purposes.

3.3.1 *The Role of MicroRNAs in Tissue Development*

It is now emerging that miRNAs function in essentially all stages of tissue development. MicroRNAs play key roles from the very onset of human development, the growth of undifferentiated embryonic stem cells into a foetus with differentiated germ layers as well as in adult tissue repair such as those found after a bone fracture.

3.3.1.1 MicroRNAs and Embryonic Development

The involvement of miRNAs in embryonic development is interesting as it illustrates their immense capacity for controlling the progression of cell differentiation; furthermore, the therapeutic use of embryonic stem cells in clinical trials has started [26]. At the earliest stages of life, the embryonic stem cells are kept undifferentiated and pluripotent by a self-regulating circuitry of key proteins including OCT4, SOX2 and KLF4 which are silenced by miR-145 upon differentiation [27]. When cells begin to differentiate into the earliest progenitor cells specific to the three germ layers, the endo-, meso- and ectoderm lineages, the cell identity is controlled, in part, by microRNAs. In mice, for example, mesodermal specialisation is blocked by microRNA members of the miR-290 cluster [28]. These miRNAs repress the expression of DKK1, an inhibitor of the WNT pathway, which plays a complex role in embryonic cell specialisation. During increased specialisation, waves of successive miRNAs are expressed. A study detailing miRNA expression at various developmental stages during embryonic stem cell to adipocyte differentiation found that the expression level of 129 individual miRNAs changed twofold or greater during differentiation [29]. At the mesodermal progenitor stage after just 5 days, the expression levels of 104 miRNAs differed from the outset. Cell specialisation continues throughout development and is activated in many tissues in the adult body when cell populations need replenishing. We shall now explore this further.

3.3.1.2 MicroRNAs and Adult Tissue Development

In the adult, stem cells are located in specialised niches in most organs. An example of adult stem cells is the neural stem cells that reside in the subventricular zone in the brain and which are capable of generating new neurons, glia and astrocytes [30]. Another example is the mesenchymal stem cells (MSCs) which are found in the stromal compartment of many tissues such as bone marrow (BMSCs) and adipose tissue (ADSCs). MSCs are, at least, capable of generating bone cells (osteoblasts), cartilage cells (chondrocytes), fat cells (adipocytes) and muscle cells (myoblasts) [31]. Adult stem cells are either activated upon tissue damage or are continually active if replacing constantly removed cells, such as in the case of hematopoietic stem cells. Activation leads to proliferation and differentiation of a subpopulation of the stem cells into new specialised cells that replace or replenish the needed cells.

The maintenance of the stem cells in their quiescent multipotent state as well as the coordinated exit from this state upon damage is controlled by microRNAs such as miR-335 [32]. The subsequent differentiation pathway taken is also controlled tightly by microRNAs. Two well-studied differentiation pathways are osteoblastic and adipocytic specialisations. Many miRNAs have been shown to be involved in these processes (summarised in Table 3.1).

Of the osteogenesis regulating miRNAs, miR-138 seems to regulate a pathway involving focal adhesion kinase and ERK 1/2 [5]. In undifferentiated MSCs, miR-138 suppresses the expression of the late osteogenic transcription factor OSX, whereas miR-138 is downregulated, leading to the expression of OSX in connection with osteogenesis. MicroRNA-20a was recently [33] found to be upregulated during osteogenesis, resulting in downregulation of adipocyte marker PPARγ and of the osteoblast antagonists BAMBI and CRIM1. These are negative regulators of BMP signalling. Hence, BMP and the early osteogenic transcription factor RUNX2 were upregulated in the presence of miR-20a compatible with the observation that this miRNA induces osteogenesis in human MSC. Co-repressing of PPARγ, BAMBI and CRIM leads to activation of the BMPs/RUNX2 signalling pathway. MicroRNA-148b, miR-27a and miR-489 may also be involved in osteogenesis [36]. MicroRNA-27a and miR-489 are downregulated during osteogenesis with grancalcin as a potential target, whereas miR-148b is upregulated.

Numerous miRNAs have also been reported as control factors for adipogenesis from human MSC. Both miR-371 and miR-369-5p reduce proliferation [52]. MiR-369-5p, however, blocks adipogenesis, whereas miR-371 increases differentiation. A putative target of miR-371 is FABP4. MicroRNA-221, miR-222 and miR-155 seem to act anti-adipogenic similar to miR-369-5p [51]. The levels of these three miRNAs decrease during differentiation, and overexpression negatively affects adipogenesis. A putative target for miR-221 and miR-222 is cyclin-dependent inhibitor CDKN1B and for miR-369-5p it is adiponectin [fatty-acid-binding protein (FABP4)]. For two other miRNAs, miR-155 and miR-371, a target has not been identified yet.

3.3.2 Vascularisation

Almost regardless of what tissue is to be generated, there is a need for implant vascularising due to oxygen diffusion limits of 100–200 μm in dense tissue [60]. While adjacent blood vessels can neo-vascularise biodegradable implants [61], the elapsed time before sufficient vascularisation has occurred may lead to necrotic zones at the cores of larger implants. Vascular tissue is composed of endothelial cells surrounded by smooth muscle cells (except in the case of capillaries). In tissue engineering the generation of these two cell types has traditionally been promoted using vascular endothelial growth factor (VEGF) [62] and transforming growth factor beta (TGF-β) [63], respectively. However, microRNAs can also be applied to promote smooth muscle and endothelial cell differentiation. Transfecting fibroblasts with miR-143 and miR-145, for example, enhance their differentiation into smooth muscle cells

Table 3.1 miRNAs involved in osteogenesis and adipogenesis

Effect	miRNA	Cell type	Species	Target	Reference
Pro-osteogenic	miR-20a	BMSC	Human	PPARγ BAMBI CRIM1	[33]
Pro-osteogenic	miR-20b	BMSC	Human	PPARγ BAMBI CRIM1	[34]
Anti-osteogenic (final stage)	miR-26a	ADSC	Human	SMAD1	[35]
Anti-osteogenic	mi-27a	BMSC	Human	GCA	[36]
Anti-osteogenic	miR-138	BMSC	Human	FAK	[5]
Pro-osteogenic	miR-146a	Periodontal ligament cells	Human	NF-κB	[37]
Pro-osteogenic	miR-148b	BMSC	Human	N/A	[36]
Pro-osteogenic	miR-196a	ADSC	Human	HOXC8	[38]
Anti-osteogenic	miR-489	BMSC	Human	GCA	[36]
Pro-osteogenic	miR-29b	Rat calvaria osteoblasts, MC3T3	Mouse rat	COL1A1 COL5A3 COL4A2 HDAC4 TGFβ3 ACVR2A CTNNBIP1 DUSP2	[39]
Anti-osteogenic	miR-125b	ST2	Mouse	ERBB2	[40]
Anti-osteogenic	miR-133	C2C12	Mouse	RUNX2	[41]
Anti-osteogenic	miR-135	C2C12	Mouse	SMAD5	[41]
Pro-osteogenic	miR-141	MC3T3-E1	Mouse	DLX5	[42]
Pro-osteogenic	miR-200a	MC3T3-E1	Mouse	DLX5	[42]
Anti-osteogenic	miR-206	C2C12, primary osteoblasts	Mouse	Connexin 43	[43]
Pro-osteogenic	miR-210	ST2	Mouse	ACVR1b	[44]
Pro-osteogenic	miR-2861	Primary osteoblasts, ST2	Mouse	HDAC5	[45]
Anti-adipogenic	miR-27b	ADSC	Human	PPARγ	[46]
Pro-adipogenic	miR-30	ADSC	Human	RUNX2	[47]
Anti-adipogenic	miR-130	Primary pre-adipocytes	Human	PPARγ	[48]
Anti-adipogenic	miR-138	ADSC	Human	EID-1	[49]
Pro-adipogenic	miR-143	Pre-adipocytes	Human	ERK5	[50]
Anti-adipogenic	miR-155	BMSC	Human	N/A	[51]
Anti-adipogenic	miR-221	BMSC	Human	CDKN1B	[51]
Anti-adipogenic	miR-222	BMSC	Human	CDKN1B	[51]
Pro-adipogenic	miR-642-3p	ADSC	Human	N/A	[47]
Anti-adipogenic	miR-369-5p	BMSC	Human	FABP4	[52]
Pro-adipogenic	miR-371	BMSC	Human	N/A	[52]
Pro-adipogenic	miR-17-92 cluster	3T3-L1	Mouse	RB2/P130	[53]
Anti-adipogenic	miR-31	C3H10T1/2	Mouse	CEBPA	[54]
Anti-adipogenic	miR-27a/b	3T3-L1, OP9	Mouse	PPARγ	[55]
Pro-adipogenic	miR-103	3T3-L1	Mouse	N/A	[56]
Pro-adipogenic	miR-200	ST2	Mouse	N/A	[57]
Pro-adipogenic	miR-204/211	ST2	Mouse	RUNX2	[58]
Pro-adipogenic	miR378/378*	ST2	Mouse	N/A	[59]

[64], while antagonists against miR-16 and miR-424 induce secretion of VEGF [65].

3.3.3 MicroRNAs and Induced Pluripotent Stem Cell Generation

Another potential use of microRNAs in regenerative medicine is to promote the generation of IPSC from somatic adult cells, a use that is related to the major role of miRNA in embryonic stem cells. The generation of IPSC cells is usually performed using gene therapy with a vector encoding the pluripotency genes OCT3/4, SOX2 and KLF4. However, a major obstacle is the P53 tumour suppressor protein, which partly through P21 counteracts IPSC formation [66]. These proteins act, in part, by increasing the levels of certain miRNAs, for example miR-34 [67], which in turn silence the expression of pluripotency genes such as NANOG, SOX2 and MYCN. Hence, the expression of these microRNAs represents a barrier for the generation of enough IPSCs for practical purposes. Such microRNAs would potentially be ideal targets for anti-miRNAs for enhancing IPSC formation. Other miRNAs, for example miR-93 and miR-106b [68], are induced during early dedifferentiation in which they silence P21 leading to further de-specialisation. Increasing the amount of these miRNAs in fibroblasts by introducing miRNA mimics can potentially be used to promote IPSC formation. The increasing possibilities for regulating key genes involved in IPSC formation and maintenance suggests that it may be possible to drive dedifferentiation with miRNA and anti-miRNA alone, thereby avoiding problematic viral vectors and the overexpression of oncogenes associated with the current methods [69].

3.3.4 Immune Modulation

The immune system is an important component in regenerative medicine; it is a primary target for strategies that seek to reduce localised inflammation and subsequent tissue destruction prior to regeneration. MicroRNA-155, for example, is upregulated in rheumatoid arthritis where it down regulates SHIP1, an inhibitor of inflammation [70]. In contrast, miRNA-146a alleviates rheumatoid arthritis by inhibiting the differentiation of monocytes and macrophages into multi-nucleated osteoclasts, which are responsible for joint destruction [3]. Incidentally miR-146a also seems to be involved in promoting de novo osteogenesis of cells [37], and it may, thus, play a dual role both reducing tissue inflammation and increasing regeneration. MicroRNA-mediated modulation of immune responses, therefore, presents a strategy to arrest further tissue degeneration before regeneration is induced.

3.3.5 Cell Survival

One of the major obstacles in any tissue engineering strategy employing transplanted cells is to ensure cell survival upon implantation. MicroRNAs play important roles in cell survival and modulating miRNA levels may, therefore, be used to promote cell survival. This is especially important with cells that display very limited proliferative capacity or post-transplantation survival such as cardiomyocytes. In one study, overexpression of a cocktail of miR-21, miR-24 and miR-221 from lentiviral vectors promoted survival of transduced cardiac progenitor cells both in vitro under serum-free condition and in vivo after intramuscular injection [4]. Injection of cardiac progenitor cells, overexpressing the miRNA cocktail, in ischemic heart myocardium led to improved function and reduced infarction size compared to injections of non-transfected cells. The microRNAs used acted partly by repressing a number of apoptotic genes, notably BIM which is a critical apoptotic activator, leading to less activated Caspase 3. Similarly, in another study, it was shown that diabetic cardiomyopathy was partly driven by the progressive loss of PIM1, whose restoration after gene therapy led to improved heart function [71]. PIM1 is repressed by miR-1, and it was found that transfecting cardiomyocytes with an anti-miR against miR1 led to increased cell survival as revealed by increased anti-apoptotic gene expression and lowered caspase activity.

3.3.6 Proliferation and Senescence

The adult body contains some 100 trillion cells [72]; therefore, growing a tissue or organ of any significant volume requires a large amount of cells. This is typically achieved by expanding explanted cells in vitro before they are re-implanted into the patient. Many ill patients, however, cannot wait long periods for a replacement organ. In order to generate a large number of cells in a short time, there is a need for increasing cell proliferation during the in vitro culture. Unfortunately, the proliferation of adult stem cells is limited by senescence. When cells are explanted and expanded, in vitro adult stem cells will experience successive telomere shortening and finally stop dividing. Previous efforts to circumvent the senescence barrier in adult stem cells have focused on the activity of the telomerase gene, which, if transduced into these cells, causes a dramatic increase in proliferation capacity [73]. However, recent results indicate that miRNAs also play a role in senescence, for example, microRNA-486-5p, which represses SIRT1, a major regulator of longevity and metabolic disorders. Blocking the function of this miRNA with anti-miRNAs was shown to inhibit senescence and promote cell proliferation [74].

3.3.7 *Potential for Oncogenic Transformation*

Any strategy that seeks to modulate cells in order to increase their proliferation and survival should carefully investigate the risk of oncogenic potential before implantation into patients. In fact, the microRNAs employed in the cardiac progenitor study mentioned [4] have all previously been found to be overexpressed in tumours and be involved in tumourigenesis. Furthermore, the earlier study [73], in which stem cells were transduced with telomerase, has demonstrated that such manipulations can lead to cells with oncogenic potential. However, the same study also demonstrated that it is possible to generate cells with enhanced proliferative capacity without apparent oncogenicity. There is probably a fine balance between promoting greater cell numbers while preventing uncontrolled expansion and transformation. In the telomerase overexpression study [73], it was concluded that conditional or temporary expression of telomerase would be a safe approach to increase cell proliferation. Since miRNAs levels are typically modulated in a transient manner, they would be ideal for this purpose.

3.4 MicroRNA Delivery

Delivery has traditionally been the bottleneck for RNAi-based drugs [75]. Apart from special cases, such as the employment of hydrodynamic pressure [76] or using very small anti-miRNA [77], naked RNA cannot be successfully delivered. MicroRNAs and anti-miRNA are large anionic molecules that do not readily cross biological membranes such as the cell membrane; therefore, delivery vectors are needed to enable the miRNA modulators to reach the cellular and subcellular target sites. Many different vectors have been utilised in regenerative medicine for the delivery of other nucleic acid species such as plasmids and siRNA [6]. Successful RNA transfection is often a complex compromise between silencing efficiency, duration and toxicity. There are several delivery routes for miRNA modulation; one can inject the miRNA with a carrier either systemically or locally, or it can be transplanted together with cells and/or implant materials. Delivery in conjunction with implantable materials such as scaffolds for the durations and timed phases needed in tissue regeneration complicates the story further, but also opens up new possibilities. Most tissues and organs are composed of different cell types organised in specific spatial locations, and each of the differentiation pathways leading to these cells relies on the expression of different miRNAs at different time points [29]. Hence, to mimic natural tissue formation, there is a need for scaffolds that can release different miRNAs at different locations at controlled time points.

Tissue destruction can be prevented and regeneration promoted by i.v. injection of the miRNA vector systemically or directly into the damaged site. One study used tail-vein injections of atelocollagen/miR-146a complexes which circulated systemically and prevented joint destruction in collagen-induced arthritis [3]. However, many microRNAs play different roles in different tissues, and a systemic delivery approach for one condition may, thus, have unforeseen consequences in other tissues. In the case of miRNA-146a, for example, its overexpression alters haematopoiesis [78]. Local injections may lead to fewer side-effects and avoid a number of delivery barriers, but serum nucleases and the local immune system still have to be considered. In one study, a cocktail of miR-1, miR-133 and miR-206, all miRNAs that have roles in muscle development, was shown to promote skeletal muscle differentiation and fibre formation in vitro [2]. Subsequent in vivo local injections of atelocollagen complexes containing the miRNA cocktail directly into damaged muscle tissue accelerated muscle regeneration.

3.4.1 MicroRNA Delivery in Combination with Implantable Cells

When combining delivery and cell therapy, the implanted cells are typically transfected prior to transplantation, allowing unbound miRNA transfection reagent to be removed prior to injection. The advantage of this approach is the lowered risk of targeting undesired cells. In one study, for example, embryonic stem cells were transfected with a plasmid-encoding miR-1, which promotes cardiomyocyte differentiation, and afterwards implanted in infarcted myocardium where the transfected cells led to improved cardiac function [79].

In another study, mesenchymal stem cells were transfected with an anti-miR directed against miR-138, a repressor of osteogenesis [5]. The transfected cells were then transferred onto a hydroxyapatite/tricalcium phosphate ceramic powder, which was implanted subcutaneously in mice. Interestingly, the anti-miR-138-transfected cells produced more bone than the cells containing a control miRNA and much more than pre-miR-138-transfected cells. In another study, when miR-148b and an antagonist against miR-489 were transfected into mesenchymal stem cells, which were encapsulated into PEG/peptide hydrogels, osteogenic differentiation of hMSCs was enhanced in vitro [80].

In many cases, however, it is beneficial to deliver the miRNA vector from the implant itself. Benefits include the following (a) the possibility of producing a cell-free bioactive implant, which may be easier to use and to get regulatory approval for; (b) pre-synthesis of miRNA-enhanced scaffolds simplifying handling by the end user to cell seeding and implantation; (c) controlled release of miRNA from an implant, prolonging its activity without resorting to viral vectors (d) tailoring of implants to deliver different miRNAs in multiple temporal phases and/or spatial locations.

3.4.2 Scaffold-Directed MicroRNA Delivery

There are two main ways in which miRNAs can be incorporated into scaffolds: They can be adsorbed onto the surface or be embedded within the scaffold polymer. The release rate of nucleic acids from implants depends on the method of incorporation. The release rate is crucial as differentiation cues need to be released during specific phases of differentiation. Surface-adsorbed drugs are released at rates that are highly dependent on the strength of drug–surface interaction (by van der Waals, electrostatic and covalent bonds), and the size of the surface area with which these interactions can take place. In contrast, embedded drugs are typically released as the implant degrades, although diffusion may play a minor role as well. To our knowledge, no studies exist on delivery of miRNA from scaffolds; however, delivery of siRNA from scaffolds has been carried out by various approaches [6]. These strategies should be directly applicable to scaffold-directed miRNA delivery [81].

Solid scaffolds containing embedded siRNA are typically prepared by adding the siRNA during implant production. Two studies showed naked siRNA and TransIT-TKO/siRNA vectors could be incorporated into electrospun fibres during their production [82, 83]. Release rates were determined by the porosity of the polymer surrounding the siRNA and the presence of a vector. Fibres releasing naked siRNA and siRNA vectors induced different degrees of GAPDH silencing in HEK293 cells (21 % and 31 %, respectively). The limited knockdown may reflect damage to the siRNA or vector from exposure to organic solvents or mechanical processing or limited release during the cell culture period. Interestingly, naked siRNA demonstrated a knockdown despite the inherent barriers; possibly the scaffold polymer polycaprolactone could act as a delivery vehicle as polycaprolactone nanoparticles have previously been used to deliver siRNA successfully [84]. It has also been demonstrated that siRNA-carrying micelles can be incorporated into polyurethane foam scaffolds by mixing the dried vector powder with a scaffold polymer prior to foaming, induced by the addition of lysine triisocyanate [85].

Drugs that are adsorbed onto a scaffold are generally released faster than embedded drugs [86]. One advantage of adsorbing drugs lies in the simpler methodology; the miRNA can be added after scaffold production, enabling a laboratory to coat the scaffold with their miRNA of interest without being involved in scaffold manufacturing. Since an adsorbed drug is added after implant production, it is not subjected to potentially detrimental chemicals or processing during scaffold production. Furthermore, miRNA can be coated onto commercial implants that are already clinically approved. In one case, chitosan/siRNA nanoparticles targeting the RHOA mRNA were adsorbed onto polymer filaments, where they could induce RHOA silencing in neurons desensitising them to myelin inhibition of neurite outgrowth [87].

Hydrogels represent an interesting alternative to solid polymer sponges. They can be loaded with siRNA vectors simply by mixing their components with the vector prior to gelling which, depending on the nature of the gelling component, makes the process less damaging to the RNA and vector. In one study, alginate or collagen was mixed with siRNA prior to $CaCl_2$-, UV-radiation- or heating-induced gelling [88]. The gels were subsequently found to silence a reporter gene in co-encapsulated cells and cells growing beneath a hydrogel. Release rates were found to depend on electrostatic interaction between vector and gel components. Finally, it was shown that transfecting cells directly from a scaffold were much more efficient than delivering a similar amount of siRNA through the medium.

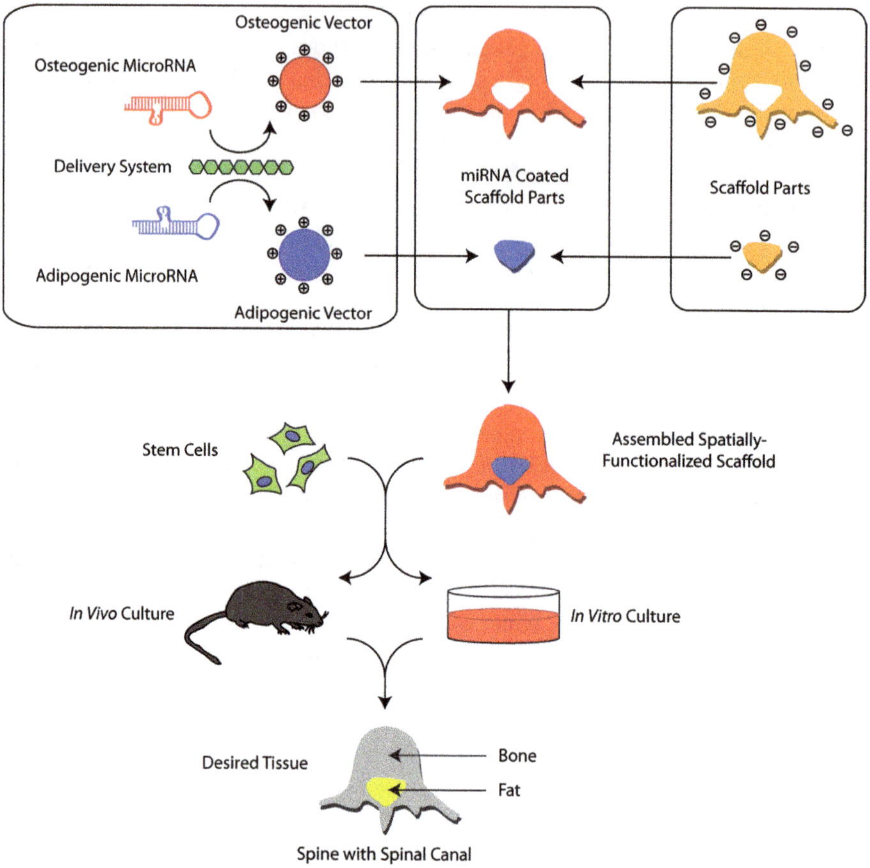

Fig. 3.3 Shows a modification of the strategy previously used to create a spatially differentiated tissue [25]. First, osteogenic and adipogenic miRNA modulators are incorporated into delivery vectors; these are then coated onto scaffold parts with the shape of the tissue section they will create. The miRNA vectors interact with the surface through van der Waals and ionic bonding, binding them to that specific scaffold part. The scaffold components are then assembled in the correct form, and stem cells added. During further in vitro or in vivo culture, the cells take up the miRNA from their local environment, and the miRNA induces them to locally create the correct tissue. Figure modified from [6]

The main advantage of implant-mediated miRNA delivery is the ability to control spatial-temporal miRNA transfection. Our lab has previously devised a method for delivering two different siRNAs from the surface of different parts of an implant [25]. Using TransIT-TKO, we produced nanoparticles containing siRNAs against BCL2L2 and TRIB2, proteins that inhibit osteogenesis and adipogenesis, respectively. These cationic nanoparticles could be adsorbed onto anionic nanoporous polycaprolactone scaffolds prior to seeding with MSCs. After implantation, the siRNAs guided the development of two distinct tissues at discrete locations. The same system could in principle be used for delivering miRNA or anti-miRNA in specific regions of an implant (Fig. 3.3). Instead of using siRNAs against BCL2L2 and TRIB2 to promote osteogenesis and adipogenesis, one could instead use miR-148b and miR-371, respectively [36, 52], to grow two component tissues such as a spinal disc.

Temporally controlled release of miRNA modulators could be accomplished by adsorbing one and encapsulating another miRNA modulator or, alternatively, by encapsulating the nucleic acids in two materials with different degradation rates. In one study, two growth factors, BMP2 and BMP7, were incorporated into a scaffold demonstrating the feasibility of the latter method [89]. Here BMP2 acts to promote early stage osteogenesis, while BMP7 accelerates late stage osteogenesis. The differential release profile was accomplished by encapsulating the proteins into microparticles made of either fast degrading poly(lactide-co-glycolide) or slow degrading poly(3-hydroxybutyrate-co-3-hydroxyvalerate), which were subsequently adsorbed onto chitosan fibre mesh scaffolds. The sequential release system, releasing first BMP2 and then BMP7, demonstrated superior bone formation over scaffolds releasing the two growth factors simultaneous or a reversed sequential manner. This approach illustrates the benefits of temporal control of drug delivery. Similarly, if one desired to differentiate embryonic stem cells to osteoblasts, one could suggest a strategy that involves rapid and delayed release of two different miRNA modulators. For example, anti-miR-290 could be released rapidly to promote mesodermal differentiation followed by a delayed release of miR-148b to promote subsequent osteoblastic differentiation of the mesodermal cells [28, 36].

3.4.3 Secreted MicroRNAs and Regenerative Medicine

Recent findings indicate that miRNAs may play an even greater role than previously thought. Traditionally, miRNAs have been regarded as endogenous regulators of the cells wherein they are expressed. However, secreted vesicular bodies (including the smaller exosomes [90] and the larger microvesicles [91]) have been shown to carry mRNA and miRNA between cells as a form of nature's own miRNA drug delivery [92]. MicroRNAs may, thus, serve as paracrine messengers. Adipocytes, for example, are known to coordinate lipid synthesis using secreted vesicular miRNA [93]. The presence of miRNA in all tested body fluids and the fact that it can be transferred between different cell types indicate that RNA may prove to represent a whole new level of endocrine signals [94]. Little research has been done on the role secreted miRNAs play in tissue development and repair. However, it is known that embryonic stem cells secrete miRNA-loaded vesicles capable of cell transfer [95], and MSCs secrete miRNAs that repress differentiation and may play a role in coordinating stem cell differentiation [96]. The first therapeutic applications of miRNA-carrying vesicles in relation to regenerative medicine have already been performed. For example, MSC-secreted microvesicles were found to carry miRNA involved in cell survival, cell differentiation and organ development [97], and intravenous injections of such microvesicles could alleviate glycerol and ischemia reperfusion injury-induced acute and chronic kidney damage [98, 99]. Furthermore, miRNA-carrying microvesicles seem to play a major role in angiogenesis and are predicted to have therapeutic roles in treating ischemic conditions [100]. The vesicles can be used as natural targeted delivery vehicles if the secreting cells are genetically modified to express modified exosomal membrane proteins that bind to specific receptors on the targeted cells [101]. After the targeted exosomes have been isolated from these cells, they can then be loaded with specific RNAs by electroporation. Such exosomes were successfully used to deliver siRNA to the brain in vivo where the target BACE1, a protein involved in Parkinson's disease, was silenced. We foresee that such vesicles will play a great role in future regenerative medicine as delivery vehicles for siRNA and miRNA modulators. However, while the direct delivery of miRNA using such vesicles has been demonstrated, we have yet to see whether such vesicles can be successfully combined with scaffolds and cell therapy.

3.5 Future Perspectives

The real advantage of miRNA-based regenerative medicine lies in the flexibility offered by the multitude of cellular functions that can be modulated once a miRNA delivery system has been devised for a particular application. Our group has, for example, used one siRNA delivery system to enhance both adipogenic and osteogenic differentiation on a scaffold [25]. MicroRNAs, however, would probably perform even better for this purpose than siRNAs since they target multiple genes and are "optimised" by nature for purposes such as differentiation. Numerous miRNAs have been characterised well enough for these purposes already, and the continued elucidation of endogenous and exogenous miRNA functions in tissue regeneration and repair will undoubtedly expand the potential miRNA can have in regenerative medicine. Currently, high-throughput methods for characterising miRNA expression, such as microarrays [5] and next generation sequencing [29], are providing detailed descriptions of the presence and role various miRNAs play in differentiation events. Ideally, such studies would be designed to provide us with the temporal and spatial information needed to grow complex tissues: For example, by detailing the sequential expression of miRNAs at successive differentiation stages so that we can place those miRNAs into scaffolds that release them in the correct order or by analysing different differentiation pathways in parallel studies to reveal which miRNAs that may be placed in spatially different parts of an implant generate tissues with multiple cell types.

At the same time, scaffold and implant development must take place to ensure that such detailed biological knowledge can be mimicked artificially. Continual research on how drugs interact with implants is necessary so that different methods of drug incorporation can be devised that give controllable release rates. This may include devising polymers or scaffold architectures with different bulk degradation rates or tailoring the delivery systems to bind the implants with different degrees of affinity and, thus, release rates. Alternatively, one could envisage a scaffold containing stimuli-responsive polymers where an externally applied stimulus such as heat or ultrasound triggers the release of drugs. Such a scaffold could be used to accommodate differences in stem cell potential and rate of differentiation, as may be found between patients, by synchronising the release of drugs with the completion of each stage of differentiation.

The best method for incorporating complex spatial information into implants is probably further development of 3D printing of scaffolds together with cells and/or drugs. However, it is necessary to further investigate the degree of spatially restricted drug function when using such approaches, and it remains to be seen whether temporally controlled release of different drugs can be achieved in a printed scaffold. Currently, too few materials are designed to incorporate the knowledge on sequential or temporal function of differentiation cues, and to our knowledge, no approach has yet to try and incorporate both temporal and spatial information on differentiation cues, such as miRNA expression, into implant design. If tissue engineering is to replicate nature's ingenious way of making organs, this needs to change.

References

1. Sen CK (2011) MicroRNAs as new maestro conducting the expanding symphony orchestra of regenerative and reparative medicine. Physiol Genomics 43(10):517–520
2. Nakasa T, Ishikawa M, Shi M, Shibuya H, Adachi N, Ochi M (2010) Acceleration of muscle regeneration by local injection of muscle-specific microRNAs in rat skeletal muscle injury model. J Cell Mol Med 14(10):2495–2505
3. Nakasa T, Shibuya H, Nagata Y, Niimoto T, Ochi M (2011) The inhibitory effect of microRNA-146a expression on bone destruction in collagen-induced arthritis. Arthritis Rheum 63(6):1582–1590
4. Hu S, Huang M, Nguyen PK, Gong Y, Li Z, Jia F, Lan F, Liu J, Nag D, Robbins RC, Wu JC (2011) Novel microRNA prosurvival cocktail for improving engraftment and function of cardiac progenitor cell transplantation. Circulation 124(11 Suppl):S27–S34
5. Eskildsen T, Taipaleenmäki H, Stenvang J, Abdallah BM, Ditzel N, Nossent AY, Bak M, Kauppinen S, Kassem M (2011) MicroRNA-138 regulates osteogenic differentiation of human stromal (mesenchymal) stem cells in vivo. Proc Natl Acad Sci U S A 108(15): 6139–6144
6. Andersen MØ, Kjems J (2011) RNA interference enhanced implants. In: Meital Z (ed) Active implants and scaffolds for tissue regeneration, vol 8, Studies in mechanobiology, tissue engineering and biomaterials. Springer, Berlin, pp 145–165
7. L'Heureux N, McAllister TN, de la Fuente LM (2007) Tissue-engineered blood vessel for adult arterial revascularization. N Engl J Med 357(14):1451–1453
8. Cebotari S, Lichtenberg A, Tudorache I, Hilfiker A, Mertsching H, Leyh R, Breymann T, Kallenbach K, Maniuc L, Batrinac A, Repin O, Maliga O, Ciubotaru A, Haverich A (2006) Clinical application of tissue engineered human heart valves using autologous progenitor cells. Circulation 114(1 Suppl):I132–I137
9. Atala A, Bauer SB, Soker S, Yoo JJ, Retik AB (2006) Tissue-engineered autologous bladders for patients needing cystoplasty. Lancet 367(9518):1241–1246
10. Macchiarini P, Jungebluth P, Go T, Asnaghi MA, Rees LE, Cogan TA, Dodson A, Martorell J, Bellini S, Parnigotto PP, Dickinson SC, Hollander AP, Mantero S, Conconi MT, Birchall MA (2008) Clinical transplantation of a tissue-engineered airway. Lancet 372(9655): 2023–2030
11. Raya-Rivera A, Esquiliano DR, Yoo JJ, Lopez-Bayghen E, Soker S, Atala A (2011) Tissue-Engineered autologous urethras for patients who need reconstruction: an observational study. Lancet 377(9772):1175–1182

12. Leeb C, Jurga M, McGuckin C, Moriggl R, Kenner L (2010) Promising new sources for pluripotent stem cells. Stem Cell Rev 6(1):15–26
13. Davidson BL, McCray PB Jr (2011) Current prospects for RNA interference-based therapies. Nat Rev Genet 12(5):329–340
14. Axtell MJ, Westholm JO, Lai EC (2011) Vive la difference: biogenesis and evolution of microRNAs in plants and animals. Genome Biol 12(4):221
15. Krol J, Loedige I, Filipowicz W (2010) The widespread regulation of microRNA biogenesis, function and decay. Nat Rev Genet 11(9):597–610
16. Zeng Y, Yi R, Cullen BR (2003) MicroRNAs and small interfering RNAs can inhibit mRNA expression by similar mechanisms. Proc Natl Acad Sci USA 100(17):9779–9784
17. Zeng Y, Wagner EJ, Cullen BR (2002) Both natural and designed micro RNAs can inhibit the expression of cognate mRNAs when expressed in human cells. Mol Cell 9(6):1327–1333
18. Rahbek UL, Howard KA, Oupicky D, Manickam DS, Dong M, Nielsen AF, Hansen TB, Besenbacher F, Kjems J (2008) Intracellular siRNA and precursor miRNA trafficking using bioresponsive copolypeptides. J Gene Med 10(1):81–93
19. Meister G, Landthaler M, Dorsett Y, Tuschl T (2004) Sequence-specific inhibition of microRNA- and siRNA-induced RNA silencing. RNA 10(3):544–550
20. Ebert MS, Neilson JR, Sharp PA (2007) MicroRNA sponges: competitive inhibitors of small RNAs in mammalian cells. Nat Methods 4(9):721–726
21. Kelly EJ, Russell SJ (2009) MicroRNAs and the regulation of vector tropism. Mol Ther 17(3):409–416
22. Martin I, Padera RF, Vunjak-Novakovic G, Freed LE (1998) In vitro differentiation of chick embryo bone marrow stromal cells into cartilaginous and bone-like tissues. J Orthop Res 16(2):181–189
23. Song I, Kim BS, Kim CS, Im GI (2011) Effects of BMP-2 and vitamin D3 on the osteogenic differentiation of adipose stem cells. Biochem Biophys Res Commun 408(1):126–131
24. Yang X, Walboomers XF, van den Dolder J, Yang F, Bian Z, Fan M, Jansen JA (2008) Non-viral bone morphogenetic protein 2 transfection of rat dental pulp stem cells using calcium phosphate nanoparticles as carriers. Tissue Eng Part A 14(1):71–81
25. Andersen MØ, Nygaard JV, Burns JS, Raarup MK, Nyengaard JR, Bünger C, Besenbacher F, Howard KA, Kassem M, Kjems J (2010) siRNA nanoparticle functionalization of nanostructured scaffolds enables controlled multilineage differentiation of stem cells. Mol Ther 18(11):2018–2027
26. Alper J (2009) Geron gets green light for human trial of ES cell-derived product. Nat Biotechnol 27(3):213–214
27. Xu N, Papagiannakopoulos T, Pan G, Thomson JA, Kosik KS (2009) MicroRNA-145 regulates OCT4, SOX2, and KLF4 and represses pluripotency in human embryonic stem cells. Cell 137(4):647–658
28. Zovoilis A, Smorag L, Pantazi A, Engel W (2009) Members of the miR-290 cluster modulate in vitro differentiation of mouse embryonic stem cells. Differentiation 78(2–3):69–78
29. Knelangen JM, van der Hoek MB, Kong WC, Owens JA, Fischer B, Santos AN (2011) MicroRNA expression profile during adipogenic differentiation in mouse embryonic stem cells. Physiol Genomics 43(10):611–620
30. Abrous DN, Koehl M, Le Moal M (2005) Adult neurogenesis: from precursors to network and physiology. Physiol Rev 85(2):523–569
31. Kassem M, Abdallah BM (2008) Human bone-marrow-derived mesenchymal stem cells: biological characteristics and potential role in therapy of degenerative diseases. Cell Tissue Res 331(1):157–163
32. Tomé M, López-Romero P, Albo C, Sepúlveda JC, Fernández-Gutiérrez B, Dopazo A, Bernad A, González MA (2011) miR-335 orchestrates cell proliferation, migration and differentiation in human mesenchymal stem cells. Cell Death Differ 18(6):985–995
33. Zhang JF, Fu WM, He ML, Xie WD, Lv Q, Wan G, Li G, Wang H, Lu G, Hu X, Jiang S, Li JN, Lin MC, Zhang YO, Kung H (2011) MiRNA-20a promotes osteogenic differentiation of human mesenchymal stem cells by co-regulating BMP signaling. RNA Biol 8(5):829–838

34. He J, Zhang JF, Yi C, Lv Q, Xie WD, Li JN, Wan G, Cui K, Kung HF, Yang J, Yang BB, Zhang Y (2010) miRNA-mediated functional changes through co-regulating function related genes. PLoS One 5(10):e13558
35. Luzi E, Marini F, Sala SC, Tognarini I, Galli G, Brandi ML (2008) Osteogenic differentiation of human adipose tissue-derived stem cells is modulated by the miR-26a targeting of the SMAD1 transcription factor. J Bone Miner Res 23(2):287–295
36. Schoolmeesters A, Eklund T, Leake D, Vermeulen A, Q S, Force Aldred S, Fedorov Y (2009) Functional profiling reveals critical role for miRNA in differentiation of human mesenchymal stem cells. PLoS One 4(5):e5605
37. Hung PS, Chen FC, Kuang SH, Kao SY, Lin SC, Chang KW (2010) miR-146a induces differentiation of periodontal ligament cells. J Dent Res 89(3):252–257
38. Kim YJ, Bae SW, Yu SS, Bae YC, Jung JS (2009) miR-196a regulates proliferation and osteogenic differentiation in mesenchymal stem cells derived from human adipose tissue. J Bone Miner Res 24(5):816–825
39. Li Z, Hassan MQ, Jafferji M, Aqeilan RI, Garzon R, Croce CM, van Wijnen AJ, Stein JL, Stein GS, Lian JB (2009) Biological functions of miR-29b contribute to positive regulation of osteoblast differentiation. J Biol Chem 284(23):15676–15684
40. Mizuno Y, Yagi K, Tokuzawa Y, Kanesaki-Yatsuka Y, Suda T, Katagiri T, Fukuda T, Maruyama M, Okuda A, Amemiya T, Kondoh Y, Tashiro H, Okazaki Y (2008) miR-125b inhibits osteoblastic differentiation by down-regulation of cell proliferation. Biochem Biophys Res Commun 368(2):267–272
41. Li Z, Hassan MQ, Volinia S, van Wijnen AJ, Stein JL, Croce CM, Lian JB, Stein GS (2008) A microRNA signature for a BMP2-induced osteoblast lineage commitment program. Proc Natl Acad Sci USA 105(37):13906–13911
42. Itoh T, Nozawa Y, Akao Y (2009) MicroRNA-141 and -200a are involved in bone morphogenetic protein-2-induced mouse pre-osteoblast differentiation by targeting distal-less homeobox 5. J Biol Chem 284(29):19272–19279
43. Inose H, Ochi H, Kimura A, Fujita K, Xu R, Sato S, Iwasaki M, Sunamura S, Takeuchi Y, Fukumoto S, Saito K, Nakamura T, Siomi H, Ito H, Arai Y, Shinomiya K, Takeda S (2009) A microRNA regulatory mechanism of osteoblast differentiation. Proc Natl Acad Sci USA 106(49):20794–20799
44. Mizuno Y, Tokuzawa Y, Ninomiya Y, Yagi K, Yatsuka-Kanesaki Y, Suda T, Fukuda T, Katagiri T, Kondoh Y, Amemiya T, Tashiro H, Okazaki Y (2009) miR-210 promotes osteoblastic differentiation through inhibition of AcvR1b. FEBS Lett 583(13):2263–2268
45. Li H, Xie H, Liu W, Hu R, Huang B, Tan YF, Xu K, Sheng ZF, Zhou HD, Wu XP, Luo XH (2009) A novel microRNA targeting HDAC5 regulates osteoblast differentiation in mice and contributes to primary osteoporosis in humans. J Clin Invest 119(12):3666–3677
46. Karbiener M, Fischer C, Nowitsch S, Opriessnig P, Papak C, Ailhaud G, Dani C, Amri EZ, Scheideler M (2009) microRNA miR-27b impairs human adipocyte differentiation and targets PPARgamma. Biochem Biophys Res Commun 390(2):247–251
47. Zaragosi LE, Wdziekonski B, Brigand KL, Villageois P, Mari B, Waldmann R, Dani C, Barbry P (2011) Small RNA sequencing reveals miR-642a-3p as a novel adipocyte-specific microRNA and miR-30 as a key regulator of human adipogenesis. Genome Biol 12(7):R64
48. Lee EK, Lee MJ, Abdelmohsen K, Kim W, Kim MM, Srikantan S, Martindale JL, Hutchison ER, Kim HH, Marasa BS, Selimyan R, Egan JM, Smith SR, Fried SK, Gorospe M (2011) miR-130 suppresses adipogenesis by inhibiting peroxisome proliferator-activated receptor gamma expression. Mol Cell Biol 31(4):626–638
49. Yang Z, Bian C, Zhou H, Huang S, Wang S, Liao L, Zhao RC (2011) MicroRNA hsa-miR-138 inhibits adipogenic differentiation of human adipose tissue-derived mesenchymal stem cells through adenovirus EID-1. Stem Cells Dev 20(2):259–267
50. Esau C, Kang X, Peralta E, Hanson E, Marcusson EG, Ravichandran LV, Sun Y, Koo S, Perera RJ, Jain R, Dean NM, Freier SM, Bennett CF, Lollo B, Griffey R (2004) MicroRNA-143 regulates adipocyte differentiation. J Biol Chem 279(50):52361–52365

51. Skårn M, Namløs HM, Noordhuis P, Wang MY, Meza-Zepeda LA, Myklebost O (2012) Adipocyte differentiation of human bone marrow-derived stromal cells is modulated by microRNA-155, microRNA-221, and microRNA-222. Stem Cells Dev 21(6):873–883
52. Bork S, Horn P, Castoldi M, Hellwig I, Ho AD, Wagner W (2011) Adipogenic differentiation of human mesenchymal stromal cells is down-regulated by microRNA-369-5p and up-regulated by microRNA-371. J Cell Physiol 226(9):2226–2234
53. Wang Q, Li YC, Wang J, Kong J, Qi Y, Quigg RJ, Li X (2008) miR-17-92 cluster accelerates adipocyte differentiation by negatively regulating tumor-suppressor Rb2/p130. Proc Natl Acad Sci USA 105(8):2889–2894
54. Sun F, Wang J, Pan Q, Yu Y, Zhang Y, Wan Y, Wang J, Li X, Hong A (2009) Characterization of function and regulation of miR-24-1 and miR-31. Biochem Biophys Res Commun 380(3):660–665
55. Lin Q, Gao Z, Alarcon RM, Ye J, Yun Z (2009) A role of miR-27 in the regulation of adipogenesis. FEBS J 276(8):2348–2358
56. Xie H, Lim B, Lodish HF (2009) MicroRNAs induced during adipogenesis that accelerate fat cell development are downregulated in obesity. Diabetes 58(5):1050–1057
57. Kennell JA, Gerin I, MacDougald OA, Cadigan KM (2008) The microRNA miR-8 is a conserved negative regulator of Wnt signaling. Proc Natl Acad Sci USA 105(40):15417–15422
58. Huang J, Zhao L, Xing L, Chen D (2010) MicroRNA-204 regulates Runx2 protein expression and mesenchymal progenitor cell differentiation. Stem Cells 28(2):357–364
59. Gerin I, Bommer GT, McCoin CS, Sousa KM, Krishnan V, MacDougald OA (2010) Roles for miRNA-378/378* in adipocyte gene expression and lipogenesis. Am J Physiol Endocrinol Metab 299(2):E198–E206
60. Radisic M, Malda J, Epping E, Geng W, Langer R, Vunjak-Novakovic G (2006) Oxygen gradients correlate with cell density and cell viability in engineered cardiac tissue. Biotechnol Bioeng 93(2):332–343
61. Wittenborn T, Nielsen T, Nygaard JV, Larsen EK, Thim T, Rydtoft LM, Vorup-Jensen T, Kjems J, Nielsen NC, Horsman MR, Falk E (2012) Ultrahigh-field DCE-MRI of angiogenesis in a novel angiogenesis mouse model. J Magn Reson Imaging 35(3):703–710
62. Zhang P, Moudgill N, Hager E, Tarola N, Dimatteo C, McIlhenny S, Tulenko T, DiMuzio PJ (2011) Endothelial differentiation of adipose-derived stem cells from elderly patients with cardiovascular disease. Stem Cells Dev 20(6):977–988
63. Narita Y, Yamawaki A, Kagami H, Ueda M, Ueda Y (2008) Effects of transforming growth factor-beta 1 and ascorbic acid on differentiation of human bone-marrow-derived mesenchymal stem cells into smooth muscle cell lineage. Cell Tissue Res 333(3):449–459
64. Cordes KR, Sheehy NT, White MP, Berry EC, Morton SU, Muth AN, Lee TH, Miano JM, Ivey KN, Srivastava D (2009) miR-145 and miR-143 regulate smooth muscle cell fate and plasticity. Nature 460(7256):705–710
65. Chamorro-Jorganes A, Araldi E, Penalva LO, Sandhu D, Fernández-Hernando C, Suárez Y (2011) MicroRNA-16 and MicroRNA-424 regulate cell-autonomous angiogenic functions in endothelial cells via targeting vascular endothelial growth factor receptor-2 and fibroblast growth factor receptor-1. Arterioscler Thromb Vasc Biol 31(11):2595–2606
66. Hong H, Takahashi K, Ichisaka T, Aoi T, Kanagawa O, Nakagawa M, Okita K, Yamanaka S (2009) Suppression of induced pluripotent stem cell generation by the p53–p21 pathway. Nature 460(7259):1132–1135
67. Choi YJ, Lin CP, Ho JJ, He X, Okada N, Bu P, Zhong Y, Kim SY, Bennett MJ, Chen C, Ozturk A, Hicks GG, Hannon GJ, He L (2011) miR-34 miRNAs provide a barrier for somatic cell reprogramming. Nat Cell Biol 13(11):1353–1360
68. Li Z, Yang CS, Nakashima K, Rana TM (2011) Small RNA-mediated regulation of iPS cell generation. EMBO J 30(5):823–834
69. Sun X, Fu X, Han W, Zhao Y, Liu H (2010) Can controlled cellular reprogramming be achieved using microRNAs? Ageing Res Rev 9(4):475–483
70. Kurowska-Stolarska M, Alivernini S, Ballantine LE, Asquith DL, Millar NL, Gilchrist DS, Reilly J, Ierna M, Fraser AR, Stolarski B, McSharry C, Hueber AJ, Baxter D, Hunter J, Gay

S, Liew FY, McInnes IB (2011) MicroRNA-155 as a proinflammatory regulator in clinical and experimental arthritis. Proc Natl Acad Sci USA 108(27):11193–11198
71. Katare R, Caporali A, Zentilin L, Avolio E, Sala-Newby G, Oikawa A, Cesselli D, Beltrami AP, Giacca M, Emanueli C, Madeddu P (2011) Intravenous gene therapy with PIM-1 via a cardiotropic viral vector halts the progression of diabetic cardiomyopathy through promotion of prosurvival signaling. Circ Res 108(10):1238–1251
72. Sipe JD (2002) Tissue engineering and reparative medicine. Ann N Y Acad Sci 961:1–9
73. Kassem M, Abdallah BM, Yu Z, Ditzel N, Burns JS (2004) The use of hTERT-immortalized cells in tissue engineering. Cytotechnology 45(1–2):39–46
74. Kim YJ, Hwang SH, Lee SY, Shin KK, Cho HH, Bae YC, Jung JS (2012) miR-486-5p induces replicative senescence of human adipose tissue-derived mesenchymal stem cells and its expression is controlled by high glucose. Stem Cells Dev 21(10):1749–1760
75. Howard KA (2009) Delivery of RNA interference therapeutics using polycation-based nanoparticles. Adv Drug Deliv Rev 61(9):710–720
76. Lewis DL, Wolff JA (2007) Systemic siRNA delivery via hydrodynamic intravascular injection. Adv Drug Deliv Rev 59(2–3):115–123
77. Obad S, dos Santos CO, Petri A, Heidenblad M, Broom O, Ruse C, Fu C, Lindow M, Stenvang J, Straarup EM, Hansen HF, Koch T, Pappin D, Hannon GJ, Kauppinen S (2011) Silencing of microRNA families by seed-targeting tiny LNAs. Nat Genet 43(4):371–378
78. Starczynowski DT, Kuchenbauer F, Wegrzyn J, Rouhi A, Petriv O, Hansen CL, Humphries RK, Karsan A (2011) MicroRNA-146a disrupts hematopoietic differentiation and survival. Exp Hematol 39(2):167.e4–178.e4
79. Glass C, Singla DK (2011) MicroRNA-1 transfected embryonic stem cells enhance cardiac myocyte differentiation and inhibit apoptosis by modulating the PTEN/Akt pathway in the infarcted heart. Am J Physiol Heart Circ Physiol 301(5):H2038–H2049
80. Mariner PD, Johannesen E, Anseth KS (2012) Manipulation of miRNA activity accelerates osteogenic differentiation of hMSCs in engineered 3D scaffolds. J Tissue Eng Regen Med 6(4):314–324
81. Monaghan M, Pandit A (2011) RNA interference therapy via functionalized scaffolds. Adv Drug Deliv Rev 63(4–5):197–208
82. Cao H, Jiang X, Chai C, Chew SY (2010) RNA interference by nanofiber-based siRNA delivery system. J Control Release 144(2):203–212
83. Rujitanaroj PO, Wang YC, Wang J, Chew SY (2011) Nanofiber-mediated controlled release of siRNA complexes for long term gene-silencing applications. Biomaterials 32(25):5915–5923
84. Bouclier C, Moine L, Hillaireau H, Marsaud V, Connault E, Opolon P, Couvreur P, Fattal E, Renoir JM (2008) Physicochemical characteristics and preliminary in vivo biological evaluation of nanocapsules loaded with siRNA targeting estrogen receptor alpha. Biomacromolecules 9(10):2881–2890
85. Nelson CE, Gupta MK, Adolph EJ, Shannon JM, Guelcher SA, Duvall CL (2012) Sustained local delivery of siRNA from an injectable scaffold. Biomaterials 33(4):1154–1161
86. Nie H, Wang CH (2007) Fabrication and characterization of PLGA/HAp composite scaffolds for delivery of BMP-2 plasmid DNA. J Control Release 120(1–2):111–121
87. Mittnacht U, Hartmann H, Hein S, Oliveira H, Dong M, Pêgo AP, Kjems J, Howard KA, Schlosshauer B (2010) Chitosan/siRNA nanoparticles biofunctionalize nerve implants and enable neurite outgrowth. Nano Lett 10(10):3933–3939
88. Krebs MD, Jeon O, Alsberg E (2009) Localized and sustained delivery of silencing RNA from macroscopic biopolymer hydrogels. J Am Chem Soc 131(26):9204–9206
89. Yilgor P, Tuzlakoglu K, Reis RL, Hasirci N, Hasirci V (2009) Incorporation of a sequential BMP-2/BMP-7 delivery system into chitosan-based scaffolds for bone tissue engineering. Biomaterials 30(21):3551–3559
90. Théry C, Zitvogel L, Amigorena S (2002) Exosomes: composition, biogenesis and function. Nat Rev Immunol 2(8):569–579
91. György B, Szabó TG, Pásztói M, Pál Z, Misják P, Aradi B, László V, Pállinger E, Pap E, Kittel A, Nagy G, Falus A, Buzás EI (2011) Membrane vesicles, current state-of-the-art: emerging role of extracellular vesicles. Cell Mol Life Sci 68(16):2667–2688

92. Valadi H, Ekström K, Bossios A, Sjöstrand M, Lee JJ, Lötvall JO (2007) Exosome-mediated transfer of mRNAs and microRNAs is a novel mechanism of genetic exchange between cells. Nat Cell Biol 9(6):654–659
93. Müller G, Schneider M, Biemer-Daub G, Wied S (2011) Microvesicles released from rat adipocytes and harboring glycosylphosphatidylinositol-anchored proteins transfer RNA stimulating lipid synthesis. Cell Signal 23(7):1207–1223
94. Ogawa R, Tanaka C, Sato M, Nagasaki H, Sugimura K, Okumura K, Nakagawa Y, Aoki N (2010) Adipocyte-derived microvesicles contain RNA that is transported into macrophages and might be secreted into blood circulation. Biochem Biophys Res Commun 398(4): 723–729
95. Yuan A, Farber EL, Rapoport AL, Tejada D, Deniskin R, Akhmedov NB, Farber DB (2009) Transfer of microRNAs by embryonic stem cell microvesicles. PLoS One 4(3):e4722
96. Koh W, Sheng CT, Tan B, Lee QY, Kuznetsov V, Kiang LS, Tanavde V (2010) Analysis of deep sequencing microRNA expression profile from human embryonic stem cells derived mesenchymal stem cells reveals possible role of let-7 microRNA family in downstream targeting of hepatic nuclear factor 4 alpha. BMC Genomics 11(Suppl 1):S6
97. Collino F, Deregibus MC, Bruno S, Sterpone L, Aghemo G, Viltono L, Tetta C, Camussi G (2010) Microvesicles derived from adult human bone marrow and tissue specific mesenchymal stem cells shuttle selected pattern of miRNAs. PLoS One 5(7):e11803
98. Bruno S, Grange C, Deregibus MC, Calogero RA, Saviozzi S, Collino F, Morando L, Busca A, Falda M, Bussolati B, Tetta C, Camussi G (2009) Mesenchymal stem cell-derived microvesicles protect against acute tubular injury. J Am Soc Nephrol 20(5):1053–1067
99. Gatti S, Bruno S, Deregibus MC, Sordi A, Cantaluppi V, Tetta C, Camussi G (2011) Microvesicles derived from human adult mesenchymal stem cells protect against ischaemia-reperfusion-induced acute and chronic kidney injury. Nephrol Dial Transplant 26(5):1474–1483
100. Martinez MC, Andriantsitohaina R (2011) Microparticles in angiogenesis: therapeutic potential. Circ Res 109(1):110–119
101. Alvarez-Erviti L, Seow Y, Yin H, Betts C, Lakhal S, Wood MJ (2011) Delivery of siRNA to the mouse brain by systemic injection of targeted exosomes. Nat Biotechnol 29(4):341–345

Chapter 4
Intracellular Delivery Considerations for RNAi Therapeutics

Yu Zhu, Jing Li, and David Oupický

Abstract Intracellular delivery of the wide range of RNAi therapeutics is a complex process that remains to be fully understood and mastered. This chapter discusses the current understanding of the process of intracellular delivery of RNAi triggers from multiple perspectives. The delivery requirements are primarily determined by the properties and intracellular site of action of the RNAi molecule. Main intracellular barriers to nonviral delivery of siRNA, shRNA, and miRNA are discussed, together with the latest methods to overcome them. The barriers covered include cellular uptake, endosomal escape, cytoplasmic trafficking, nuclear entry, and therapeutic cargo release from delivery vectors.

4.1 Introduction

Since RNA interference (RNAi)-based gene silencing by double-stranded RNA was first described in 1998 [1], the field of RNAi-mediated gene therapy has been quickly developing. However, the clinical application of RNAi-based therapeutics is hindered by the lack of adequate delivery methods. Not only are nucleic acids subject to degradation by a host of nucleases [2], their polyanionic nature makes translocation across cellular membranes difficult. Further compromising RNAi efficacy is the often-inadequate delivery to the specific subcellular compartments and incorporation into the RNAi machinery. The application of RNAi triggers as a potential clinical agent requires proper design and development of effective and safe delivery vectors. This chapter focuses on the intracellular trafficking of different RNAi triggers and recent developments in the design of delivery vectors capable of breaking down intracellular delivery barriers.

Y. Zhu • J. Li • D. Oupický (✉)
Department of Pharmaceutical Sciences, Wayne State University,
259 Mack Avenue, Detroit, MI, USA
e-mail: oupicky@wayne.edu

K.A. Howard (ed.), *RNA Interference from Biology to Therapeutics*,
Advances in Delivery Science and Technology, DOI 10.1007/978-1-4614-4744-3_4,

4.2 Delivery Considerations for Different RNAi Molecules

There are three major types of RNAi triggers: small interfering RNA (siRNA), small hairpin RNA (shRNA), and microRNA (miRNA). These molecules differ in the way they are processed and the RNAi machinery involved at different intracellular locations (Fig. 4.1). We will first consider individual requirements for delivery of the three types of RNAi molecules.

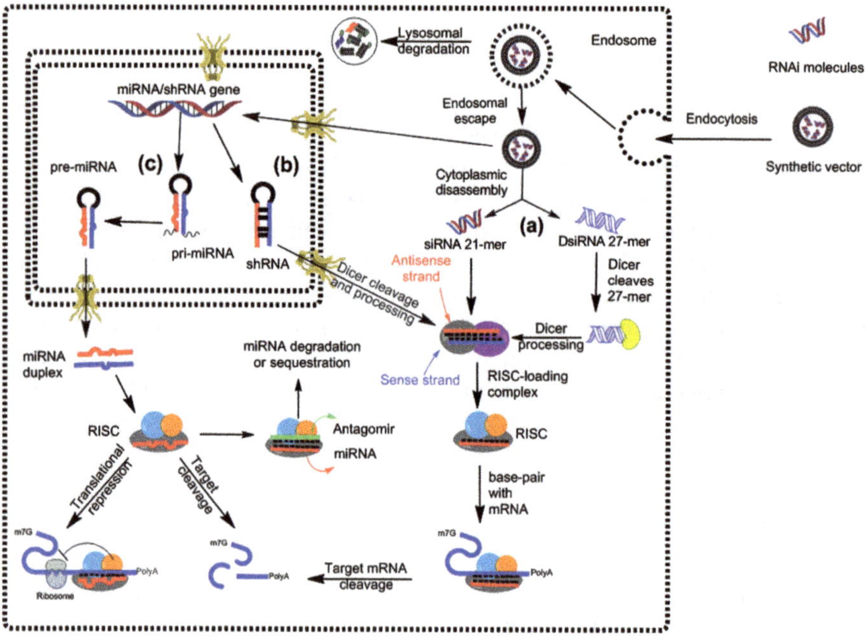

Fig. 4.1 Intracellular trafficking of RNAi delivery systems. RNAi triggers are delivered using either viral or synthetic delivery systems. Synthetic delivery vectors enter the cell mainly by endocytosis; some escape from the endosomes into the cytoplasm, while the rest are routed for degradation in lysosomes. Complexes located in the cytoplasm undergo partial or full disassembly, and the released RNAi triggers can be degraded by cytosolic nucleases or transported to the nucleus. (**a**) siRNA molecules require cytoplasmic delivery. Conventional siRNA molecules (21-mer) are incorporated into RISC, and the sense strand is excluded through an unwinding process. This complex is directed to the targeting mRNA sequence through complementary recognition of the mRNA and antisense strand and then triggers the cleavage of the mRNA. DsiRNA (27-mer), on the other hand, first enters the Dicer pathway where it is cleaved into standard siRNA strand followed by being processed to RISC. (**b**) After entering the nucleus, the gene encoding for shRNA is transcribed to shRNA molecule, which is then exported into the cytoplasm. Subsequently, shRNA is cleaved by the endoribonuclease Dicer into dsRNA, which is then loaded into the RISC-loading complex and triggers the mRNA cleavage in the same way described in (**a**). (**c**) miRNA can be delivered either to the nucleus as the encoding gene or to the cytoplasm as mature miRNA molecule. It then pairs with the targeting mRNA and leads to either target cleavage or suppression of translation process. Antagomirs, after entering the cytoplasm, complementary bind to mature miRNA and lead to either miRNA degradation or sequestration, thus blocks the function of miRNA

4.2.1 siRNA

siRNA is a class of double-stranded RNA (dsRNA) molecules that are 20–25 nucleotides in length. siRNA conducts its silencing function in the cytoplasm by first incorporating into RNA-induced silencing complex (RISC). siRNA duplex then undergoes unwinding, and the sense strand is excluded from the RISC. RISC then further pairs with the complementary mRNA and mRNA cleavage facilitated by complementary base pairing with the antisense strand results in post-transcriptional gene silencing. It is worth noting that a class of longer siRNA, the 27-mer Dicer-substrate siRNA (DsiRNA), has been reported to exhibit more potent silencing than standard 21-mer siRNA [3, 4]. In the cytoplasm, the 27-mer DsiRNA is first cleaved by endonuclease Dicer to form standard 21-mer siRNA, which then proceeds through the same silencing pathway as stated above. One possible explanation for the greater potency of DsiRNA is that the process of Dicer-mediated cleavage may facilitate the loading of siRNA into RISC [5], thus enhancing gene silencing. While it is generally accepted that delivering siRNA into the cytoplasm is optimal, recent findings suggest that siRNA can also induce transcriptional gene silencing through DNA methylation and knockdown of transcripts restricted to the nucleus [6–10]. siRNA that engages in the nuclear RNAi pathways would require siRNA delivery into the nucleus. This notion is supported by the observed increase in the transcriptional gene silencing efficiency with delivery vectors containing nuclear localization sequences [11].

4.2.2 miRNA

The active mature miRNA is a single-stranded RNA generated during endogenous transcription [12]. It is ~22 nucleotides in length, firstly transcribed in the nucleus as long primary transcripts (pri-miRNA) with local hairpin structure and then processed into precursor miRNAs (pre-miRNA) by Drosha enzyme, followed by export to the cytoplasm where it is processed to form mature miRNA by Dicer [13]. Unlike siRNA, which forms a perfect duplex with its target mRNA and directs the RISC-mediated mRNA degradation, the action of miRNA depends on the level of complementarity. It may cause mRNA cleavage, or it may bind imperfectly with the untranslated regions of mRNA, leading to translational repression. The endogenous processing pathway of miRNA improves the silencing efficacy and avoids activation of the interferon system associated with externally introduced synthetic siRNA [14]. Depending on the type of miRNA, the desired intracellular site of delivery varies from cytoplasm to the nucleus. Commercially available miRNA mimics are small dsRNA molecules mimicking endogenous mature miRNA and thus should be delivered directly into the cytoplasm. Antagomir, a single-stranded oligonucleotide, binds endogenous mature miRNA and blocks the function of miRNA-mediated silencing [15, 16]. As the cytoplasm is the site for antagomir–miRNA binding, antagomir success requires delivery into cytoplasm. In contrast, nuclear delivery is necessary for the RNA transcripts of miRNAs, such as pri-miRNA precursors. The requirement for enhancement of miRNA expression or downregulation depends on the role of the miRNA in disease.

4.2.3 shRNA

The shRNA is a class of RNAi agents with tight hairpin turn that can be cleaved by Dicer enzyme to generate siRNA with 21–23 nucleotides. shRNA can be introduced into host cells via plasmid or viral vectors and be further integrated into the host genome. This offers the advantage of the silencing effect not necessarily being lost or diluted during cell division. When compared with siRNA, shRNA is used more frequently in the laboratory to achieve permanent knockdown of a specific mRNA, typically by using a virus as the delivery vector. Unlike siRNA, shRNA is usually delivered as the corresponding gene inserted into a plasmid DNA, which undergoes transcription process to form shRNA. Therefore, the sequence encoding for shRNA has to enter the nucleus, and the delivery considerations fall into the DNA delivery category [17].

4.3 Main Intracellular Barriers for RNAi Therapeutic Delivery

Viral vectors and synthetic vectors are the two major classes of delivery systems used for nucleic acid delivery. In this chapter, we will focus on nanoparticle-based synthetic vectors used in RNAi delivery. In most cases, RNAi delivery vectors are internalized into cells via endocytosis. Available evidence suggests that multiple endocytic pathways are involved in the delivery, including clathrin-mediated pathway, caveolar pathway, macropinocytosis, and phagocytosis. If the vectors contain specific ligands that bind to cell surface receptors, cells can internalize the vectors by receptor-mediated endocytosis [18].

Initial uptake of the RNAi delivery vectors proceeds through a sequence of trafficking vesicles including early endosomes, late endosomes, and lysosomes. A fraction of the endocytosed vectors can be trafficked into the cytoplasm, while the rest are routed for degradation into lysosomes. If the vectors successfully escape lysosomal degradation and enter the cytoplasm, siRNA and miRNA require release from the vectors in order to be incorporated into RISC and to trigger the target mRNA degradation. In contrast, shRNA and pri- and pre-miRNA need to be further transported into the nucleus for processing in order to perform their silencing function. shRNA and miRNA released from the vectors in the cytoplasm can be either degraded by cytosolic nucleases or be further transported to the nucleus. RNAi vectors that do not disassemble in the cytoplasm are transported into the nucleus either passively by association with nuclear material during breakdown of the nuclear envelope during cell division or actively through nuclear pore complexes. RNAi vectors that release their cargo in the nucleus provide the transcription apparatus access to shRNA and miRNA.

Table 4.1 Intracellular barriers and methods to improve RNAi therapeutic delivery

Intracellular barrier	Delivery solution	References
Cell uptake and targeting	Folate-conjugated block copolymers	[19]
	Aptamer–siRNA conjugates	[20, 21]
	siRNA complex with protamine–antibody fusion protein	[22]
	siRNA–antibody conjugate	[23]
	Cell-penetrating peptides	[24–27]
Endosomal escape	Direct membrane translocation	[28]
	Abrogate immune activation	[29, 30]
	Proton sponge vectors	[31–40]
	pH-sensitive vectors	[41, 42]
	Temperature-responsive pluronics	[43]
	Photosensitizers	[44]
	Fusogenic lipids	[45]
	Influenza-derived fusogenic peptide diINF-7	[46]
	Membranolytic peptides	[47]
Nuclear transport	Reducible copolypeptide containing nuclear localization signal (NLS) to deliver siRNA and pri-miRNA	[11]
RNAi release from vectors	Biodegradable poly(β-amino ester)s	[48, 49]
	Degradable poly(lactide-*co*-glycolide)	[50–52]
	Carbamate-linked oligoamine–polyglycerol	[53]
	Chitosan	[54]
	Bioreducible polycations	[55–59]

The following sections will describe the main intracellular barriers encountered by the RNAi vectors during intracellular delivery. We will discuss current strategies applied to overcome the individual barriers and to control the RNAi trafficking (Table 4.1).

4.3.1 Cell Uptake and Targeting

Cell uptake is a key step required for successful delivery of RNAi therapeutics into the target cells. Cell membrane consists of a hydrophobic phospholipid bilayer embedded with various surface proteins. The physicochemical properties of this biological barrier prevent direct translocation of the hydrophilic and negatively charged RNAi molecules. Electrostatic complexation of RNAi molecules with various cationic lipids and polymers is often employed to facilitate cell uptake. The complexation results in a formation of particles with a net positive charge, which helps to facilitate interaction with the negatively charged cell membranes and to trigger internalization of the RNAi-containing particles via endocytosis.

Selective delivery of RNAi into specific cells can be achieved by direct binding of the vectors to a variety of cell-type-specific receptors. For example, cholesterol–siRNA conjugates can be efficiently internalized via receptor-mediated process by specific cells such as hepatocytes [60]. Recent studies have demonstrated that cholesterol promotes the fusion with cell membrane and leads to improved intracellular trafficking and transfection efficiency [61].

Synthetic RNAi vectors can be easily modified with various cell targeting ligands, including monoclonal antibodies, peptides, small-molecule ligands, and aptamers, capable of binding to specific cell surface receptors [62]. Folate is among the most widely investigated targeting ligands in RNAi therapeutic delivery due to the overexpression of the folate receptor in cancer cells [60]. York et al. reported cell-specific delivery of siRNA using multivalent folate-conjugated block copolymers to obtain enhanced receptor-mediated endocytosis and cell-specific delivery [19].

Aptamers are nucleic acids that can specifically bind to target receptors in specific cells. This approach is popular for selective delivery of siRNA to target cells [60]. It is reported that aptamer–siRNA conjugates are able to selectively enter cells expressing prostate-specific membrane antigens [63]. Zhou et al. reported the use of anti-gp120 aptamers–siRNA conjugates for in vitro silencing of HIV viral replication to suppress HIV infection [20, 21].

Protamine, a short cationic peptide, has been used to link siRNA to F_{ab} fragment of an HIV-1 envelope antibody to selectively silence gene expression in HIV-infected cells [22]. Another reported example of antibody targeting of RNAi includes scFv antibody used to deliver siRNA to T-cell CD7 receptor [23].

Various cell-penetrating peptides (CPPs) such as TAT transactivator protein, penetratin, and transportan have been incorporated into the RNAi delivery vectors to promote cell uptake and membrane translocation [24–26]. Arthanari et al. utilized a fusion peptide of TAT with a membranolytic peptide (LK15) to deliver therapeutic shRNA and siRNA targeting bcr–abl fusion gene in chronic myeloid leukemia K562 cells [27].

4.3.2 Endosomal Escape

It has been suggested that CPPs avoid sequestration of RNAi therapeutics in the endo/lysosomal pathway and facilitate RNAi cargo delivery directly into the cytosol [64]. The endosomal-independent membrane translocation is presumably driven by transmembrane potential [65, 66], with acylation of arginine-rich CPPs being able to promote direct penetration of the delivery system through the plasma membrane [67]. Rydström et al. suggested a CPP-based siRNA delivery system, named CADY, with some evidence of direct membrane translocation across the plasma membrane [28].

Despite the continuing desire and attempts to avoid endo/lysosomal trafficking, the vast majority of RNAi delivery vectors are internalized by cells via endocytosis. The intracellular trafficking thus starts in early endosomes and proceeds through the stage of late endosomes, which are acidified by a vacuolar-type H^+ ATPase membrane proton pump [68]. In the absence of an escaping mechanism, RNAi vectors remain trapped in the endosomes and are subsequently routed to the lysosomes where they are exposed to lysosomal enzymes and likely degraded [69]. It is imperative that RNAi vectors either escape or facilitate release of free RNAi therapeutics from the endosomes before lysosomal degradation.

In addition to avoiding lysosomal degradation, finding alternatives to endocytosis as means of internalizing RNAi therapeutics might be desirable also because of the presence of Toll-like receptors TLR 7 and TLR 8 in the endosomes, which recognize exogenous siRNA and potentiate innate immune response [70, 71]. Chemical modification (i.e., 2′-*O*-methyl) of the 2′-ribose site of siRNA, however, has been shown to block TLR-mediated immune activation [29, 30, 70, 72].

Endosomal escape is one of the most significant barriers encountered during delivery of RNAi therapeutics, and multiple strategies have been devised that enhance the efficiency of endosomal escape.

4.3.2.1 Proton Sponge Effect

So-called proton sponge polycations, i.e., polycations with high buffering capacity in the endo/lysosomal pH range, represent the most widely employed design principle in the development of nucleic acid delivery vectors. These polycations are thought to enhance endosomolysis by absorbing protons and preventing the acidification of the endosomes, ultimately resulting in the release of the delivery vectors into the cytoplasm. The proposed mechanism of action relies on elevated influx of the protons to endosomes mediated by the H+-ATPase. The increased concentration of the protons and the counter ions in the endosomal vesicles lead to increasing osmotic pressure, causing endosomal swelling, membrane rupture, and finally the release of the RNAi vectors into the cytosol [73]. Despite growing evidence questioning the validity of the proton sponge hypothesis, polycations that were designed following this hypothesis, nevertheless, have become among the most successful ones in nucleic acid delivery.

Polyethyleneimine (PEI) was the first and remains the most wildly used polycation based on the proton sponge hypothesis [31–33]. The mix of primary, secondary, and tertiary amines in the PEI chain not only provide multiple positive charges at physiological conditions to bind with the negatively charged siRNA but they also act as the proton sponge to facilitate the endosomal escape [34]. Unfortunately, the high cationic charge density also contributes to high cytotoxicity of PEI, which has impeded its further applications in vivo [74, 75]. Efforts have been made to improve this “gold standard” polycation by conjugating with PEG or by introducing biodegradable moieties into the structure to decrease the toxicity [35, 36].

PAMAM dendrimers are another group of RNAi delivery vectors with high buffering capacity [37–39]. These treelike polymers have highly controlled and defined sizes and architecture with surface primary amines and central tertiary amines. These multiple amino groups provide high buffering capacity and proton absorbing ability, which improve the endosomal escape of the vectors [40].

4.3.2.2 Stimuli-Responsive Vectors

Additional approaches to improving the endosomal escape of RNAi therapeutics rely on stimulus-responsive lipids and polymers. The main categories include pH-sensitive vectors, temperature-sensitive vectors, and photosensitive vectors [76].

Langer et al. used titratable pH-sensitive lipids to design polymer-protected cationic liposomes with minimized binding of serum proteins. The liposomes were then able to shed the protecting polymer and restore their fusogenic ability after internalization into the acidic environment in the endosomes [77]. Carmona et al. developed liposomes stabilized by PEG attached to cholesterol by pH-sensitive oxime linkage, whose hydrolysis then facilitates pH-triggered escape of the liposomes from the endosomes [41]. Andaloussi et al. utilized pH titratable trifluoromethylquinoline moieties and covalently linked them to CPP, which resulted in rapid endosomal release [42].

Pluronic block copolymers exhibit a thermally reversible swelling/deswelling behavior, which triggers endosomal escape of RNAi delivery vectors. Nanocapsules prepared from PEI and Pluronic were able to efficiently deliver siRNA, in part, due to improved endosomal escape [43].

Photosensitive vectors utilize photosensitizers for endosomal disruption. Upon exposure to light, photosensitizers are stimulated to form reactive oxygen species leading to the damage of endosomal membranes, causing release of the RNAi vectors in the cytoplasm. Oliveira et al. have shown that by incorporating photosensitizers into siRNA delivery vectors which target epidermal growth factor receptor (EGFR), endosomal escape efficiency of siRNA vectors improved significantly, resulting in a tenfold increase in the knockdown efficacy of EGFR [44].

4.3.2.3 Fusogenic Lipids and Peptides

Lipid vectors with non-lamellar structure adhere to anionic vesicles without inducing fusion event [68]. In contrast, lipids with inverted hexagonal structure fuse with anionic membranes and cause the release of nucleic acids from endosomes [78]. Thus, incorporation of fusogenic lipids such as 1,2-dioleoyl-sn-glycero-3-phosphoethanolamine (DOPE) into the delivery vectors enhances endosomal escape of siRNA [45].

An alternative to fusogenic lipids is to use short peptides that mimic the fusion process of viruses with endosomal membranes during infection. Several synthetic fusogenic peptides mimicking the fusion domain of the influenza virus have been developed. Oliveira et al. reported that by incorporating influenza-derived fusogenic peptide diINF-7 into the delivery system, gene silencing efficiency of siRNA targeting EGFR and K-ras oncogenes is significantly enhanced [46].

Another alternative approach is to introduce pore-forming ability of viroporins into the delivery vectors. Viroporins are a group of hydrophobic proteins that participate in several steps during viral infection, including the release of viral particles from cells, altering membrane permeability, creating channels, and facilitating ion flow across membranes [79]. Kwon and coworkers have chosen a membranolytic peptide from the endodomain of HIV gp41 and covalently linked it to PEI. The peptide-modified polyplexes exhibited significantly enhanced siRNA delivery efficiency compared to unmodified control polyplexes. Confocal microscopy imaging of intracellular polyplex distribution suggested that the enhanced transfection was the result of increased endosomal escape [47].

4.3.3 Transport Through the Cytoplasm

Different types of RNAi triggers require delivery to different intracellular compartments. Mechanism of action of pri-miRNA, pre-miRNA, and shRNA demands delivery to the nucleus. Even siRNA may benefit from controlled localization within the cytoplasm, as evidenced by higher activity of siRNA delivered to the perinuclear region of the cell. The cytoplasmic environment is composed of densely packed soluble proteins and cytoskeleton protein structures such as actin filaments and microtubules. This dense packing creates a substantial barrier for trafficking of large delivery vectors because their passive diffusion is severely limited. The trafficking of the delivery vectors, and endosomes containing the vectors, mainly relies on microtubules and kinesin/dynein motor proteins [80, 81]. Disrupting the microtubule network using nocodazole to depolymerize tubulin significantly decreases the cytoplasmic transport of the delivery vectors. In contrast, applying cyclic mechanical stretch to the cells to reorganize and stabilize the microtubules leads to tubulin acetylation and enhanced endosomal trafficking leading to improved transfection efficiency [82].

4.3.4 Nuclear Entry

Early research from the 1980s to 1990s showed that microinjection of exogenous genes into the nucleus produced strong gene expression, while injection into the cytoplasm resulted in poor expression [83–85]. These findings clearly demonstrated that the nuclear envelope is a crucial barrier for delivery of nucleic acids to the nucleus. Successful delivery of shRNA and pri- and pre-miRNA vectors requires efficient entry to the nucleus, and the challenges thus mirror those encountered in gene delivery. This problem is somewhat minimized in rapidly dividing cells because delivery vectors can gain access to the nucleus during the process of mitosis when the nuclear envelope breaks down and becomes permeable to the material present in the cytoplasm [86]. However, in non-dividing cells, the nuclear envelope is a major barrier for nonviral delivery vectors. Significant effort has been made in improving nuclear delivery of DNA in the past, but not until recently have similar approaches been applied to delivery of RNAi triggers.

First discovered in the SV40 large T-antigen [87], nuclear localization signals (NLS) have become a convenient tool for targeting delivery systems to the nucleus. The NLS sequence when conjugated to the delivery vector provides the ability to guide through the nuclear pore complex into the nucleus. Rahbek et al. have designed delivery vectors based on reducible copolypeptides of endosomolytic peptide and an NLS peptide to deliver siRNA and pri-miRNA. The inclusion of NLS promoted nuclear localization of the RNAi molecules, allowing for transcriptional silencing of EF1A and Drosha- and Dicer-dependent expression of mature miRNA. The authors observed an NLS-dependent effect on the production of mature miRNA. An optimized content of NLS was necessary to achieve maximum levels of mature miRNA when delivering pri-miRNA using the described delivery vectors [11].

4.3.5 RNAi Therapeutic Release from Delivery Vectors

RNAi triggers can be delivered using a variety of vectors in which they are either directly conjugated by a covalent bond to a carrier molecule or noncovalently complexed or encapsulated in lipid or polymeric materials. Regardless of the delivery method, RNAi molecules have to be released from the delivery vectors to become biologically active. Most of the delivery vectors are designed based on the assumption that it is ideal to release the RNAi molecules at the sites of their action rather than relying on endogenous trafficking processes of naked RNAi molecules within the cells [88]. At present, it is not clearly understood whether intracellular release of RNAi triggers represents a significant barrier to delivery; vectors that permit spatial control of RNAi release have been developed. Hydrolytic and reductive degradation are the two main approaches to control intracellular cargo release. The rationale for developing biodegradable delivery vectors is based on the requirement to control RNAi molecule release and to lower the vector toxicity by degradation into less toxic low molecular weight by-products that avoid intracellular accumulation of the vector material.

4.3.5.1 Hydrolytically Degradable Delivery Vectors

Hydrolytically degradable vectors are usually based on polymers with degradable ester or, less frequently, amide bonds [89]. For example, poly(β-amino esters) (PBAE) have received much attention due to their favorable degradation profile [90]. PBAE undergo hydrolysis in acidic environment to yield low molecular weight β-amino acids and dialcohols [48]. Jere et al. used PBAE to deliver siRNA and shRNA to lung cancer cells and achieved a nearly 1.5 times greater silencing in addition to improving the safety profile of the delivery vector when compared with control nondegradable PEI [49]. Saltzman and coworkers deployed a well-known degradable polymer, poly(lactide-*co*-glycolide), for local delivery of siRNA to vaginal mucosa [50]. This biocompatible and biodegradable polymer with favorable sustained release characteristics showed promising properties for controlled siRNA delivery [51, 52].

Fischer et al. reported another interesting design of a hydrolytically degradable vector. The authors synthesized several polycations with core-shell architecture that contained a shell of amines attached via a hydrolyzable carbamate bond to a polyglycerol core. It was suggested that enzymatic hydrolysis of the carbamates improved intracellular release of siRNA and contributed to enhanced silencing activity [53].

Natural polymers such as chitosan and atelocollagen represent another type of degradable carriers for siRNA delivery [91–93]. Chitosan is degraded enzymatically by lysozyme and *N*-acetyl-glucosaminidase [94] and, is thus, expected that the degradation and release of RNAi cargo may be triggered following intracellular delivery of the chitosan-based vectors [95]. For example, Howard et al. described a chitosan-based siRNA delivery system, which mediated efficient knockdown of endogenous enhanced green fluorescent protein in H1299 human lung carcinoma cells and murine peritoneal macrophages (78% and 89% knockdown of the fluorescence signal, respectively) [54].

4.3.5.2 Reductively Degradable Delivery Vectors

Hydrolytically degradable delivery vectors offer only a limited spatial control of degradation and release of RNAi molecules. Reductively degradable vectors have emerged as more suitable alternatives because of the existence of robust and reliable redox potential gradient that localizes their degradation predominantly to the intracellular space (cytoplasm and nucleus) thus making them more favorable for RNAi delivery [96]. Bioreducible delivery vectors with incorporated disulfide bonds into the vector structure exhibit the capability of responding to the redox potential gradients between the oxidative extracellular environment and the reducing intracellular space. The existence of this redox potential gradient is determined by small redox molecules like glutathione (GSH) with the assistance of redox enzymes [97–100]. The low GSH concentration (1–2 μM) in the plasma allows the nucleic acids to remain condensed in the delivery vector, while high intracellular GSH level (>1 mM in the cytoplasm and ~20 mM in the nucleus) [101] triggers the reduction of the disulfide bonds, followed by fast disassembly of the polyplexes and release of the RNAi molecules. Introducing reducible disulfide bonds into the backbone of the polycations or lipids allows nanoparticles to undergo intracellular thiol–disulfide exchange reactions, leading to the cleavage of the disulfide bonds and fragmentation of the polycations or lipids. The increased disassembly rate of the RNAi delivery vesicles ultimately enhances the knockdown efficiency and decreases cytotoxicity of the vectors [55–59].

4.4 Future Perspectives

In order to design effective RNAi delivery systems and to advance RNAi therapy into clinic, countless efforts are being made to improve our understanding of the intracellular trafficking of the delivery vectors. To design efficient and safe vectors for RNAi delivery, each and every intracellular barrier including cell uptake, endosomal escape, vector disassembly, and, where applicable, nuclear localization should be taken into consideration. Off-target effects and immune responses should also be taken into account especially for siRNA delivery. Viral RNAi delivery methods suffer from several safety and preparation issues and are less feasible for modifications. On the other hand, synthetic vectors are more versatile in designing, which allows specific modifications to overcome each barrier. The synthetic delivery systems have shown great potential, and they promise to not only become useful research tools but also be applied in future clinical trials.

An important consideration is the balance between the extracellular stability and intracellular disassembly to allow RNAi trigger release and subsequent interaction with the RNAi machinery.

Tremendous progress has been made in our understanding of intracellular trafficking of RNAi vectors over the past several years. As increasing numbers of researchers join the field of RNAi delivery, we will undoubtedly see further progress, which will translate better understanding of the underlying processes to designing more efficacious delivery vectors. The improvements in delivery vectors will be closely tied with the rapidly developing field of nanotechnology, especially in areas of multifunctional delivery systems that will permit improved integration of multiple functionalities into RNAi delivery vectors. Major shortcoming of the current research on intracellular trafficking of RNAi delivery vectors has been its almost exclusive focus on in vitro experiments. Considering how poorly translatable and predictive these findings have so far been to in vivo conditions, it is important that future research focuses on enhancing our understanding of intracellular trafficking of RNAi vectors in vivo.

References

1. Fire A, Xu SQ, Montgomery MK, Kostas SA, Driver SE, Mello CC (1998) Potent and specific genetic interference by double-stranded RNA in Caenorhabditis elegans. Nature 391: 806–811
2. Rao NM (2010) Cationic lipid-mediated nucleic acid delivery: beyond being cationic. Chem Phys Lipids 163:245–252
3. Siolas D, Lerner C, Burchard J, Ge W, Linsley PS, Paddison PJ, Hannon GJ, Cleary MA (2005) Synthetic shRNAs as potent RNAi triggers. Nat Biotechnol 23:227–231
4. Kim DH, Behlke MA, Rose SD, Chang MS, Choi S, Rossi JJ (2005) Synthetic dsRNA Dicer substrates enhance RNAi potency and efficacy. Nat Biotechnol 23:222–226
5. Rose SD, Kim DH, Amarzguioui M, Heidel JD, Collingwood MA, Davis ME, Rossi JJ, Behlke MA (2005) Functional polarity is introduced by Dicer processing of short substrate RNAs. Nucleic Acids Res 33:4140–4156
6. Weinberg MS, Villeneuve LM, Ehsani A, Amarzguioui M, Aagaard L, Chen ZX, Riggs AD, Rossi JJ, Morris KV (2006) The antisense strand of small interfering RNAs directs histone methylation and transcriptional gene silencing in human cells. RNA 12:256–262
7. Robb GB, Brown KM, Khurana J, Rana TM (2005) Specific and potent RNAi in the nucleus of human cells. Nat Struct Mol Biol 12:133–137
8. Langlois MA, Boniface C, Wang G, Alluin J, Salvaterra PM, Puymirat J, Rossi JJ, Lee NS (2005) Cytoplasmic and nuclear retained DMPK mRNAs are targets for RNA interference in myotonic dystrophy cells. J Biol Chem 280:16949–16954
9. Haussecker D, Proudfoot NJ (2005) Dicer-dependent turnover of intergenic transcripts from the human beta-globin gene cluster. Mol Cell Biol 25:9724–9733
10. Morris KV, Chan SW, Jacobsen SE, Looney DJ (2004) Small interfering RNA-induced transcriptional gene silencing in human cells. Science 305:1289–1292
11. Rahbek UL, Howard KA, Oupicky D, Manickam DS, Dong M, Nielsen AF, Hansen TB, Besenbacher F, Kjems J (2008) Intracellular siRNA and precursor miRNA trafficking using bioresponsive copolypeptides. J Gene Med 10:81–93
12. He L, Hannon GJ (2004) MicroRNAs: small RNAs with a big role in gene regulation. Nat Rev Genet 5:522–531

13. Lee Y, Jeon K, Lee JT, Kim S, Kim VN (2002) MicroRNA maturation: stepwise processing and subcellular localization. EMBO J 21:4663–4670
14. Boden D, Pusch O, Silbermann R, Lee F, Tucker L, Ramratnam B (2004) Enhanced gene silencing of HIV-1 specific siRNA using microRNA designed hairpins. Nucleic Acids Res 32:1154–1158
15. Krutzfeldt J, Rajewsky N, Braich R, Rajeev KG, Tuschl T, Manoharan M, Stoffel M (2005) Silencing of microRNAs in vivo with 'antagomirs'. Nature 438:685–689
16. Yang M, Mattes J (2008) Discovery, biology and therapeutic potential of RNA interference, microRNA and antagomirs. Pharmacol Ther 117:94–104
17. Tokatlian T, Segura T (2010) siRNA applications in nanomedicine. Wiley Interdiscip Rev Nanomed Nanobiotechnol 2:305–315
18. Juliano R, Alam MR, Dixit V, Kang H (2008) Mechanisms and strategies for effective delivery of antisense and siRNA oligonucleotides. Nucleic Acids Res 36:4158–4171
19. York AW, Zhang Y, Holley AC, Guo Y, Huang F, McCormick CL (2009) Facile synthesis of multivalent folate-block copolymer conjugates via aqueous RAFT polymerization: targeted delivery of siRNA and subsequent gene suppression. Biomacromolecules 10:936–943
20. Zhou J, Li H, Li S, Zaia J, Rossi JJ (2008) Novel dual inhibitory function aptamer-siRNA delivery system for HIV-1 therapy. Mol Ther 16:1481–1489
21. Zhou J, Li H, Zhang J, Piotr S, Rossi J (2011) Development of cell-type specific anti-HIV gp120 aptamers for siRNA delivery. J Vis Exp. doi:10.3791/2954
22. Song E, Zhu P, Lee SK, Chowdhury D, Kussman S, Dykxhoorn DM, Feng Y, Palliser D, Weiner DB, Shankar P, Marasco WA, Lieberman J (2005) Antibody mediated in vivo delivery of small interfering RNAs via cell-surface receptors. Nat Biotechnol 23:709–717
23. Kumar P, Ban HS, Kim SS, Wu H, Pearson T, Greiner DL, Laouar A, Yao J, Haridas V, Habiro K, Yang YG, Jeong JH, Lee KY, Kim YH, Kim SW, Peipp M, Fey GH, Manjunath N, Shultz LD, Lee SK, Shankar P (2008) T cell-specific siRNA delivery suppresses HIV-1 infection in humanized mice. Cell 134:577–586
24. Nam HY, Kim J, Kim S, Yockman JW, Kim SW, Bull DA (2011) Cell penetrating peptide conjugated bioreducible polymer for siRNA delivery. Biomaterials 32:5213–5222
25. Ishihara T, Goto M, Kodera K, Kanazawa H, Murakami Y, Mizushima Y, Higaki M (2009) Intracellular delivery of siRNA by cell-penetrating peptides modified with cationic oligopeptides. Drug Deliv 16:153–159
26. Turner JJ, Jones S, Fabani MM, Ivanova G, Arzumanov AA, Gait MJ (2007) RNA targeting with peptide conjugates of oligonucleotides, siRNA and PNA. Blood Cells Mol Dis 38:1–7
27. Arthanari Y, Pluen A, Rajendran R, Aojula H, Demonacos C (2010) Delivery of therapeutic shRNA and siRNA by Tat fusion peptide targeting BCR-ABL fusion gene in Chronic Myeloid Leukemia cells. J Control Release 145:272–280
28. Rydstrom A, Deshayes S, Konate K, Crombez L, Padari K, Boukhaddaoui H, Aldrian G, Pooga M, Divita G (2011) Direct translocation as major cellular uptake for CADY self-assembling peptide-based nanoparticles. PLoS One 6:e25924
29. Sioud M, Furset G, Cekaite L (2007) Suppression of immunostimulatory siRNA-driven innate immune activation by 2′-modified RNAs. Biochem Biophys Res Commun 361: 122–126
30. Ocampo SM, Romero C, Avino A, Burgueno J, Gassull MA, Bermudez J, Eritja R, Fernandez E, Perales JC (2012) Functionally enhanced siRNA targeting TNFalpha attenuates DSS-induced colitis and TLR-mediated immunostimulation in mice. Mol Ther 20:382–390
31. Baker A, Saltik M, Lehrmann H, Killisch I, Mautner V, Lamm G, Christofori G, Cotten M (1997) Polyethylenimine (PEI) is a simple, inexpensive and effective reagent for condensing and linking plasmid DNA to adenovirus for gene delivery. Gene Ther 4:773–782
32. Zou SM, Erbacher P, Remy JS, Behr JP (2000) Systemic linear polyethylenimine (L-PEI)-mediated gene delivery in the mouse. J Gene Med 2:128–134
33. Urban-Klein B, Werth S, Abuharbeid S, Czubayko F, Aigner A (2005) RNAi-mediated gene-targeting through systemic application of polyethylenimine (PEI)-complexed siRNA in vivo. Gene Ther 12:461–466

34. Goyal R, Tripathi SK, Tyagi S, Sharma A, Ram KR, Chowdhuri DK, Shukla Y, Kumar P, Gupta KC (2012) Linear PEI nanoparticles: efficient pDNA/siRNA carriers in vitro and in vivo. Nanomedicine 8:167–175
35. Malek A, Merkel O, Fink L, Czubayko F, Kissel T, Aigner A (2009) In vivo pharmacokinetics, tissue distribution and underlying mechanisms of various PEI(-PEG)/siRNA complexes. Toxicol Appl Pharmacol 236:97–108
36. Xia W, Wang P, Lin C, Li Z, Gao X, Wang G, Zhao X (2011) Bioreducible polyethylenimine-delivered siRNA targeting human telomerase reverse transcriptase inhibits HepG2 cell growth in vitro and in vivo. J Control Release 157:427–436
37. Kukowska-Latallo JF, Bielinska AU, Johnson J, Spindler R, Tomalia DA, Baker JR Jr (1996) Efficient transfer of genetic material into mammalian cells using Starburst polyamidoamine dendrimers. Proc Natl Acad Sci USA 93:4897–4902
38. Tsutsumi T, Hirayama F, Uekama K, Arima H (2007) Evaluation of polyamidoamine dendrimer/alpha-cyclodextrin conjugate (generation 3, G3) as a novel carrier for small interfering RNA (siRNA). J Control Release 119:349–359
39. Arima H, Tsutsumi T, Yoshimatsu A, Ikeda H, Motoyama K, Higashi T, Hirayama F, Uekama K (2011) Inhibitory effect of siRNA complexes with polyamidoamine dendrimer/alpha-cyclodextrin conjugate (generation 3, G3) on endogenous gene expression. Eur J Pharm Sci 44:375–384
40. Patil ML, Zhang M, Minko T (2011) Multifunctional triblock Nanocarrier (PAMAM-PEG-PLL) for the efficient intracellular siRNA delivery and gene silencing. ACS Nano 5:1877–1887
41. Carmona S, Jorgensen MR, Kolli S, Crowther C, Salazar FH, Marion PL, Fujino M, Natori Y, Thanou M, Arbuthnot P, Miller AD (2009) Controlling HBV replication in vivo by intravenous administration of triggered PEGylated siRNA-nanoparticles. Mol Pharm 6:706–717
42. Andaloussi SE, Lehto T, Mager I, Rosenthal-Aizman K, Oprea II, Simonson OE, Sork H, Ezzat K, Copolovici DM, Kurrikoff K, Viola JR, Zaghloul EM, Sillard R, Johansson HJ, Said Hassane F, Guterstam P, Suhorutsenko J, Moreno PM, Oskolkov N, Halldin J, Tedebark U, Metspalu A, Lebleu B, Lehtio J, Smith CI, Langel U (2011) Design of a peptide-based vector, PepFect6, for efficient delivery of siRNA in cell culture and systemically in vivo. Nucleic Acids Res 39:3972–3987
43. Lee SH, Choi SH, Kim SH, Park TG (2008) Thermally sensitive cationic polymer nanocapsules for specific cytosolic delivery and efficient gene silencing of siRNA: swelling induced physical disruption of endosome by cold shock. J Control Release 125:25–32
44. Oliveira S, Fretz MM, Hogset A, Storm G, Schiffelers RM (2007) Photochemical internalization enhances silencing of epidermal growth factor receptor through improved endosomal escape of siRNA. Biochim Biophys Acta 1768:1211–1217
45. Hassani Z, Lemkine GF, Erbacher P, Palmier K, Alfama G, Giovannangeli C, Behr JP, Demeneix BA (2005) Lipid-mediated siRNA delivery down-regulates exogenous gene expression in the mouse brain at picomolar levels. J Gene Med 7:198–207
46. Oliveira S, van Rooy I, Kranenburg O, Storm G, Schiffelers R (2007) Fusogenic peptides enhance endosomal escape improving siRNA-induced silencing of oncogenes. Int J Pharm 331:211–214
47. Kwon EJ, Bergen JM, Pun SH (2008) Application of an HIV gp41-derived peptide for enhanced intracellular trafficking of synthetic gene and siRNA delivery vehicles. Bioconjug Chem 19:920–927
48. Vandenbroucke RE, De Geest BG, Bonne S, Vinken M, Van Haecke T, Heimberg H, Wagner E, Rogiers V, De Smedt SC, Demeester J, Sanders NN (2008) Prolonged gene silencing in hepatoma cells and primary hepatocytes after small interfering RNA delivery with biodegradable poly(beta-amino esters). J Gene Med 10:783–794
49. Jere D, Xu CX, Arote R, Yun CH, Cho MH, Cho CS (2008) Poly(beta-amino ester) as a carrier for si/shRNA delivery in lung cancer cells. Biomaterials 29:2535–2547
50. Woodrow KA, Cu Y, Booth CJ, Saucier-Sawyer JK, Wood MJ, Saltzman WM (2009) Intravaginal gene silencing using biodegradable polymer nanoparticles densely loaded with small-interfering RNA. Nat Mater 8:526–533

51. Cun D, Jensen DK, Maltesen MJ, Bunker M, Whiteside P, Scurr D, Foged C, Nielsen HM (2011) High loading efficiency and sustained release of siRNA encapsulated in PLGA nanoparticles: quality by design optimization and characterization. Eur J Pharm Biopharm 77:26–35
52. Zhou J, Patel TR, Fu M, Bertram JP, Saltzman WM (2012) Octa-functional PLGA nanoparticles for targeted and efficient siRNA delivery to tumors. Biomaterials 33:583–591
53. Fischer W, Calderon M, Schulz A, Andreou I, Weber M, Haag R (2010) Dendritic polyglycerols with oligoamine shells show low toxicity and high siRNA transfection efficiency in vitro. Bioconjug Chem 21:1744–1752
54. Howard KA, Rahbek UL, Liu X, Damgaard CK, Glud SZ, Andersen MO, Hovgaard MB, Schmitz A, Nyengaard JR, Besenbacher F, Kjems J (2006) RNA interference in vitro and in vivo using a novel chitosan/siRNA nanoparticle system. Mol Ther 14:476–484
55. Manickam DS, Li J, Putt DA, Zhou QH, Wu C, Lash LH, Oupicky D (2010) Effect of innate glutathione levels on activity of redox-responsive gene delivery vectors. J Control Release 141:77–84
56. Read ML, Bremner KH, Oupicky D, Green NK, Searle PF, Seymour LW (2003) Vectors based on reducible polycations facilitate intracellular release of nucleic acids. J Gene Med 5:232–245
57. Read ML, Singh S, Ahmed Z, Stevenson M, Briggs SS, Oupicky D, Barrett LB, Spice R, Kendall M, Berry M, Preece JA, Logan A, Seymour LW (2005) A versatile reducible polycation-based system for efficient delivery of a broad range of nucleic acids. Nucleic Acids Res 33:e86
58. Beloor J, Choi CS, Nam HY, Park M, Kim SH, Jackson A, Lee KY, Kim SW, Kumar P, Lee SK (2012) Arginine-engrafted biodegradable polymer for the systemic delivery of therapeutic siRNA. Biomaterials 33:1640–1650
59. Won YW, Yoon SM, Lee KM, Kim YH (2011) Poly(oligo-D-arginine) with internal disulfide linkages as a cytoplasm-sensitive carrier for siRNA delivery. Mol Ther 19:372–380
60. Shim MS, Kwon YJ (2010) Efficient and targeted delivery of siRNA in vivo. FEBS J 277:4814–4827
61. Zidovska A, Evans HM, Ahmad A, Ewert KK, Safinya CR (2009) The role of cholesterol and structurally related molecules in enhancing transfection of cationic liposome-DNA complexes. J Phys Chem B 113:5208–5216
62. Phalon C, Rao DD, Nemunaitis J (2010) Potential use of RNA interference in cancer therapy. Expert Rev Mol Med 12:e26
63. McNamara JO II, Andrechek ER, Wang Y, Viles KD, Rempel RE, Gilboa E, Sullenger BA, Giangrande PH (2006) Cell type-specific delivery of siRNAs with aptamer-siRNA chimeras. Nat Biotechnol 24:1005–1015
64. Morris MC, Deshayes S, Heitz F, Divita G (2008) Cell-penetrating peptides: from molecular mechanisms to therapeutics. Biol Cell 100:201–217
65. Terrone D, Sang SL, Roudaia L, Silvius JR (2003) Penetratin and related cell-penetrating cationic peptides can translocate across lipid bilayers in the presence of a transbilayer potential. Biochemistry 42:13787–13799
66. Thoren PE, Persson D, Isakson P, Goksor M, Onfelt A, Norden B (2003) Uptake of analogs of penetratin, Tat(48-60) and oligoarginine in live cells. Biochem Biophys Res Commun 307:100–107
67. Katayama S, Hirose H, Takayama K, Nakase I, Futaki S (2011) Acylation of octaarginine: implication to the use of intracellular delivery vectors. J Control Release 149:29–35
68. Dominska M, Dykxhoorn DM (2010) Breaking down the barriers: siRNA delivery and endosome escape. J Cell Sci 123:1183–1189
69. Di Guglielmo GM, Le Roy C, Goodfellow AF, Wrana JL (2003) Distinct endocytic pathways regulate TGF-beta receptor signalling and turnover. Nat Cell Biol 5:410–421
70. Sioud M (2006) Innate sensing of self and non-self RNAs by Toll-like receptors. Trends Mol Med 12:167–176
71. Sioud M (2006) RNA interference below the immune radar. Nat Biotechnol 24:521–522

72. Lares MR, Rossi JJ, Ouellet DL (2010) RNAi and small interfering RNAs in human disease therapeutic applications. Trends Biotechnol 28:570–579
73. Sonawane ND, Szoka FC Jr, Verkman AS (2003) Chloride accumulation and swelling in endosomes enhances DNA transfer by polyamine-DNA polyplexes. J Biol Chem 278:44826–44831
74. Florea BI, Meaney C, Junginger HE, Borchard G (2002) Transfection efficiency and toxicity of polyethylenimine in differentiated Calu-3 and nondifferentiated COS-1 cell cultures. AAPS PharmSci 4:E12
75. Aravindan L, Bicknell KA, Brooks G, Khutoryanskiy VV, Williams AC (2009) Effect of acyl chain length on transfection efficiency and toxicity of polyethylenimine. Int J Pharm 378:201–210
76. Pecot CV, Calin GA, Coleman RL, Lopez-Berestein G, Sood AK (2011) RNA interference in the clinic: challenges and future directions. Nat Rev Cancer 11:59–67
77. Auguste DT, Furman K, Wong A, Fuller J, Armes SP, Deming TJ, Langer R (2008) Triggered release of siRNA from poly(ethylene glycol)-protected, pH-dependent liposomes. J Control Release 130:266–274
78. Koltover I, Salditt T, Radler JO, Safinya CR (1998) An inverted hexagonal phase of cationic liposome-DNA complexes related to DNA release and delivery. Science 281:78–81
79. Gonzalez ME, Carrasco L (2003) Viroporins. FEBS Lett 552:28–34
80. Kulkarni RP, Castelino K, Majumdar A, Fraser SE (2006) Intracellular transport dynamics of endosomes containing DNA polyplexes along the microtubule network. Biophys J 90:L42–L44
81. Leopold PL, Kreitzer G, Miyazawa N, Rempel S, Pfister KK, Rodriguez-Boulan E, Crystal RG (2000) Dynein- and microtubule-mediated translocation of adenovirus serotype 5 occurs after endosomal lysis. Hum Gene Ther 11:151–165
82. Geiger RC, Kaufman CD, Lam AP, Budinger GR, Dean DA (2009) Tubulin acetylation and histone deacetylase 6 activity in the lung under cyclic load. Am J Respir Cell Mol Biol 40:76–82
83. Capecchi MR (1980) High efficiency transformation by direct microinjection of DNA into cultured mammalian cells. Cell 22:479–488
84. Graessmann M, Menne J, Liebler M, Graeber I, Graessmann A (1989) Helper activity for gene expression, a novel function of the SV40 enhancer. Nucleic Acids Res 17:6603–6612
85. Zabner J, Fasbender AJ, Moninger T, Poellinger KA, Welsh MJ (1995) Cellular and molecular barriers to gene transfer by a cationic lipid. J Biol Chem 270:18997–19007
86. Dean DA, Strong DD, Zimmer WE (2005) Nuclear entry of nonviral vectors. Gene Ther 12:881–890
87. Kalderon D, Roberts BL, Richardson WD, Smith AE (1984) A short amino acid sequence able to specify nuclear location. Cell 39:499–509
88. Howard KA (2009) Delivery of RNA interference therapeutics using polycation-based nanoparticles. Adv Drug Deliv Rev 61:710–720
89. Nair LS, Laurencin CT (2007) Biodegradable polymers as biomaterials. Prog Polym Sci 32:762–798
90. Schaffert D, Wagner E (2008) Gene therapy progress and prospects: synthetic polymer-based systems. Gene Ther 15:1131–1138
91. Rudzinski WE, Aminabhavi TM (2010) Chitosan as a carrier for targeted delivery of small interfering RNA. Int J Pharm 399:1–11
92. Forootan SS, Bao ZZ, Forootan FS, Kamalian L, Zhang Y, Bee A, Foster CS, Ke Y (2010) Atelocollagen-delivered siRNA targeting the FABP5 gene as an experimental therapy for prostate cancer in mouse xenografts. Int J Oncol 36:69–76
93. Minakuchi Y, Takeshita F, Kosaka N, Sasaki H, Yamamoto Y, Kouno M, Honma K, Nagahara S, Hanai K, Sano A, Kato T, Terada M, Ochiya T (2004) Atelocollagen-mediated synthetic small interfering RNA delivery for effective gene silencing in vitro and in vivo. Nucleic Acids Res 32:e109

94. Nordtveit RJ, Varum KM, Smidsrod O (1996) Degradation of partially N-acetylated chitosans with hen egg white and human lysozyme. Carbohydr Polym 29:163–167
95. Koping-Hoggard M, Tubulekas I, Guan H, Edwards K, Nilsson M, Varum KM, Artursson P (2001) Chitosan as a nonviral gene delivery system. Structure-property relationships and characteristics compared with polyethylenimine in vitro and after lung administration in vivo. Gene Ther 8:1108–1121
96. Du FS, Wang Y, Zhang R, Li ZC (2010) Intelligent nucleic acid delivery systems based on stimuli-responsive polymers. Soft Matter 6:835–848
97. Jiang XM, Fitzgerald M, Grant CM, Hogg PJ (1999) Redox control of exofacial protein thiols/disulfides by protein disulfide isomerase. J Biol Chem 274:2416–2423
98. Donoghue N, Hogg PJ (2002) Characterization of redox-active proteins on cell surface. Methods Enzymol 348:76–86
99. Gilbert HF (1997) Protein disulfide isomerase and assisted protein folding. J Biol Chem 272:29399–29402
100. Hansen RE, Roth D, Winther JR (2009) Quantifying the global cellular thiol-disulfide status. Proc Natl Acad Sci USA 106:422–427
101. Soundara Manickam D, Oupicky D (2006) Polyplex gene delivery modulated by redox potential gradients. J Drug Target 14:519–526

Chapter 5
Mucosal Delivery of RNAi Therapeutics

Borja Ballarín González, Ebbe Bech Nielsen, Troels Bo Thomsen, and Kenneth A. Howard

Abstract The effectiveness of RNA interference-based drugs is dependent on accumulation at the target site in therapeutically relevant amounts. Local administration to the mucosal surfaces lining the respiratory, gastrointestinal and genitourinary tracts allows access into diseased areas without the necessity to overcome serum nuclease degradation, rapid renal and hepatic clearance and non-specific tissue accumulation associated with systemic delivery. This work describes RNAi therapeutics focused on pulmonary, oral, rectal and intravaginal routes of administration. Mucosal barrier components including site variations and delivery considerations are addressed in order to design an effective mucosal delivery strategy.

5.1 Introduction

Regulation of cellular gene expression by harnessing the natural process of RNA interference (RNAi) offers an exciting gene medicine approach [1, 2]. Post-transcriptional silencing occurs by mRNA engagement with small interfering RNA (siRNA) or microRNA (miRNA) facilitated by complementary base pairing [3]. Gene specificity coupled with the capability for externally introduced synthetic siRNA and miRNA to be recruited into the cellular RNAi pathway provides the rational for RNAi drug development. A greater understanding of the molecular mechanism of RNAi has resulted in a wide repertoire of potential RNAi drugs involved in the RNAi pathway cascade that offers diverse therapeutic options.

B.Ballarin González • E.B. Nielsen • T.B. Thomsen • K.A. Howard, Ph.D. (✉)
Department of Molecular Biology and Genetics, Interdisciplinary Nanoscience Center, Aarhus University, Gustav Wieds Vej 14, Building 1593, 8000 Aarhus C, Denmark
e-mail: kenh@inano.au.dk

K.A. Howard (ed.), *RNA Interference from Biology to Therapeutics*, Advances in Delivery Science and Technology, DOI 10.1007/978-1-4614-4744-3_5,

The clinical potential of nucleic acid-based drugs is restricted by the susceptibility to serum nuclease degradation, rapid renal clearance and non-specific tissue accumulation [4]. Furthermore, the macromolecular and polyanionic nature reduces interaction and uptake across the cellular membrane required for recruitment into the intracellular RNAi machinery. Improvements in both extracellular and intracellular delivery are key to the therapeutic success of RNAi therapeutics. Chemical modification [5], conjugation [6] and incorporation into nanoparticle-based delivery systems [7, 8] are common strategies that have been employed to maximise delivery [9, 10].

The route of administration is an important determinant for successful RNA-based silencing therapeutics. The administration route dictates both migratory pathway and biological barriers the drug must undertake in order to reach its target. Local administration to the mucosal surfaces lining the respiratory, gastrointestinal and genitourinary tract is an attractive alternative to the intravenous route [11–13]. It is a non-invasive method that avoids hepatic and renal clearance associated with the systemic route and allows direct access to regions that are the main portal of entry and pathogenesis for many pathogens, inflammation and cancer.

Recent Phase II clinical trials with RNAi therapeutics delivered directly to the lung [14] highlight the potential and support the use of the mucosal route.

This work describes pulmonary, oral, rectal and intravaginal delivery of RNAi therapeutics focused on nanoparticle-based delivery of synthetic siRNA. Attention will be given to the biological and physical barriers occurring at the mucosal surfaces that restrict uptake of luminal material. Strategies to improve mucosal penetration will be discussed with a view to better design of mucosal delivery systems.

5.2 Mucosal Barriers

RNAi-based therapeutics must overcome the physical barrier of the mucus gel layer and tightly packed epithelial cells combined with mucus capture and consequent active clearance mechanism. Understanding these barriers and their evolutionary differences can provide guidelines for siRNA-based therapy targeted at specific mucosal sites. This section focuses on mucus and epithelial components relevant to naked siRNA and nanoparticle-based siRNA delivery.

5.2.1 Mucus

Mucus is a hydrated protein gel which overlays the luminal surface at mucosal sites and serves as a barrier between the external environment and the underlying tissue. It lines the respiratory, gastrointestinal and genitourinary tracts, and eyes (Fig. 5.1). Its role is to serve as a first line of defence against various pathogens [15] and toxins

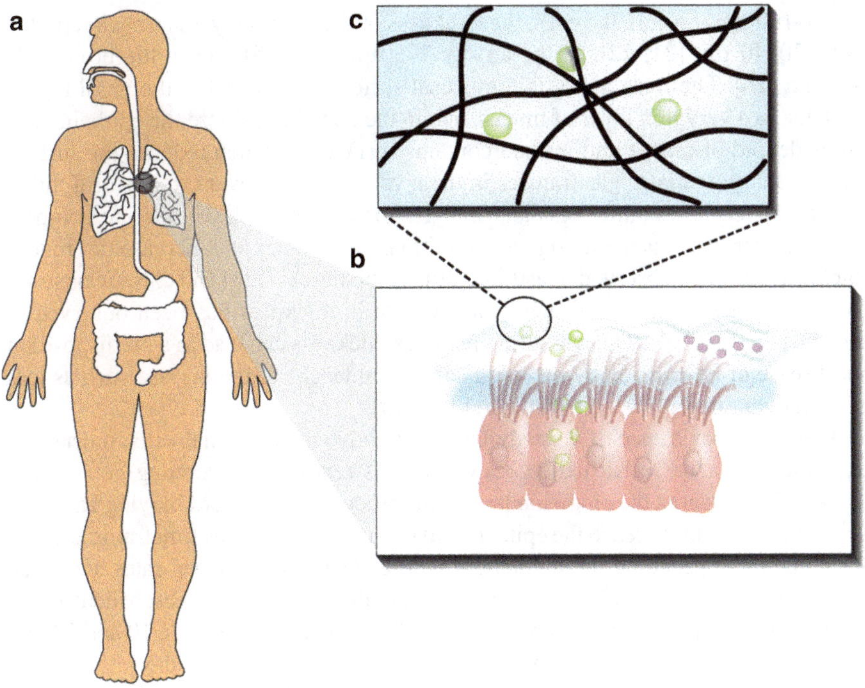

Fig. 5.1 Schematic representation of nanoparticle uptake across mucosal epithelium. (**a**) The mucosal surfaces lining the respiratory, gastrointestinal and genitourinary tracts and eyes. (**b**) The mucosal epithelial border restricts penetration of luminal particulates (*right*) through a combination of overlying mucus composed of a lower steady layer (dark) and an upper layer (light), tight cellular junctions and ciliary clearance. Exploitation of mucoadhesive and mucopermeable particles allows cellular uptake across mucosal surfaces (*left*). (**c**) Particles diffusing through the network of mucin fibres

[16] and to facilitate continuous exchange of nutrient, water and gases. Mucus has macroscopic properties of a gel and exhibits non-Newtonian rheological behaviour [17, 18]. It is composed of ions, glycoproteins (termed mucins), proteins, lipids, DNA and cellular debris [18]. The mucins are extended 0.5–40-MDa molecules that are produced and secreted by goblet cells. It has a two-layer composition composed of a lower steady-state layer in contact with the epithelium and a mobile outer layer. Mucus site variations have evolved to suit the role performed at particular sites.

5.2.1.1 Site Variations

Mucus thickness, rate of renewal and pH are properties that can vary between tissues. The mucus layer thickness determines the accessibility to the underlying epithelium and depends on both luminal conditions and functional requirements of the underlying tissue. The layer thickness varies along the human gastrointestinal tract with the thickest layer in the stomach (~50–450 μm) [19] and in the colon

(~110–160 μm) [20]. In the eyes, the thickness of the mucus layer has been reported to be 30–40 μm [21, 22], in the airway 7–30 μm [23–25] and in the bronchiole ~55 μm [26]. The most accessible mucosal surfaces are found in the nasal region [27] due to a very thin layer of mucus, and in the deep lung where the epithelial lining is devoid of mucus and instead contains surfactants which reduces the surface tension and potentiates gas transfer in the alveoli [28]. Relevant to mucosal therapeutic delivery is the thickness and integrity of the mucus layer may be compromised under various pathological conditions. For example increase in mucus thickness has been observed in asthma [29], cystic fibrosis [26] and chronic obstructive pulmonary disease [30], whilst a decrease in thickness has been observed in ulcerative colitis [20]. This variation in mucus thickness can lead to a predisposition for disease or be induced by pathological consequences of the disease such as loss of epithelial integrity as seen in ulcerative colitis.

Despite the existence of a steady layer just above the epithelial cells, mucus is a dynamic substance which undergoes continuous renewal by secreting goblet cells dispersed throughout the epithelial layer. This produces an outward moving barrier to any entity that aims to reach the epithelial layer and determines the timeframe allowed for particles to penetrate the epithelium before clearance. Renewal rates are tissue dependent which has implications for designing therapeutic strategies. Mucus in the nasal cavity is replaced approximately every 20 min [31] and between 10 and 20 min in the respiratory tract [32] compared to a clearance rate of 4–6 h in the gastrointestinal tract [33] as found in rats but values have not been fully determined in humans.

In addition to thickness and rate of renewal, the pH value varies at different sites, with the lung and nasal mucus being nearly neutral [34, 35], the eye possessing a weak basic pH [21] and the mucus of the stomach having a pH gradient from pH 1–2 at the luminal side to approximately pH 7 at the surface of the epithelial cells underlying the mucus [36, 37]. These pH variations could be utilised for pH-responsive delivery systems which release their cargo at specific mucosal sites.

5.2.1.2 Mucus Penetration

The protective properties of mucus pose a barrier to RNAi-based therapeutics both in naked or nanoparticle form. Mucus constituents such as the glycoproteins (mucins), cellular debris and lipids form a heterogeneous environment through which drugs and/or drug carriers need to diffuse to reach their target [38]. Glycosylated domains of the mucin fibres possess a negative charge under physiological conditions, and hence mucus selectively controls the diffusion of particles not only through particle physical parameters such as size, but also by their chemical surface properties. Based on the net negative charge of mucins, one could speculate that naked siRNA might be repelled by the mucin fibres. Multivalent interactions between particles and the mucus network are main determinants of particle diffusion. Both electrostatic and hydrophobic interactions occur, and the possibility of making large numbers of non-specific hydrophobic interactions together with the more thermal stable electrostatic interactions enables mucus to trap particles. Several studies have demonstrated the efficacy of hydrophobic interactions to immobilise particles within mucus [38–41] and Ribbeck and co-workers have shown that particle surface charge, density and pH of mucin hydrogels can alter particle diffusion [42]. Particles with a PEGylated surface possessing a neutral charge had a greater diffusion rate compared to its negative or positive charged counterpart [42]. Two approaches govern the design of particles for mucosal delivery (1) the mucoadhesive and (2) mucus penetrative approach.

Regarding the mucoadhesive strategy, much attention has been given to the design of particles which associate with the mucus barrier and hence lower the clearance rate. The essence of this strategy lies with the previously mentioned bi-layered structure of mucus. Association of particles with the lower undisturbed layer will avoid clearance and enhance the bioavailability of the bioactive entity. Another positive effect from mucus association is increased viscosity due to greater cross-linking of mucus fibres, which in turn may lower the clearance rate. Materials of choice have been the so-called mucoadhesive materials. A common characteristic of these materials is their adherence to mucus through various forces. A widely used mucoadhesive polymer is chitosan [43], which have been utilised to form particles with siRNA and exhibit mucosal silencing [7]. Thiolation of polymers have been shown to enhance mucus interaction through formation of disulfide linkages with mucins [44] and various thiolated chitosan's have been synthesised [45, 46]. Not only does the mucoadhesiveness prolong the bioavailability of particles at mucosal surfaces, but it can also alter the structure of mucus so that it becomes more permeable to siRNA-loaded particles [47]. As an alternative, mucus-penetrating particle with limited interaction with mucus and increased diffusion rates is an exciting approach. Coating with PEG has been demonstrated to mediate such surface properties on latex particles (200–500 nm) [48] and nanoparticles composed of a biodegradable di-block copolymer of poly (sebacic acid) and PEG [49]. It has been further shown that both molecular weight and degree of surface coverage determined the mobility of the coated particles [50]. Reports, however, have previously classified PEG as a mucoadhesive polymer [51–53]. As speculated by Lai et al. the contradicting reports might be attributable to variations in type of PEGylation used, but the use of PEG-coated particles have yet to demonstrate intracellular delivery of nucleic acids to the respiratory or gastrointestinal tract. Interestingly, the design of mucus-penetrating particles builds on lessons learned from nature where viruses with an equal surface density of positive and negative charges readily penetrate mucus barriers [54]. Thus, surface chemistry together with size appear to be determining factors for mucus penetration and evidence points towards neutral surfaces for effective diffusion.

5.2.2 *Epithelial Cell Barrier*

An ordered array of closely packed epithelial cells overlaying a basement membrane constitutes the mucosal epithelium. Cell type, morphology and arrangement differ dependent on site and function. For example the small intestine comprises of single-layered enterocytes assembled into structured villi that increase the adsorptive surface area. Whilst in the upper respiratory tract, movement of apical cilia on pseudostratified epithelium restricts interactions at the luminal surface. Common to all mucosal epithelium is the close packing of adjacent cells separated by tight junctions.

Material uptake across the epithelium can occur by transcellular or paracellular pathways that are determined by the physicochemical characteristics of the material. The main transcellular mechanism for nanoparticle transport across the epithelium is adsorptive or cell-mediated endocytosis. Modification of material properties or targeting specific sites can be used to maximise delivery across the mucosa. It is generally accepted that the tight junctions restrict paracellular transport of micro-nanoparticles; however, mucopenetration enhancers have been used to facilitate transient opening of the junctions and mediate paracellular movement of small molecules.

The migration of particles from local to systemic tissue relies on translocation through the lymphatic or vasculature system and is dependent on the physicochemical properties of the material and the site. Disease pathogenesis dictates whether there is a necessity for local and/or systemic delivery and should determine the therapeutic strategy adopted.

5.3 Pulmonary Delivery

Direct access to a vast array of lung-associated diseases makes the lung an ideal target for RNAi-based therapies. With a total surface area of 140 m^2 [54], the pulmonary route offers an attractive alternative to the invasive nature of intravenous injections. The future clinical potential for pulmonary RNAi therapeutics holds promise based on the large number of current inhalable traditional drugs and established pulmonary delivery technologies provided by pressurised metered dose inhalers (pMDIs) or dry powder inhalers (DPIs) [55]. The respiratory system also provides an opportunity for drugs to reach the systemic circulation by uptake across the thin epithelium of the alveoli.

Lung-associated diseases such as influenza and respiratory syncytial virus (RSV) infection are prime candidates for siRNA therapy. The transient nature of gene silencing is sufficient for acute viral disease treatment. Moreover, silencing of host factors or conserved genes involved in viral replication could overcome the necessity for seasonal drugs directed towards surface proteins that are susceptible to mutational changes. Several host factors critical for viral replication [56] have now been identified in influenza which provides a selection of novel targets for RNAi-based therapies.

5.3.1 Considerations for Pulmonary Delivery of siRNA

The anatomy, physiology and immunology of the lung present a challenge to delivery of nanoparticle-based or naked siRNA. The lung is composed of the conducting and respiratory regions in which site variations in both structure and cell composition pose specific regional challenges to siRNA delivery.

The main role of the upper respiratory tract is to filter and conduct air to the lower respiratory segment. As a consequence, its anatomical and cellular features restrict material adsorption that includes naked siRNA or nanoparticle-based delivery systems. The trachea divides into the two primary bronchi at the carina, after which, heavy branching from the lobular bronchi occurs and then onto continuously narrowing tubes to the respiratory segment beginning with the respiratory bronchioles and the alveolar ducts and sacs. Delivery of siRNA to the lungs, whether naked or incorporated in a particle, needs to address the branching of airways and the mucus layer covering the conducting segments. Ciliated cells are abundant in the nasal cavity and trachea, with apical cilia working in coordinated sweeps to transport mucus along with trapped material towards the oesophagus. The constant removal of mucus by the ciliated cells, termed the "mucociliary escalator" plays a critical role in preventing inhaled particulates and pathogens from residing within the trachea and the upper bronchiolar tree. The respiratory mucus consists of an outer luminal layer and an inner layer (termed periciliary liquid) in direct contact with the cilia. Under normal physiological conditions, the luminal mucus layer is refreshed every 10–20 min, whereas renewal of the underlying layer is cleared much slower [38]. The mucus layer is swept away and replenished continually requiring trapped material or nanoparticles to diffuse across a current gradient in order to reach the epithelial surface [17].

Deeper into the lung, the mucus layer diminishes, but the passageways narrow. This restricts the transit of particle-based delivery systems into the alveoli. If administered as an aerosol, inertia determines whether or not particles will impact on the epithelium walls, in which case they will be cleared by the mucociliary escalator. Furthermore, surfactant that covers the deeper regions to prevent collapse of the respiratory sections during exhalation may interfere with particle integrity [57], leaving the siRNA exposed to enzymatic degradation. Alveoli macrophages that compensate for the lack of mucus protection are able to scavenge foreign material by extending processes into the lumen of the alveolus. This could limit the effectiveness of nanoparticle-based RNAi therapeutics; however, subsequent macrophage migration may offer a mechanism for systemic delivery of the nanoparticles.

5.3.2 Naked siRNA Delivery

There is an ongoing discussion on whether non-formulated naked siRNA is sufficient or a particle formulation is needed for effective pulmonary siRNA delivery. Both approaches have been used (Table 5.1).

Non-formulated siRNA administered by intranasal or intratracheal instillation have been able to mediate a reduction in target gene expression [58–62, 74] or viral titres [63, 65, 76] in mice and non-human primates [64]. In an interesting study from 2005, Bitko et al. demonstrated that naked phosphoprotein-specific anti-RSV siRNA (70 μg single dose) performed near equally as siRNA complexed with the commercial

Table 5.1 Pulmonary delivery of siRNA

Formulation	Route/animal	Molecular target/model	Effect (dosage)	Ref./year
Naked siRNA				
Naked	Intranasal C57BL/6 mice	PAI-1/bleomycin induced pulmonary fibrosis model	Suppresion of PAI-1 resulted in prevention of fibrosis (multiple doses, 2 μM in 50 μl)	[58]/2010
Naked	Intranasal C57BL/6J mice	HO-1/ischaemia reperfusion injury	Administration of HO-1 siRNA increased apoptosis in lung (1 × 2 mg/kg)	[59]/2004
Naked	Intratracheal C3H/HeN mice	KC and MIP-2/acute lung injury	~40% reduction of KC and MIP-2 mRNA (1 × 75 μg)	[60]/2005
Naked	Intratracheal C57BL/6 mice	XCL-1/*M. tuberculosis*	50% reduction in xcl1 mRNA levels and 40% reduction in protein levels (1 × 5–15 μg)	[61]/2009
Naked	Intratracheal C3H/HeN mice	Fas and caspase-8/acute lung injury	Reduction of Fas and caspase-8 mRNA (1 × 75 μg)	[62]/2005
Naked/Mirus TKO	Intranasal BALB/c mice	Phosphoroprotein/RSV	Several log reduction of viral titres (1 × 70 μg)	[63]/2005
Naked	Intranasal BALB/c mice	Nucleocapsid mRNA /RSV	2.5–3 log reduction in RSV lung concentration (single or multiple doses, 40–120 μg)	[76]/2009
Naked/Lipofectamine	Intranasal BALB/c mice	Ori and glycoprotein B/EHV-1	Antiviral effect observed 1 × 62.5 pmol)	[65]/2009
Naked	Intranasal rhesus macaque	SARS Corona virus/replicase, transcriptase and structural proteins	Anti-SARS effect by prophylactic or therapeutic regimens (1 × 30 μg)	[64]/2005
Naked	Nasal spray human	Nucleocapsid mRNA/RSV	38% reduction in experimentally infected patients (150 mg/day for 4 days)	[14]/2010
Naked LNA modified	I.V. C57BL/6-Yg mice	EGFP/EGFP transgenic mice	55% reduction of EGFP (5 × 50 μg)	[66]/2009
Polymers				
Chitosan	Intranasal C57BL/6-Yg mice	EGFP/EGFP transgenic mice	~40% reduction in EGFP expressing bronchoepithelium cells (5 × 30 μg)	[7]/2006

Chitosan	Intratracheal B6;129P2-RAGE tm1.1 mice	EGFP/EGFP transgenic mice	68% silencing of EGFP (3×0.26 μg)	[13]/2010
Chitosan/imidazole-PEG modified chitosan	I.V./intranasal BALB/c-C57BL/6 mice	GAPDH	40–50% silencing of GAPDH after I.V. or intranasal administration (1–3×0.5–1 mg)	[67]/2010
PEI	Retroorbital injection—C57BL/J mice	Nucleocapsid protein influenza A	10- to 1,000-fold reduction in viral titres (1×120 μg)	[68]/2004
Fully deacetylated PEI	Retroorbital injection—C57BL/J mice	Nucleocapsid protein influenza A	94% drop in viral titre (1×120 μg)	[69]/2005
PEG-PEI	Intratracheal C57BL/6-Tg (CAG-EGFP) 1 Osb/J mice	EGFP DsiRNA	42% knockdown of EGFP (1×50 μg)	[57]/2009
Fatty acid modified PEG-PEI	Intratracheal C57BL/6J-Tg mice	EGFP	69% knockdown of EGFP (1×35 μg)	[70]/2011
Lipid-based vectors				
Oligofectamine/naked siRNA	Intranasal /hydrodynamic injection BALB/cAnNCR mice	Nucleoprotein, acidic polymerase/influenza A	63-fold reduction of viral titres (1 × 50 μg + 20 μg)	[71]/2004
DharmaFECT	Intratracheal C57BL/6 mice	SPARC/bleomycin induced lung fibrosis	58% reduced collagen content in lung (3×3 μg)	[72]/2010
GL67	Intranasal BALB/c mice	lacZ/β-galactosidase	33% lower β-galactosidase mRNA levels (1×40 μg)	
Cholesterol/cell penetrating peptides	Intratracheal BALB/c mice	p38 MAP kinase	45% knockdown of p38 MAP kinase mRNA, no change in protein levels (1×10 nmol)	[73]/2007
AtuPLEX	I.V. C57BL/6N mice	E-cadherin	~50% reduction of VE-cadherin mRNA (4×50 μg)	[74]/2010
Infasurf	Intranasal C57BL/6J mice	GAPDH	50–67% lowered lung concentration of GAPDH protein at 24 h and 7 days (1×10 μg)	[75]/2004

transfection agent Mirus TKO in a RSV mouse model. In this study, viral titres were reduced several logs after intranasal administration with no adverse or immunostimulatory effects observed [63]. Alvarez et al. likewise, was able to reduce RSV titres (2.5–3 log reduction, 100 μg single dose) using the naked Alnylam siRNA against RSV nucleocapsid gene (ALN-RSV01) after intranasal delivery in mice. By RACE analysis of the ALN-RSV01 cleavage product, it was also confirmed that the reduction of the viral titre was in fact an RNAi-mediated effect [76]. Recently, the results from an Alnylam phase II clinical trial was published showing a 38% reduction of experimentally RSV infected test subjects receiving ALN-RSV01 (150 mg) compared to subjects receiving a placebo [14]. A nasal spray was used to deliver the PBS/ALN-RSV01 solution. As mentioned by the authors of the report, the induced RSV infection in the study resulted in a mild to moderate upper respiratory tract illness in the region the nasal spray is likely to reach. Studies are currently underway to evaluate the effect of ALN-RSV01 in naturally infected patients, and these are likely to use aerosolised delivery methods in order to reach both the upper and lower respiratory segments simultaneously. The success achieved in studies using non-formulated siRNA is unexpected when one considers the polyionic nature of the siRNA molecule that restricts cellular uptake. A possible explanation could be loss of epithelial integrity due to infection that might allow entry of naked siRNA. Nonetheless, at the time of writing, the Alnylam RSV programme is one of the most advanced RNAi clinical trial programmes and the simplistic naked siRNA approach could fulfil the clinical requirement of cost-effectiveness.

Whilst direct administration of naked siRNA to the mucosa has been extensively used, the susceptibility of the duplex to serum nucleases makes intravenous (i.v.) delivery less attractive. Modification of the siRNA backbone, however, is now standard to reduce serum degradation [77]. In a recent study, serum stability and silencing of enhanced green fluorescent protein (EGFP) in the bronchoepithelium of mice have been demonstrated after i.v. administration of naked LNA modified siRNAs [66]. Intravenous injections of naked LNA modified siRNA (five doses of 50 μg siRNA) resulted in comparable reduction of EGFP (55% reduction) in the bronchoepithelium as animals dosed intranasally with chitosan/siRNA particles (single 30 μg dose). Naked modified siRNA was less effective after intranasal dosing. The authors suggest that the success of the naked modified siRNA to reach the lung epithelium after i.v. injection might result from increased serum stability, allowing for longer circulation time compared to unmodified siRNAs.

5.3.3 Nanoparticle Delivery

It is generally accepted that nanoparticle-based systems are needed to improve the therapeutic potential of the siRNA despite the success of naked siRNA. The ability to package high levels of siRNA into nanoscale carriers with a predisposition to enter cells has promoted their use. Two prominent classes are polyplexes and lipoplexes formed by self-assembly of polycations or cationic lipids with siRNA resulting from ionic interaction between cationic amines and siRNA-bearing anionic phosphates [78]. The net positive charge facilitates cellular uptake, and incorporation of mucopenetrative components into the design promotes use for mucosal siRNA delivery applications.

5.3.3.1 Polymer-Based Systems

Chitosan-Based Systems

The polysaccharide chitosan has been used extensively for the mucosal delivery of drugs. It is a deacetylated derivative of the natural polymer chitin and is composed of randomly distributed repeating units of β (1,4)-*N*-acetyl-D-glucosamine and β (1,4)-D-glucosamine and is non-toxic, biocompatible and biodegradable [79]. The cationic glucosamine component facilitates mucoadhesion, mucopermeation and polyplex formation. It is involved in transient opening of epithelial tight junctions improving paracellular drug transport [80, 81]. Moreover, it adheres to the mucus layer by interaction with sialic acid in mucus glycoproteins that increases viscosity, leading to decreased mucociliary clearance and prolonged residence time [43, 82]. The cationic amine has been utilised for entropy-driven formation of sub-micron particles with polyanionic DNA [83] and siRNA [84]. Chitosan has demonstrated excellent transfection abilities and several in vivo studies have revealed the ability of chitosan to enhance respiratory delivery of siRNAs and DNA. A study by Köping-Höggård et al. achieved expression of β-galactosidase after intratracheal delivery of chitosan/DNA polyplexes [85] and another study managed to partly immunise mice from RSV by intranasal application of chitosan/DNA particles coding for RSV epitopes [86].

Chitosan-based nanoparticle delivery of siRNA was first introduced by Howard and co-workers [7]. Parameters such as high molecular weight (~100 kDa) and highly deacetylated (>80%) chitosan at N:P (amine:phosphate ratio) >30 showed improved formation, stability and knockdown in vitro [84]. It is proposed that excess chitosan at high N:P ratio may improve mucosal properties. Silencing (~40%) of enhanced green fluorescent protein (EGFP) was observed in the bronchiolar epithelium in transgenic mice after intranasal administration (30 μg of siRNA per dose) over 5 consecutive days of the chitosan/siRNA polyplexes [7]. Intranasal administration suffers from particle adsorption in the mucus layers of the nasal cavity and the amount of drug reaching the lung can, therefore, only be estimated.

In a recent study by the same group, the airway deposition of the chitosan/siRNA particle system was improved with an aerosolised formulation using a nebulising catheter (AeroProbe™, Trudell Medical Instruments) inserted directly into the trachea of the mouse. Silencing of the target gene (EGFP) was accomplished with a very low dose of siRNA (three doses of 0.26 μg) [13]. The low dose is a significant step towards reduction of potential off-target and immunological side effects [87].

Chitosan has predominately been used for local delivery; however, a recent report showing chitosan/siRNA particles accumulated in the kidneys after i.v.

administration [4] suggests circulatory properties. Furthermore, modification of chitosan with an imidazole group and PEG has been used for intravenous delivery of siRNA in mice resulting in a 49% reduction of mRNA levels of GAPDH in the lungs, suggesting that chitosan/copolymers might be useful as an intravenous delivery vector [67].

PEI-Based Systems

Since the introduction of polyethylenimine (PEI) as a gene transfer reagent in 1995 [99], this cationic polymer has been studied extensively for both DNA and siRNA delivery [100]. Effective polyplex formation, protection from nucleases and endosomolytic properties attributed to its high charge amino density have promoted its use. PEI has been used for systemic delivery of siRNA to a number of tissues including the lungs in mice [68, 69]. Ge et al. and Thomas et al. used PEI/siRNA polyplexes (N:P 5) against the influenza nucleocapsid protein (120 μg single dose) administered by retroorbital injection. The study by Thomas et al. expanded on the previous study by Ge et al., by evaluating the ability of various high molecular weight PEI polymers to enhance the delivery of siRNA and in both studies, a significant reduction of viral titres was observed (10- to 1,000-fold reduction).

The mechanism of antiviral effects from the studies by Thomas and Ge, however, has been brought to question in a landmark paper from Robbins et al. [101]. In this work, several published siRNA sequences, including the nucleocapsid sequence used by Ge et al. and Thomas et al., were tested for an ability to stimulate the innate immune system ascribed to intracellular Toll-like receptor activation. Remarkably, it was shown that the control GFP sequence used in several in vivo studies [60, 68, 69, 71] did not elicit an immune response, whereas the nucleocapsid sequence (among others) stimulated the production of interferon α suggested to be largely responsible for the observed antiviral effect.

The ability of systemic PEI-based siRNA systems to reach the lungs [102, 103] could result from serum-induced aggregation and its consequent entrapment within the lung capillary beds. This, however, could result in lung embolisms and restricts the likelihood for clinical translation.

PEI, unlike chitosan, is not generally thought to be a mucoadhesive polymer, but it is reasonable to speculate that some amino-mediated interaction with mucins can occur if delivered locally. *Hitherto*, PEI interactions and effects on mucus have not been studied in detail. Two recent studies have demonstrated pulmonary EGFP silencing in transgenic mice. Using intratracheal administration of PEG-PEI/siRNA polyplexes, Merkel et al. showed a 42% reduction of EGFP expression compared to luciferase siRNA control (single 50 μg siRNA dose) [57]. Moderate inflammation was seen by analysis of cytokine levels, but no histological abnormality was observed. Beyerle et al. used a fatty acid modified PEG-PEI/siRNA polyplex to achieve a 69% reduction of EGFP expression compared to untreated controls (35 μg single dose) [70]. As in the previous study by Merkel et al., PEGylation increased inflammation, whilst at the same time also decreased cytotoxicity. These findings seem to contradict the usual perception that PEG limits interaction with the immune system.

Whilst PEGylated PEG-PEI polymers appear less cytotoxic than non-modified PEI [104], they and the fatty acid modified PEI may have a higher proinflammatory potential that is of concern in a clinical setting. Although PEI has been used extensively for several years in animal studies, safety concerns still restrict its use in the clinic.

5.3.3.2 Lipid-Based Systems

Lipid vectors have been widely used for in vitro and in vivo delivery of siRNA [105], most based on cationic lipids that form lipoplexes with siRNA. Mirus TKO, a cationic lipid/polymer formulation, has been used by Bitko et al. to deliver siRNA against RSV [63]. 70 μg siRNA delivered intranasally with Mirus TKO was able to reduce viral titres in mice without inducing an interferon response, an effect shown to be a 20–30% improvement on naked siRNA. In a mouse influenza model, the animals received hydrodynamic injections (3.78 nmol) of naked siRNA followed 16–24 h later by intranasal delivered oligofectamine/siRNA (1.51 nmol) complexes against the viral nucleoprotein and acidic polymerase to reduce viral titres in the lung (63-fold compared to EGFP siRNA) [71]. Whilst interferon levels were investigated and found not to be upregulated, concerns remain for the use of the EGFP sequence as a negative control due to its non-stimulatory uniqueness [101]. A third commercial lipid-based transfection reagent, DharmaFECT, has been used as a pulmonary delivery vector in a bleomycin-induced lung fibrosis mouse model [72]. siRNA against SPARC, a matricellular protein overexpressed in fibrotic diseases, markedly reduced collagen content in the lung (58%) compared to the bleomycin-only group after intratracheal dosing (3×3 μg siRNA).

The cationic lipid from Genzyme, GL67, has been used in k18-lacZ mice which express β-galactosidase in airway epithelial cells [106]. A 33% reduction of mRNA level (but no change in protein levels) was observed after intranasal administration of lacZ siRNA (40 μg siRNA). Histological analysis showed that the GL67/siRNA lipoplexes were mainly associated with pulmonary macrophages which could explain the lack of change in protein levels.

Direct conjugation of siRNA to cholesterol has been explored by Moschos et al. [73]. A single intratracheal administration of siRNA–cholesterol conjugates (10 nmol) facilitated a 45% knockdown of p38 MAP kinase mRNA in mouse lungs after 12 h compared to vehicle-only controls. The effect appeared to be transient as the detected mRNA levels were almost back to normal after 24 h which was attributed to poor stability of the siRNA. It was suggested that chemical modification of the backbone might increase the silencing effect.

The respiratory vasculature in mice can be targeted by systemic delivery of cationic lipoplexes (AtuPLEX) [74]. A ~50% reduction of the endothelial cell-specific protein VE-cadherin was achieved in the lungs after intravenous injection of 50 μg of siRNA on 4 consecutive days compared to a luciferase-specific siRNA. The lipoplexes were also administered intratracheally, but only a 21% reduction of epithelial

E-cadherin was observed compared to a luciferase control, suggesting better suitability for systemic delivery. Capture within the lung microvasculature and subsequent endothelial uptake were proposed by the authors as the mechanism of delivery.

5.3.4 Aerosolised Formulations

It is anticipated that clinical translation will require inhalation technology based on aerosols of dry powder formulation or solutions. Aerosols are by definition a gaseous suspension of fine solid particles or liquid droplets. The size and weight of these particles or droplets determine their ability to follow the flow of inhaled air through the airways.

The main parameter for linking particle or droplet size and weight in regard to lung deposition is the aerodynamic diameter. This parameter takes into account shape, roughness and porosity of the particles or droplets in an aerosol. The aerodynamic diameter is the diameter of a unit density (1 g/cm^3) sphere having the same gravitational settling velocity as the particle being investigated. The mass median aerodynamic diameter (MMAD) is the diameter at which 50% of the particle/droplet distribution by mass will have a larger or smaller diameter. In other words, if deposition at a specific airway depth is required and is achieved at a given aerodynamic diameter (e.g. 5 μm), then if the MMAD of an aerosol is 5 μm, then 50% of the total aerosol mass will in principle deposit above the selected depth and 50% will deposit below. This restricts nanoparticle diameter to a narrow size distribution if deposition at a certain depth is required.

Investigation of MMADs of aerosols is typically carried out on cascade impactors mimicking different airway depths. Particles with an aerodynamic diameter between 1 and 5 μm are likely to reach the pulmonary regions, whereas larger particles will be deposited in the upper airways [55]. However, if particles become too small, they are prone to being exhaled before depositing. This means solid nanoparticles, per definition, are in principle too small to be effectively deposited in the lungs, and a large portion of these particles may end up leaving the lung again after inhalation. There are two solutions to this problem. Either the particles are kept in solution or they are attached to a carrier formulation which will facilitate deposition at the required depth. Nanoparticles such as those consisting of polymers and siRNAs are formed in solution, but subsequent drying by either spray drying [107] or freeze drying [108] can produce particles retaining their silencing ability which in terms of storage and stability of a therapeutic agent might be preferable compared to a solution-based formulation.

Intratracheal administration in animal models has provided preclinical evaluation of aerosolised siRNA formulations and is more cost-effective than inhalation chambers. Nebulisers developed specifically for delivery of aerosols to animals such as the "AeroProbe" from Trudell Medical and the "Microsprayer" from Penncentury are examples of devices used for particle delivery directly to the mucosal surfaces of the respiratory tract. These devices overcome the difficult nature of the mouse breathing pattern and anatomy [109] and allow dose-response studies to be conducted.

Substantial clinical evaluation of dry powder-based siRNA formulations is lacking, although Alnylam (http://www.alnylam.com, 2011) has used a handheld battery-driven nebulising system (http://www.paripharma.com, 2010) in their current phase II RSV clinical trial with naked siRNAs. The promise of nanotechnology and the advances with surface and particle engineering combined with recent advances in inhaler technology hold promise for future inhalable siRNA-based particles.

5.4 Oral Delivery

Oral administration of therapeutics is considered the most favourable in terms of cost-effectiveness, ease of administration and patient compliance. This route potentially provides rapid systemic distribution of the drug [110] due to the enormous adsorptive surface area (~200 m^2). Utilisation of this route depends on overcoming the challenges of enzymatic degradation, mucus and epithelial penetration. Oral administration of RNAi-based drugs offers great potential for both the treatment of diseases occurring locally within the gastrointestinal (GI) tract, such as inflammatory bowel disease (IBD), and to combat systemic pathologic conditions.

5.4.1 Considerations for Oral Delivery of siRNA

The GI tract possesses a specialised epithelium involved in the degradation of macromolecules and assimilation of the obtained products while restricting the transport of pathogens. Unfortunately, these processes often compromise the integrity and absorption of therapeutics. In this respect, exposure to a highly active enzymatic environment, extreme pH conditions and the existence of a selective-permeability epithelial barrier are the main challenges for oral delivery of RNAi therapeutics. Nucleases, highly abundant in pancreatic secretions, constitute the main enzymatic barrier to nucleic acids. Moreover, the delivery system itself may be susceptible to degradation by other enzymes present in the lumen (e.g. lipases, glycosidases or proteases) or the microvillus (e.g. P450).

pH extremes along the GI tract ranging from 1 to 2 in the stomach to >7 at the terminal part of the small intestine and colon may affect acid- or base-labile components of the delivery system, although increased stability of nucleic acids under these pH conditions can be achieved by chemical modifications [111, 112] or incorporation into delivery systems. Exploitation of the localised pH conditions could offer an exciting strategy for site-specific release of siRNA using pH-sensitive carriers. Luminal pH determines the drug's ionisation degree that affects transcellular passive diffusion and/or the interactions between the formulation components [84].

Since the GI tract epithelium is covered by mucus, drugs must diffuse through this lubricant and protective layer in order to reach the absorptive surface. Therefore, the uptake of a therapeutic compound will depend upon the interactions of the drugs with the mucus components as well as the thickness of this layer, which in experimental animal models has been shown to vary along the gastrointestinal tract [113, 114].

Whilst paracellular transport across the GI epithelium is limited to ions and small hydrophilic molecules that can diffuse across tight junctions, the hydrophobic nature of the cell membranes impede the diffusion of most polar and charged molecules [110, 115]. Consequently, macromolecular siRNA absorption across the epithelia is restricted, although binding of specific ligands may facilitate uptake as demonstrated in different cell types [6, 116–118]. The capability to attach different chemical components by simple nucleic acid chemistry could promote this approach.

Nanoparticle-based carriers are taken up by adsorptive or receptor-mediated endocytosis across enterocytes dependent on surface moieties. The level of uptake is thought to be low, although penetration enhancers may potentiate paracellular delivery. An important consideration is delivery and breakdown in the liver due to the first-pass effect commonly encountered by absorbed drugs. An alternative route through the gut-associated lymphoid tissue (GALT) has been exploited for the delivery of micro and nanoparticles [119–122]. The overlying follicle-associated epithelium (FAE) contains specialised cells termed M-cells [115] that are anatomically designed to sample luminal particles as part of the mucosal immune response. The lymphoid follicle domes are highly populated with macrophages, which have been shown to capture material. Systemic dissemination of these macrophages has been proposed as a mechanism of transport to peripheral tissue. Although particle transport through M-cells may be augmented by increasing particle-surface hydrophobicity or attachment of specific targeting ligands [123, 124], it is important to bear in mind when utilising this route for intestinal absorption that GALT only constitutes a small fraction of the GI tract, with the numbers deceasing with age. Recent attention has focused on transport across the epithelial barrier directly mediated by dendritic cells [125]. These phagocytic cells, widespread throughout the epithelia, have been shown to disrupt tight junction and sample luminal content through the projection of dendrites, providing an exciting opportunity for the design of oral vaccines.

Inter-species differences exist between humans and the animal models commonly used for the in vivo evaluation of oral drug administration. For example in contrast to humans, mice and rats exhibit a less acidic stomach pH (~4 vs. ~1.7) and lower mean intestinal pH [126]. These are important considerations when assessing clinical translation.

5.4.2 *Oral Studies*

A number of studies have used the oral route for siRNA delivery (Table 5.2). A high profile study was reported in 2009 by Aouadi et al. [88]. In this study, porous β-1, 3-D-glucan shells were loaded with siRNA targeting expression of tumour necrosis factor-alpha (TNF-α) or mitogen-activated protein kinase 4 (Map4k4) in mice. The

Table 5.2 Oral, rectal and intravaginal delivery of siRNA

Formulation	Route/animal	Molecular target/model	Effect (dosage)	Ref./year
β-1, 3-D-glucan shells	Oral C57BL6/J mice	TNF-α, Map4k4 Untreated animals and LPS lethality test	Knock-down of target genes. Protection from LPS-induced lethality (~0.4 μg/dose for 8 consecutive days)	[88]/2009
Thioketal nanoparticles	Oral C57BL/6 mice	TNF-α DSS-induced colitis	TNF-α Knock-down. Protection from colitis (~46 μg/dose or 4.6 μg/dose for 5 consecutive days)	[89]/2010
NiMOS	Oral Balb/c mice	TNF-α DSS-induced colitis	TNF-α knock-down. Milder colitis symptoms (~24 μg/dose for 3 alternated days)	[90]/2011
Stabilised unilamellar vesicles β7 integrin-targeted	Intravenous C57BL/6 mice	CyD1 DSS-induced colitis	CyD1 knock-down. Alleviated colitis symptoms (~50 μg/dose for 4 alternated days)	[91]/2008
Lipoplex (Lipofectamine)	Rectal C57BL/6 mice	TNF-α DSS-induced colitis	TNF-α knock-down in descending colon. Mild or moderate inflammation (~53 μg/dose for 2 alternated days)	[92]/2006
Lipoplex (DOTAP) chemical modified siRNA	Rectal Swiss-Webster mice	None detection of fluorescent-labelled siRNA	Uptake in spleen, colon and bone marrow (single dose of ~20 μg)	[93]/2007
Lipoplex (Oligofectamine)	Intravaginal Balb/c mice	UL27 &UL29 viral genes HSV-2 lethal challenge	Protection from HSV-2 lethal infection (two doses of ~7 μg)	[94]/2006
Naked (cholesterol conjugated chemical stabilised siRNA)	Intravaginal Balb/c mice	Viral (UL29) & host (nectin-1) genes HSV-2 lethal challenge	Protection from HSV-2 lethal infection (~27 μg/dose for 2 consecutive days)	[95]/2009
PLGA nanoparticles	Intravaginal FVB Cg-Tg (GFPU)5 Nagy mice	GFP transgenic GFP mice	GFP gene silencing throughout reproductive tract (single dose of ~7 μg)	[96]/2009
Lipoplex (Lipofectamine)	Intravaginal C57BL/6 mice	Lamin A/C, CCR5	Knock-down of target genes (single dose of ~53 μg)	[92]/2006
PEGylated Lipoplexes entrapped in alginate scaffold	Intravaginal C57BL/6 and K14E7 mice	Lamin A/C	Lamin A/C knock-down (two doses of ~8 μg)	[97]/2011
Naked (CD4 aptamer-siRNA chimera)	Intravaginal NOD/SCID-BLT & NSG-BLT mice	Viral (*gag,vif*) and host (*CCR5*) genes Humanised mice	Protection against HIV vaginal transmission (~4 μg in complex dosing regimen)	[98]/2011

internal element of the 2–4 μm particles contained a tRNA-core coated with consecutive layers of PEI and siRNA. Daily particle administration (~0.4 μg siRNA/dose) by oral gavage over an 8-day period resulted in reduced mRNA levels of Map4k4 (~60–70%) or TNF-α (~40–60%) in peritoneal exudate cells (PECs) compared to animals receiving scrambled-siRNA containing particles. Map4k4 downregulation elicited a concomitant reduction in TNF-α expression that suggests a role for Map4k4-mediated control of TNF-α. In addition extended knockdown duration of ~8 days was observed after the final dose. Interestingly, the authors proposed that siRNA release from the glucan shell was triggered by the acidic environment in phagosomes; this, however, could compromise particle integrity at low pH within the GI tract. Notably, no unspecific interferon-γ response was detected even though non-modified siRNA was used. Map4k4 and TNF-α silencing in macrophage-enriched cells isolated from spleen, liver and lung tissues was observed ascribed to particle uptake across GALT and subsequent dissemination in migrating macrophages. No direct evidence for M-cell uptake or adsorption levels was provided; however, this study does suggest the possibility for systemic silencing via the oral route.

In certain pathologies, such as IBD, a localised rather than systemic effect is more desirable. IBD encompass a group of complex autoimmune diseases, which are broadly categorised as Crohn's disease or ulcerative colitis found in the small intestine and colon respectively. An attractive pathological condition that could be exploited for improved siRNA-based therapeutic delivery is mucosa integrity loss in IBD. Despite the numerous targets investigated that includes IL-12, IL-23, IL-17 and IL6 [127], the current biologic treatment of IBD is based on anti-TNF-α molecules [128]. TNF-α has been the preferred target for oral-based siRNA therapies in two recent studies employing orally delivered siRNA for the prevention/treatment of dextran sodium sulphate (DSS)-induced ulcerative colitis in mice. In the first of these studies, Wilson et al. [89] used thioketal (poly-1, 4-phenyleneacetone dimethylene thioketal) nanoparticles (TKNs) designed for triggered anti-TNF-α siRNA release in response to raised levels of reactive oxygen species (ROS) common to inflamed regions. A tenfold specific decrease in colonic mRNA levels of TNF-α and other proinflammatory cytokines (IFN-γ, IL-6 and IL-1) was detected after oral administration of anti-TNF-α TKNs (~46 μg siRNA/dose) over 5 consecutive days during colitis induction. Furthermore, the authors also demonstrated by histological and weight analysis that a ten times lower siRNA dose (~4.6 μg siRNA/dose) was sufficient to protect mice from DSS-induced colitis.

An alternative approach for site-specific delivery was reported by Kriegel et al. [90]. The nanoparticles-in-microsphere oral system (NiMOS) is based on lipase-mediated intestinal degradation of poly epsilon-caprolactone microspheres for triggered release of gelatin nanoparticles containing TNF-α-specific siRNA. Administration by oral gavage at days 2, 4 and 6 after DSS treatment of anti-TNF-α siRNA-loaded NiMOS (~24 μg siRNA/dose) resulted in reduced intestinal mRNA and protein levels compared with controls. In addition, ELISA showed decreased levels of several proinflammatory cytokines (IFN-γ, IL-1b, IL-2, IL-5, IL-6 and

IL-12p70); however, some non-specific silencing intrinsic to the formulation was observed. Colitis protection (moderate intestinal inflammation and healthy colon morphology) was only evident in the anti-TNF-α siRNA-treated mice.

An alternative strategy based on a systemic RNAi-based IBD treatment was revealed by the elegant study by Peer et al. [91]. In this study, i.v. administration of particles targeting a specific leucocyte subset (β7 integrin expressing gut mononuclear leucocytes) was utilised for the treatment of DSS-induced colitis. The design of the system, termed β7-I-tsNPs, has a protamine/siRNA core complex coated within a unilamellar vesicle decorated with an anti-integrin β-7 antibody. Administration of CyD1-specific siRNA (~50 μg/dose) β7-I-tsNPs at days 0, 2, 4 and 6 resulted in reduced intestinal mRNA levels of this cell cycle regulatory molecule and simultaneous mRNA reduction of the proinflammatory cytokines TNF-α and IL-12. This resulted in significantly less severe lesions at the intestinal tissue and the reversal of clinical and pathological characteristics associated with the onset of the DSS-induced colitis. The observed local effect may be attributed to the CyD1 silencing of peripheral blood and spleen leucocytes prior their recruitment to the inflamed gut.

We are currently evaluating the potential of siRNA nanoparticles formulated with the non-toxic, biodegradable and mucoadhesive polymer chitosan for the reduction of proinflammatory cytokines after oral administration. Encouraging results have been recently obtained in animal experiments, suggesting strong nuclease-protection and high gastrointestinal siRNA deposition provided by this system (unpublished results).

5.5 Rectal Delivery

Rectal administration is an attractive route for siRNA delivery as it circumvents the low stomach pH, is an established route for traditional drugs and the colon presents a low enzymatic milieu. In addition, direct access to the site of several diseases such as colorectal carcinoma or ulcerative colitis further promotes this route.

Zhang et al. [92] demonstrated that rectal administration of lipoplexes containing anti-TNF-α siRNA (two doses of ~53 μg siRNA) significantly reduced the upregulation of TNF-α mRNA in a DSS-induced ulcerative colitis mouse model. Reduced perirectal TNF-α mRNA levels were associated with mild or moderated inflammation at the mucosa of the descending colon compared to the severe inflammation observed in the controls. Interestingly, despite toxicity previously reported with similar liposomal formulations, no increase in proinflammatory cytokines (IL-1, IL-10, TNF-α) or interferon responses was found. In a later study [93], fluorescent and chemically modified siRNA contained within DOTAP liposomes was detected in the spleen, bone marrow, colon and liver after rectal administration in mice. This supports the capability for nanoparticles to migrate into the systemic circulation that could be exploited for both local and systemic gene silencing.

5.6 Intravaginal Delivery

Advantages such as an established therapeutic route, marketed products for sustained drug release, low enzymatic activity and possible avoidance of the first-pass effect [129] promote vaginal administration of siRNA. Poor systemic absorption of polar high molecular weight molecules across the epithelium, however, seemingly restricts siRNA-based therapies to local vaginal treatments.

5.6.1 Vaginal Studies

A number of studies have utilised acute infection of mice with herpes simplex virus 2 (HSV-2) as a model for the development of antiviral siRNA-based therapeutics. In this setting, the capacity of the treatment to inhibit viral spread across the genital mucosa after HSV-2 challenge is evaluated. In 2006, a study by Palliser et al. [94] assessed the protection provided by lipid-complexed siRNA (~7 μg/dose) targeting essential HSV-2 viral genes. Two siRNA (UL27.2 and UL29.2) conferred significant protection with considerable reduction in the lethality and severity of the lesions, when administrated in a double regime 2 h prior and 4 h after an otherwise lethal HSV-2 vaginal challenge. The protection was, however, transient and a post-exposure treatment (3 and 6 h after the viral challenge) was only effective when both siRNA were administrated in combination but not individually. No inflammatory response or interferon induction was detected by histological and expression analysis respectively in this study. More exhaustive follow-up studies have revealed, however, several undesirable features and toxic side effects related to lipid formulations [95, 96].

A chemically modified siRNA approach has been also used for the treatment of HSV-2 [95]. The cholesterol (Chol)–siRNA conjugate, stabilised with phosphorothioate residues, was used to knockdown viral and host gene expression. Consistent with previous results [94], targeting an essential viral gene (UL29) exclusively conferred protection when the siRNA was administrated within a few hours of the viral challenge. Interestingly, this protection could not be replicated if a too high siRNA dose (~135 μg) was employed, a matter that requires further investigation. In contrast, targeting of nectin-1, the receptor used by HSV-2 to penetrate in the cells, conferred protection only when administrated 1–7 days prior, but not immediately before or after the HSV-2 challenge. Treatment of the mice with two doses (~27 μg/dose) of Chol–siRNA combining the targeting of nectin-1 and viral genes provided significant protection for 1 week irrespective of the time of challenge.

Woodrow et al. [96] developed a delivery system based on a siRNA/polyamine (spermidine) core encapsulated into PLGA nanoparticles. A single dose (~7 μg) of these particles induced sustained GFP mRNA silencing throughout the female reproductive tract for at least 14 days in a transgenic GFP mouse model. Reduction of fluorescence was maximal at day 10 in the vaginal tract, with only 30–40% of the siRNA dose (~2.8 μg) released due to the slow degradation rate of the nanoparticles.

In the aforementioned studies, thorough cleaning of the vaginal tract and/or progesterone treatment of the animals prior to particle administration was performed [94–96]. Whilst mucus removal eliminates one of the main barriers for vaginal epithelial transfection, the hormone treatment arrests the oestrous cycle in the dioestrus phase in which the epithelium is thin and porous that most probably contributed to higher drug absorption [130]. In addition this treatment has been associated with a reduced immune response in the vagina, which may mask potential undesirable side effects of the evaluated drugs.

In a model more closely resembling the normal physiological conditions, Zhang et al. [92] reported liposome-mediated transfection of the squamous epithelia layer and submucosa. A single dose of siRNA (~53 μg siRNA) was sufficient to induce a significant and consistent knockdown of the targeted gene (lamin A/C or CCR5) over a 7-day period. Analysis of proinflammatory cytokines (IL-1, IL-10, TNF-α) and interferon-related genes did not detect any significant changes in the treated animals compared to controls.

In contrast, Wu et al. [97] suggested that vaginal epithelium transfection in physiological conditions with conventional lipoplexes was unlikely, most probably due to the combination of poor drug retention at the vaginal cavity and an inefficient transport across the mucus layer. In order to overcome these limitations and achieve sustained release of the entrapped therapeutic, the authors developed and characterised a system based on a biodegradable alginate scaffold. Upon exposure to sodium ions, a common element of cells and body fluids, scaffold degradation occurs, resulting in the slow release of incorporated PEGylated lipoplexes. PEGylated, but not conventional, liposomes were capable of mucosal diffusion and induce siRNA-mediated gene knockdown at the vaginal epithelium. Intra-vaginal administration of the scaffold over 2 consecutive days (daily dose of 8 μg/animal) resulted in an 85% knockdown of lamin A/C mRNA. In this report, evaluation of proinflammatory cytokine levels and unspecific interferon activation was not reported.

Encouraging results have been recently reported by Wheeler et al. [98], who by targeting viral (*gag* and *vif*) and host genes (*CCR5*) could inhibit HIV vaginal transmission in a humanised mouse model. Macrophage and CD4+ T-cell-specific targeting was achieved by the fusion of the siRNAs to a CD4 receptor-specific aptamer. The observed protection is probably due to the combination of selective gene knockdown by the siRNAs (*CCR5*, *gag* and *vif*) and a viral–aptamer competition for CD4 receptor binding. Despite the apparent absence of cellular toxicity or lymphocyte activation, caution should be taken with molecules interacting with the CD4 receptor due to its role in the host immune response and susceptibility for HIV infections in activated T lymphocytes.

5.7 Future Perspectives

Mucosal delivery of RNAi therapeutics is an exciting approach that is set to progress rapidly building on encouraging clinical studies. The relative ease of access to surfaces common to pathogen, cancer and inflammatory disease promotes their use. Local delivery avoids the necessity to install "stealth" characteristics required for systemic delivery that reduces the complexity of design that has manufacturing, cost and clinical approval benefits. The restricted entry, however, encountered by macromolecules across mucosal barriers still requires delivery strategies to improve penetration. In this context, nanoparticles rather than naked forms seem the most promising. Research to identify surface characteristics that promote mucus penetration including hydrophilic coats is set to continue, whilst coatings that mimic pathogens evolved to penetrate the mucosa is an interesting approach. Detailed studies of nanoparticle penetration in mucus and changes in the mucus morphology in response to mucopenetrative materials are a future trend. Variations in mucus characteristics at different sites and disease states are an important consideration in the design of the delivery strategy. The polyplex systems composed of siRNA and cationic polymers such as chitosan could proceed rapidly into clinical trials due to their simplistic design and mucoadhesive and mucopermeable properties. A current trend is to identify new biopolymers to improve mucosal delivery and expand the selection of available materials.

Recent attention has been directed towards oral formulations focused on treatment of inflammatory diseases of the gastrointestinal tract. Anti-inflammatory effects in IBD preclinical models using particle formulations suggest that IBD will be a primary candidate for clinical translation. The development of bioresponsive particles or coatings composed of pH-sensitive materials, employed for other drug types, will offer the possibility for localised site-specific delivery utilising the differing pH found throughout the GI tract. The necessity for particle disassembly needed for siRNA incorporation into the cellular RNAi machinery calls for intracellular release mechanisms [131, 132]. Reducible disulphide links that are cleaved in the cytoplasm is a strategy, but cost may preclude clinical translation. In contrast to the necessity for stable particles in the circulatory environment, mucosal delivery allows the use of less stable systems that could facilitate siRNA release.

The ability of nanoparticles to translocate the mucosa and enter systemic circulation is set to be exploited to elicit local and systemic silencing effects often needed to match pathogenesis. This, however, will require further modifications to avoid serum-induced aggregation and hepatic clearance. The success of this approach will depend on the technologies currently pursued for systemic nanoparticle delivery. To this end, improving nanoparticle delivery across lymphoid tissue as a route for systemic delivery is set to continue with identification of new targeting approaches running in parallel.

In addition to improved delivery systems, siRNA design is an important consideration relevant to all routes of administration. Some of the initial siRNA-mediated antiviral effects were seemingly attributed to non-specific induction of innate immune responses due to Toll-like receptor (TLR) engagement. Proinflammatory responses are highly detrimental, particularly in inflammatory disorders. The endocytic pathway undertaken by particles can increase delivery into a TLR-rich environment that can inadvertently potentiate the response. Fortunately, these TLR-dependent (through TLR-3, -7 and -8 signalling) or independent (through RIG-1 and PKR activation) adverse side effects can be avoided by siRNA structure and sequence modification such as 2′-*O*-methyl substitutions. Induction of non-specific immune responses is particularly pertinent in the mucosal immune system rich in immunocompetent sites evolved to recognise and protect against foreign luminal material. Evaluation of immune responses to both siRNA and carrier needs to be adequately addressed going forward.

In the foreseeable future, clinical trials are set to seemingly follow the lead towards treatment of pulmonary diseases such as RSV infection. Established pulmonary delivery technologies used for traditional inhalable drugs should allow rapid clinical translation. Identification of novel targets will push the field forward. An interesting approach is targeting host factors required for viral replication such as influenza rather than viral-specific targets [56].

There is a general shift in the RNAi field from conventional siRNA to miRNA-based agents that is set to follow for mucosal RNAi therapeutics. Deep sequencing technologies are set to be used for rapid identification of mucosal miRNA targets.

Mucosal delivery holds many advantages over the systemic approach and is now showing promise for delivery of siRNA that could lead to rapid clinical translation of mucosal-based RNAi therapeutics.

References

1. Davidson BL, McCray PB (2011) Current prospects for RNA interference-based therapies. Nat Rev Genet 12(5):329–340
2. de Fougerolles A, Vornlocher H-P, Maraganore J, Lieberman J (2007) Interfering with disease: a progress report on siRNA-based therapeutics. Nat Rev Drug Discov 6(6):443–453
3. Wu L, Belasco JG (2008) Let me count the ways: mechanisms of gene regulation by miRNAs and siRNAs. Mol Cell 29(1):1–7
4. Gao S, Dagnaes-Hansen F, Nielsen EJB, Wengel J, Besenbacher F, Howard KA, Kjems J (2009) The effect of chemical modification and nanoparticle formulation on stability and biodistribution of siRNA in mice. Mol Ther 17(7):1225–1233
5. Bramsen JB, Kjems J (2011) Chemical modification of small interfering RNA. Methods Mol Biol 721:77–103
6. Soutschek J, Akinc A, Bramlage B et al (2004) Therapeutic silencing of an endogenous gene by systemic administration of modified siRNAs. Nature 432(7014):173–178
7. Howard KA, Rahbek UL, Liu X et al (2006) RNA interference in vitro and in vivo using a chitosan/siRNA nanoparticle system. Mol Ther 14(4):476–484
8. Akinc A, Zumbuehl A, Goldberg M et al (2008) A combinatorial library of lipid-like materials for delivery of RNAi therapeutics. Nat Biotechnol 26(5):561–569
9. Rettig GR, Behlke MA (2011) Progress toward in vivo use of siRNAs-II. Mol Ther 20(3):483–512. doi:10.1038/mt.2011.263

10. Howard KA (2009) Delivery of RNA interference therapeutics using polycation-based nanoparticles. Adv Drug Deliv Rev 61(9):710–720
11. de Fougerolles A, Novobrantseva T (2008) siRNA and the lung: research tool or therapeutic drug? Curr Opin Pharmacol 8(3):280–285
12. Merkel OM, Kissel T (2012) Nonviral pulmonary delivery of siRNA. Acc Chem Res 45(7):961–70. doi:10.1021/ar200110p
13. Nielsen EJB, Nielsen JM, Becker D et al (2010) Pulmonary gene silencing in transgenic EGFP mice using aerosolised chitosan/siRNA nanoparticles. Pharm Res 27(12):2520–2527
14. DeVincenzo J, Lambkin-Williams R, Wilkinson T, Cehelsky J, Nochur S, Walsh E, Meyers R, Gollob J, Vaishnaw A (2010) A randomized, double-blind, placebo-controlled study of an RNAi-based therapy directed against respiratory syncytial virus. Proc Natl Acad Sci USA 107(19):8800–8805
15. McAuley JL, Linden SK, Png CW, King RM, Pennington HL, Gendler SJ, Florin TH, Hill GR, Korolik V, McGuckin MA (2007) MUC1 cell surface mucin is a critical element of the mucosal barrier to infection. J Clin Invest 117(8):2313–2324
16. Strombeck DR, Harrold D (1974) Binding of cholera toxin to mucins and inhibition by gastric mucin. Infect Immun 10(6):1266–1272
17. Cone RA (2009) Barrier properties of mucus. Adv Drug Deliv Rev 61(2):75–85
18. Lai SK, Wang Y-Y, Wirtz D, Hanes J (2009) Micro- and macrorheology of mucus. Adv Drug Deliv Rev 61(2):86–100
19. Copeman M, Matuz J, Leonard AJ, Pearson JP, Dettmar PW, Allen A (1994) The gastroduodenal mucus barrier and its role in protection against luminal pepsins: the effect of 16,16 dimethyl prostaglandin E2, carbopol-polyacrylate, sucralfate and bismuth subsalicylate. J Gastroenterol Hepatol 9(Suppl 1):S55–S59
20. Pullan RD, Thomas GA, Rhodes M, Newcombe RG, Williams GT, Allen A, Rhodes J (1994) Thickness of adherent mucus gel on colonic mucosa in humans and its relevance to colitis. Gut 35(3):353–359
21. Greaves JL, Wilson CG (1993) Treatment of diseases of the eye with mucoadhesive delivery systems. Adv Drug Deliv Rev 11(3):349–383
22. Prydal JI, Artal P, Woon H, Campbell FW (1992) Study of human precorneal tear film thickness and structure using laser interferometry. Invest Ophthalmol Vis Sci 33(6):2006–2011
23. Clunes MT, Boucher RC (2007) Cystic fibrosis: the mechanisms of pathogenesis of an inherited lung disorder. Drug Discov Today Dis Mech 4(2):63–72
24. Matsui H, Grubb BR, Tarran R, Randell SH, Gatzy JT, Davis CW, Boucher RC (1998) Evidence for periciliary liquid layer depletion, not abnormal ion composition, in the pathogenesis of cystic fibrosis airways disease. Cell 95(7):1005–1015
25. Tarran R, Grubb BR, Gatzy JT, Davis CW, Boucher RC (2001) The relative roles of passive surface forces and active ion transport in the modulation of airway surface liquid volume and composition. J Gen Physiol 118(2):223–236
26. Verkman AS, Song Y, Thiagarajah JR (2003) Role of airway surface liquid and submucosal glands in cystic fibrosis lung disease. Am J Physiol Cell Physiol 284(1):C2–C15
27. Rosen H, Abribat T (2005) The rise and rise of drug delivery. Nat Rev Drug Discov 4(5):381–385
28. Mrsny RJ (2009) Lessons from nature: "Pathogen-Mimetic" systems for mucosal nanomedicines. Adv Drug Deliv Rev 61(2):172–192
29. Cohn L (2006) Mucus in chronic airway diseases: sorting out the sticky details. J Clin Invest 116(2):306–308
30. Hogg JC, Chu F, Utokaparch S et al (2004) The nature of small-airway obstruction in chronic obstructive pulmonary disease. N Engl J Med 350(26):2645–2653
31. Merkus V, Schipper M (1998) Nasal mucociliary clearance as a factor in nasal drug delivery. Adv Drug Deliv Rev 29(1–2):13–38
32. Ali MS, Pearson JP (2007) Upper airway mucin gene expression: a review. Laryngoscope 117(5):932–938
33. Lehr C-M, Poelma FGJ, Junginger HE, Tukker JJ (1991) An estimate of turnover time of intestinal mucus gel layer in the rat in situ loop. Int J Pharm 70(3):235–240

34. Hehar SS (1999) Twenty-four hour ambulatory nasal pH monitoring. Clin Otolaryngol Allied Sci 24(1):24–25
35. Jayaraman S, Joo NS, Reitz B, Wine JJ, Verkman AS (2001) Submucosal gland secretions in airways from cystic fibrosis patients have normal [Na-] and pH but elevated viscosity. Proc Natl Acad Sci USA 98(14):8119–8123
36. Schreiber SP (1997) Gastric mucus of the guinea pig: proton carrier and diffusion barrier. Am J Physiol 272(1 Pt 1):G63–G70
37. Bahari HM, Ross IN, Turnberg LA (1982) Demonstration of a pH gradient across the mucus layer on the surface of human gastric mucosa in vitro. Gut 23(6):513–516
38. Lai SK, Wang Y-Y, Hanes J (2009) Mucus-penetrating nanoparticles for drug and gene delivery to mucosal tissues. Adv Drug Deliv Rev 61(2):158–171
39. Mantle M (1989) Binding of Yersinia enterocolitica to rabbit intestinal brush border membranes, mucus, and mucin. Infect Immun 57(11):3292–3299
40. Olmsted SS, Padgett JL, Yudin AI, Whaley KJ, Moench TR, Cone RA (2001) Diffusion of macromolecules and virus-like particles in human cervical mucus. Biophys J 81(4): 1930–1937
41. Sajjan U, Reisman J, Doig P, Irvin RT, Forstner G, Forstner J (1992) Binding of nonmucoid Pseudomonas aeruginosa to normal human intestinal mucin and respiratory mucin from patients with cystic fibrosis. J Clin Invest 89(2):657–665
42. Lieleg O, Vladescu I, Ribbeck K (2010) Characterization of particle translocation through mucin hydrogels. Biophys J 98(9):1782–1789
43. Soane RJ (1999) Evaluation of the clearance characteristics of bioadhesive systems in humans. Int J Pharm 178(1):55–65
44. Leitner VM, Walker GF, Bernkop-Schnürch A (2003) Thiolated polymers: evidence for the formation of disulphide bonds with mucus glycoproteins. Eur J Pharm Biopharm 56(2):207–214
45. Bernkop-Schnürch A (2003) Thiolated polymers—thiomers: synthesis and in vitro evaluation of chitosan–2-iminothiolane conjugates. Int J Pharm 260(2):229–237
46. Schmitz T, Grabovac V, Palmberger TF, Hoffer MH, Bernkop-Schnürch A (2008) Synthesis and characterization of a chitosan-N-acetyl cysteine conjugate. Int J Pharm 347(1–2):79–85
47. Wang Y-Y, Lai SK, So C, Schneider C, Cone R, Hanes J (2011) Mucoadhesive nanoparticles may disrupt the protective human mucus barrier by altering its microstructure. PLoS One 6(6):e21547
48. Lai SK, O'Hanlon DE, Harrold S, Man ST, Wang YY, Cone R, Hanes J (2007) Rapid transport of large polymeric nanoparticles in fresh undiluted human mucus. Proc Natl Acad Sci 104(5):1482–1487
49. Tang BC, Dawson M, Lai SK, Wang Y-Y, Suk JS, Yang M, Zeitlin P, Boyle MP, Fu J, Hanes J (2009) Biodegradable polymer nanoparticles that rapidly penetrate the human mucus barrier. Proc Natl Acad Sci USA 106(46):19268–19273
50. Wang Y-Y, Lai SK, Suk JS, Pace A, Cone R, Hanes J (2008) Addressing the PEG mucoadhesivity paradox to engineer nanoparticles that "slip" through the human mucus barrier. Angew Chem Int Ed 47(50):9726–9729
51. Fresta M, Fontana G, Bucolo C, Cavallaro G, Giammona G, Puglisi G (2001) Ocular tolerability and in vivo bioavailability of poly(ethylene glycol) (PEG)-coated polyethyl-2-cyanoacrylate nanosphere-encapsulated acyclovir. J Pharm Sci 90(3):288–297
52. Huang Y (2000) Molecular aspects of muco- and bioadhesion: tethered structures and site-specific surfaces. J Control Release 65:63–71
53. Serra L, Doménech J, Peppas NA (2006) Design of poly(ethylene glycol)-tethered copolymers as novel mucoadhesive drug delivery systems. Eur J Pharm Biopharm 63(1):11–18
54. Groneberg DA, Witt C, Wagner U, Chung KF, Fischer A (2003) Fundamentals of pulmonary drug delivery. Respir Med 97(4):382–387
55. Yang W, Peters JI, Williams RO (2008) Inhaled nanoparticles–a current review. Int J Pharm 356(1–2):239–247
56. Karlas A, Machuy N, Shin Y et al (2010) Genome-wide RNAi screen identifies human host factors crucial for influenza virus replication. Nature 463(7282):818–822

57. Merkel OM, Beyerle A, Librizzi D, Pfestroff A, Behr TM, Sproat B, Barth PJ, Kissel T (2009) Nonviral siRNA delivery to the lung: investigation of PEG-PEI polyplexes and their in vivo performance. Mol Pharm 6(4):1246–1260
58. Senoo T, Hattori N, Tanimoto T, Furonaka M, Ishikawa N, Fujitaka K, Haruta Y, Murai H, Yokoyama A, Kohno N (2010) Suppression of plasminogen activator inhibitor-1 by RNA interference attenuates pulmonary fibrosis. Thorax 65(4):334–340
59. Zhang X, Shan P, Jiang D, Noble PW, Abraham NG, Kappas A, Lee PJ (2004) Small interfering RNA targeting heme oxygenase-1 enhances ischemia-reperfusion-induced lung apoptosis. J Biol Chem 279(11):10677–10684
60. Lomas-Neira JL, Chung C-S, Wesche DE, Perl M, Ayala A (2005) In vivo gene silencing (with siRNA) of pulmonary expression of MIP-2 versus KC results in divergent effects on hemorrhage-induced, neutrophil-mediated septic acute lung injury. J Leukoc Biol 77(6):846–853
61. Rosas-Taraco AG, Higgins DM, Sánchez-Campillo J, Lee EJ, Orme IM, González-Juarrero M (2009) Intrapulmonary delivery of XCL1-targeting small interfering RNA in mice chronically infected with Mycobacterium tuberculosis. Am J Respir Cell Mol Biol 41(2):136–145
62. Perl M, Chung C-S, Lomas-Neira J, Rachel T-M, Biffl WL, Cioffi WG, Ayala A (2005) Silencing of Fas, but not caspase-8, in lung epithelial cells ameliorates pulmonary apoptosis, inflammation, and neutrophil influx after hemorrhagic shock and sepsis. Am J Pathol 167(6):1545–1559
63. Bitko V, Musiyenko A, Shulyayeva O, Barik S (2005) Inhibition of respiratory viruses by nasally administered siRNA. Nat Med 11(1):50–55
64. Li B, Tang Q, Cheng D et al (2005) Using siRNA in prophylactic and therapeutic regimens against SARS coronavirus in Rhesus macaque. Nat Med 11(9):944–951
65. Fulton A, Peters ST, Perkins GA, Jarosinski KW, Damiani A, Brosnahan M, Buckles EL, Osterrieder N, Van de Walle GR (2009) Effective treatment of respiratory alphaherpesvirus infection using RNA interference. PLoS One 4(1):e4118
66. Glud SZ, Bramsen JB, Dagnaes-Hansen F, Wengel J, Howard KA, Nyengaard JR, Kjems J (2009) Naked siLNA-mediated gene silencing of lung bronchoepithelium EGFP expression after intravenous administration. Oligonucleotides 19(2):163–168
67. Ghosn B, Singh A, Li M, Vlassov AV, Burnett C, Puri N et al. (2010) Efficient gene silencing in lungs and liver using imidazole-modified chitosan as a nanocarrier for small interfering RNA. Oligonucleotides 20(3):163–72
68. Ge Q, Filip L, Bai A, Nguyen T, Eisen HN, Chen J (2004) Inhibition of influenza virus production in virus-infected mice by RNA interference. Proc Natl Acad Sci USA 101(23):8676–8681
69. Thomas M, Lu JJ, Ge Q, Zhang C, Chen J, Klibanov AM (2005) Full deacylation of polyethylenimine dramatically boosts its gene delivery efficiency and specificity to mouse lung. Proc Natl Acad Sci USA 102(16):5679–5684
70. Beyerle A, Braun A, Merkel O, Koch F, Kissel T, Stoeger T (2011) Comparative in vivo study of poly(ethylene imine)/siRNA complexes for pulmonary delivery in mice. J Control Release 151(1):51–56
71. Tompkins SM, Lo C-Y, Tumpey TM, Epstein SL (2004) Protection against lethal influenza virus challenge by RNA interference in vivo. Proc Natl Acad Sci USA 101(23):8682–8686
72. Wang J-C, Lai S, Guo X, Zhang X, de Crombrugghe B, Sonnylal S, Arnett FC, Zhou X (2010) Attenuation of fibrosis in vitro and in vivo with SPARC siRNA. Arthritis Res Ther 12(2):R60
73. Moschos SA, Jones SW, Perry MM, Williams AE, Erjefalt JS, Turner JJ, Barnes PJ, Sproat BS, Gait MJ, Lindsay MA (2007) Lung delivery studies using siRNA conjugated to TAT(48-60) and penetratin reveal peptide induced reduction in gene expression and induction of innate immunity. Bioconjug Chem 18(5):1450–1459
74. Gutbier B, Kube SM, Reppe K, Santel A, Lange C, Kaufmann J, Suttorp N, Witzenrath M (2010) RNAi-mediated suppression of constitutive pulmonary gene expression by small interfering RNA in mice. Pulm Pharmacol Ther 23(4):334–344
75. Massaro D, Massaro GD, Clerch LB (2004) Noninvasive delivery of small inhibitory RNA and other reagents to pulmonary alveoli in mice. Am J Physiol Lung Cell Mol Physiol 287(5):L1066–L1070

76. Alvarez R, Elbashir S, Borland T et al (2009) RNA interference-mediated silencing of the respiratory syncytial virus nucleocapsid defines a potent antiviral strategy. Antimicrob Agents Chemother 53(9):3952–3962
77. Behlke MA (2006) Progress towards in vivo use of siRNAs. Mol Ther 13(4):644–670
78. Whitehead KA, Langer R, Anderson DG (2009) Knocking down barriers: advances in siRNA delivery. Nat Rev Drug Discov 8(2):129–138
79. Illum L (1998) Chitosan and its use as a pharmaceutical excipient. Pharm Res 15(9): 1326–1331
80. Artursson P, Lindmark T, Davis SS, Illum L (1994) Effect of chitosan on the permeability of monolayers of intestinal epithelial cells (caco-2). Pharm Res 11(9):1358–1361
81. Borchard G, Lueßen HL, de Boer AG, Verhoef JC, Lehr C-M, Junginger HE (1996) The potential of mucoadhesive polymers in enhancing intestinal peptide drug absorption. III: effects of chitosan-glutamate and carbomer on epithelial tight junctions in vitro. J Control Release 39(2–3):131–138
82. Aspden TJ, Mason JD, Jones NS, Lowe J, Skaugrud O, Illum L (1997) Chitosan as a nasal delivery system: the effect of chitosan solutions on in vitro and in vivo mucociliary transport rates in human turbinates and volunteers. J Pharm Sci 86(4):509–513
83. Köping-Höggård M, Vårum KM, Issa M, Danielsen S, Christensen BE, Stokke BT, Artursson P (2004) Improved chitosan-mediated gene delivery based on easily dissociated chitosan polyplexes of highly defined chitosan oligomers. Gene Ther 11(19):1441–1452
84. Liu X, Howard KA, Dong M, Andersen MØ, Rahbek UL, Johnsen MG, Hansen OC, Besenbacher F, Kjems J (2007) The influence of polymeric properties on chitosan/siRNA nanoparticle formulation and gene silencing. Biomaterials 28(6):1280–1288
85. Köping-Höggård M, Tubulekas I, Guan H, Edwards K, Nilsson M, Vårum KM, Artursson P (2001) Chitosan as a nonviral gene delivery system Structure-property relationships and characteristics compared with polyethylenimine in vitro and after lung administration in vivo. Gene Ther 8(14):1108–1121
86. Iqbal M, Lin W, Jabbal-Gill I, Davis SS, Steward MW, Illum L (2003) Nasal delivery of chitosan-DNA plasmid expressing epitopes of respiratory syncytial virus (RSV) induces protective CTL responses in BALB/c mice. Vaccine 21(13–14):1478–1485
87. Robbins M, Judge A, MacLachlan I (2009) siRNA and innate immunity. Oligonucleotides 19(2):89–102
88. Aouadi M, Tesz GJ, Nicoloro SM, Wang M, Chouinard M, Soto E, Ostroff GR, Czech MP (2009) Orally delivered siRNA targeting macrophage Map4k4 suppresses systemic inflammation. Nature 458(7242):1180–1184
89. Wilson DS, Dalmasso G, Wang L, Sitaraman SV, Merlin D, Murthy N (2010) Orally delivered thioketal nanoparticles loaded with TNF-α–siRNA target inflammation and inhibit gene expression in the intestines. Nat Mater 9(11):923–928
90. Kriegel C, Amiji M (2011) Oral TNF-[alpha] gene silencing using a polymeric microsphere-based delivery system for the treatment of inflammatory bowel disease. J Control Release 150(1):77–86
91. Peer D, Park EJ, Morishita Y, Carman CV, Shimaoka M (2008) Systemic leukocyte-directed siRNA delivery revealing cyclin D1 as an anti-inflammatory target. Science 319(5863):627–630
92. Zhang Y, Cristofaro P, Silbermann R et al (2006) Engineering mucosal RNA interference in vivo. Mol Ther 14(3):336–342
93. Larson SD, Jackson LN, Chen LA, Rychahou PG, Evers BM (2007) Effectiveness of siRNA uptake in target tissues by various delivery methods. Surgery 142(2):262–269
94. Palliser D, Chowdhury D, Wang Q-Y, Lee SJ, Bronson RT, Knipe DM, Lieberman J (2006) An siRNA-based microbicide protects mice from lethal herpes simplex virus 2 infection. Nature 439(7072):89–94
95. Wu Y, Navarro F, Lal A, Basar E, Pandey RK, Manoharan M, Feng Y, Lee SJ, Lieberman J, Palliser D (2009) Durable protection from herpes simplex virus-2 transmission following intravaginal application of siRNAs targeting both a viral and host gene. Cell Host Microbe 5(1):84–94

96. Woodrow KA, Cu Y, Booth CJ, Saucier-Sawyer JK, Wood MJ, Saltzman WM (2009) Intravaginal gene silencing using biodegradable polymer nanoparticles densely loaded with small-interfering RNA. Nat Mater 8(6):526–533
97. Wu SY, Chang H-I, Burgess M, McMillan NAJ (2011) Vaginal delivery of siRNA using a novel PEGylated lipoplex-entrapped alginate scaffold system. J Control Release 155(3): 418–426
98. Wheeler LA, Trifonova R, Vrbanac V et al (2011) Inhibition of HIV transmission in human cervicovaginal explants and humanized mice using CD4 aptamer-siRNA chimeras. J Clin Invest 121(6):2401–2412
99. Boussif O, Lezoualc'h F, Zanta MA, Mergny MD, Scherman D, Demeneix B, Behr JP (1995) A versatile vector for gene and oligonucleotide transfer into cells in culture and in vivo: polyethylenimine. Proc Natl Acad Sci USA 92(16):7297–7301
100. Pack DW, Hoffman AS, Pun S, Stayton PS (2005) Design and development of polymers for gene delivery. Nat Rev Drug Discov 4(7):581–593
101. Robbins M, Judge A, Ambegia E, Choi C, Yaworski E, Palmer L, McClintock K, MacLachlan I (2008) Misinterpreting the therapeutic effects of small interfering RNA caused by immune stimulation. Hum Gene Ther 19(10):991–999
102. Kircheis R, Schüller S, Brunner S, Ogris M, Heider KH, Zauner W, Wagner E (1999) Polycation-based DNA complexes for tumor-targeted gene delivery in vivo. J Gene Med 1(2):111–120
103. Ogris M, Brunner S, Schüller S, Kircheis R, Wagner E (1999) PEGylated DNA/transferrin-PEI complexes: reduced interaction with blood components, extended circulation in blood and potential for systemic gene delivery. Gene Ther 6(4):595–605
104. Beyerle A, Merkel O, Stoeger T, Kissel T (2010) PEGylation affects cytotoxicity and cell-compatibility of poly(ethylene imine) for lung application: structure-function relationships. Toxicol Appl Pharmacol 242(2):146–154
105. Wu SY, McMillan NAJ (2009) Lipidic systems for in vivo siRNA delivery. AAPS J 11(4):639–652
106. Griesenbach U, Kitson C, Garcia SE et al (2006) Inefficient cationic lipid-mediated siRNA and antisense oligonucleotide transfer to airway epithelial cells in vivo. Respir Res 7:26
107. Jensen DMK, Cun D, Maltesen MJ, Frokjaer S, Nielsen HM, Foged C (2010) Spray drying of siRNA-containing PLGA nanoparticles intended for inhalation. J Control Release 142(1):138–145
108. Andersen MØ, Howard KA, Paludan SR, Besenbacher F, Kjems J (2008) Delivery of siRNA from lyophilized polymeric surfaces. Biomaterials 29(4):506–512
109. Phalen RF, Oldham MJ, Wolff RK (2008) The relevance of animal models for aerosol studies. J Aerosol Med Pulm Drug Deliv 21(1):113–124
110. Hillery AM, Lloyd AW, Swarbrick J (2002) Drug delivery and targeting: for pharmacists and pharmaceutical scientists, 1st edn. CRC, Boca Raton, FL
111. Oivanen M, Kuusela S, Lönnberg H (1998) Kinetics and mechanisms for the cleavage and isomerization of the phosphodiester bonds of RNA by brønsted acids and bases. Chem Rev 98(3):961–990
112. Watts JK, Katolik A, Viladoms J, Damha MJ (2009) Studies on the hydrolytic stability of 2′-fluoroarabinonucleic acid (2′F-ANA). Org Biomol Chem 7(9):1904–1910
113. Atuma C, Strugala V, Allen A, Holm L (2001) The adherent gastrointestinal mucus gel layer: thickness and physical state in vivo. Am J Physiol Gastrointest Liver Physiol 280(5):G922–G929
114. Varum FJO, Veiga F, Sousa JS, Basit AW (2010) An investigation into the role of mucus thickness on mucoadhesion in the gastrointestinal tract of pig. Eur J Pharm Sci 40(4):335–341
115. Junqueira LC, Carneiro J (2005) Basic histology: text & atlas: text and atlas, 11th edn. McGraw-Hill Medical, New York
116. Kumar P, Wu H, McBride JL, Jung K-E, Hee Kim M, Davidson BL, Kyung Lee S, Shankar P, Manjunath N (2007) Transvascular delivery of small interfering RNA to the central nervous system. Nature 448(7149):39–43

117. Nishina K, Unno T, Uno Y, Kubodera T, Kanouchi T, Mizusawa H, Yokota T (2008) Efficient in vivo delivery of siRNA to the liver by conjugation of [alpha]-tocopherol. Mol Ther 16(4):734–740
118. Wolfrum C, Shi S, Jayaprakash KN et al (2007) Mechanisms and optimization of in vivo delivery of lipophilic siRNAs. Nat Biotech 25(10):1149–1157
119. Howard KA, Thomas NW, Jenkins PG, Davis SS, O'Hagan DT (1994) The absorption of microparticles into Peyer's patches of the rabbit and rat. Pharm Sci Commun 4(4):207–216
120. Jepson MA, Clark MA, Foster N, Mason CM, Bennett MK, Simmons NL, Hirst BH (1996) Targeting to intestinal M cells. J Anat 189(Pt 3):507–516
121. Malik B, Goyal AK, Markandeywar TS, Rath G, Zakir F, Vyas SP (2011) Microfold-cell targeted surface engineered polymeric nanoparticles for oral immunization. J Drug Target: 20(1):76–84. doi:10.3109/1061186x.2011.611516
122. Rajapaksa TE, Stover-Hamer M, Fernandez X, Eckelhoefer HA, Lo DD (2010) Claudin 4-targeted protein incorporated into PLGA nanoparticles can mediate M cell targeted delivery. J Control Release 142(2):196–205
123. Jepson MA, Clark MA, Hirst BH (2004) M cell targeting by lectins: a strategy for mucosal vaccination and drug delivery. Adv Drug Deliv Rev 56(4):511–525
124. Kim S-H, Seo K-W, Kim J, Lee K-Y, Jang Y-S (2010) The M cell-targeting ligand promotes antigen delivery and induces antigen-specific immune responses in mucosal vaccination. J Immunol 185(10):5787–5795
125. Rescigno M, Urbano M, Valzasina B, Francolini M, Rotta G, Bonasio R, Granucci F, Kraehenbuhl J-P, Ricciardi-Castagnoli P (2001) Dendritic cells express tight junction proteins and penetrate gut epithelial monolayers to sample bacteria. Nat Immunol 2(4):361–367
126. McConnell EL, Basit AW, Murdan S (2008) Measurements of rat and mouse gastrointestinal pH, fluid and lymphoid tissue, and implications for in-vivo experiments. J Pharm Pharmacol 60(1):63–70
127. Plevy SE, Targan SR (2011) Future therapeutic approaches for inflammatory bowel diseases. Gastroenterology 140(6):1838–1846
128. Triantafillidis JK, Merikas E, Georgopoulos F (2011) Current and emerging drugs for the treatment of inflammatory bowel disease. Drug Des Devel Ther 5:185–210
129. Alexander NJ, Baker E, Kaptein M, Karck U, Miller L, Zampaglione E (2004) Why consider vaginal drug administration? Fertil Steril 82(1):1–12
130. Petta CA, Faundes A, Dunson TR, Ramos M, Delucio M, Faundes D, Bahamondes L (1998) Timing of onset of contraceptive effectiveness in Depo-Provera users: part I. Changes in cervical mucus. Fertil Steril 69(2):252–257
131. Rahbek UL, Nielsen AF, Dong M, You Y, Chauchereau A, Oupicky D, Besenbacher F, Kjems J, Howard KA (2010) Bioresponsive hyperbranched polymers for siRNA and miRNA delivery. J Drug Target 18(10):812–820
132. Rahbek UL, Howard KA, Oupicky D, Manickam DS, Dong M, Nielsen AF, Hansen TB, Besenbacher F, Kjems J (2008) Intracellular siRNA and precursor miRNA trafficking using bioresponsive copolypeptides. J Gene Med 10(1):81–93

Chapter 6
Nanomedicines for Systemic Delivery of RNAi Therapeutics

Dan Peer

Abstract The discovery of RNA interference (RNAi), a ubiquitous cellular pathway of gene regulation that is dysregulated in many types of diseases, provides an exciting opportunity for relatively rapid and revolutionary approaches to drug design. Small RNAs that harness the RNAi machinery may become the next class of drugs for treating a variety of diseases. However, targeting of RNAi molecules into specific tissues and cells is still a hurdle. Major attempts are being made to develop carriers that could overcome systemic and cellular barriers. This chapter will present the recent progress in this emerging field, focusing on strategies for systemic cellular targeting using different types of nano-sized vectors. Selective cellular targeting of RNAi is considered as the major hurdle translating this new class of drugs into clinical practice.

6.1 Introduction

RNA interference (RNAi) is a natural cellular mechanism for RNA-guided regulation of gene expression. This regulation is carried out by double-stranded ribonucleic acid (dsRNA) that suppress the expression of specific genes with complementary nucleotide sequences either by degrading specific messenger RNA (mRNA) or by blocking mRNA translation. RNAi can be activated by expressing short hairpin RNA (shRNA) or by introducing synthetic small interfering RNAs (siRNAs) directly into the cell cytoplasm with viral and nonviral vectors [1, 2].

D. Peer (✉)
Laboratory of Nanomedicine, Department of Cell Research and Immunology,
George S. Wise Faculty of Life Science, Tel Aviv University, Tel Aviv, Israel

Center for Nanoscience and Nanotechnology, Tel Aviv University, Tel Aviv, Israel
e-mail: peer@tauex.tau.ac.il

K.A. Howard (ed.), *RNA Interference from Biology to Therapeutics*,
Advances in Delivery Science and Technology, DOI 10.1007/978-1-4614-4744-3_6,

siRNA is a chemically synthesised dsRNA of 19–23 base pairs with 2 nucleotides unpaired in the 5′-phosphorylated ends and unphosphorylated 3′-ends [3, 4]. Within the cell cytoplasm, siRNAs are incorporated into RNA-induced silencing complex (RISC), a protein–RNA complex that separates the strands of the RNA duplex and discards the sense strand. The antisense RNA strand then guides RISC to anneal and cleave the homologous target mRNA or block its translation [2]. The RISC complex incorporates the antisense strand and may show a therapeutic effect for up to 7 days in dividing cells and for several weeks in nondividing cells. Furthermore, repeated administration of siRNA can result in stable silencing of its target [5].

The combination of knockdown in every gene of interest and the ability to treat various diseases by addressing otherwise "undruggable" targets (i.e. molecules without ligand-binding domains or those that have a structural homology with other important molecules in the cell), and reduced toxicity, emphasises the potential of siRNAs to serve as a new platform for therapy in personalised medicine, where an individual genome will be sequenced and RNAi molecules could potentially be designed to inhibit a specific defective mRNA.

Despite this promise, utilising siRNA as therapeutics is not a trivial task. For example low cellular uptake across the plasma membrane associated with the high molecular weight (~13 kD) and the net negative of naked siRNA is found [2, 6]. When injected intravenously (one of the main routes of systemic administration), in addition to rapid renal clearance and susceptibility to degradation by RNAses, unmodified naked siRNAs are recognised by Toll-like receptors (TLRs). This often stimulates the immune system provoking an interferon response, complement activation, cytokine induction and coagulation cascades. Beside the undesired immune activation, those effects can globally suppress gene expression, generate off-target effects and misinterpreted outcomes [6, 7]. There is a clear need, therefore, for appropriate systems for extracellular delivery of siRNAs in addition to intracellular mechanisms for internalisation, release (from the carriers) and escape (from the endosomes), in addition to accumulation of siRNAs in the cell cytoplasm and RISC activation.

Silencing of gene expression in vitro is a great tool for functional and validation studies. Most of the methods commonly used for in vitro or ex vivo delivery of RNAi molecules are conventional transfection methods. Studies with purely physical methods such as microinjection and electroporation [8–11], as well as the use of calcium co-precipitation [12], commercial cationic polymers and lipids [3, 13–18] and cell penetrating peptides [19–23], have demonstrated effective knockdown of desired genes. Except for the physical methods (in which the cell is subjected to an injection of small volumes of siRNAs directly into the cell cytoplasm or to a burst of electricity that causes pores in the membrane, hence elevating the ability of extracellular material to enter into the cell), all the methods share a main feature—a positive (cationic) charge that enables complexation of the siRNAs and interact with the negatively charged plasma membrane. In this manner, it is important to note that there are evidences for toxicities of some commercial cationic lipids and polymers ([24] and reviewed in [25]). This emphasises the usefulness of the cell penetrating proteins (such as natural ligands, antibodies and their fragments), which are less toxic and have the potential to target specific cells. Studying gene expression in a disease model by validating the specific role of the gene in vivo along with the potential to induce therapeutic gene silencing opens new avenues for utilising RNAi as a novel therapeutic modality. This review will present the recent progress in this emerging field, focusing on the in vivo applications with special emphasis on the strategies for RNAi delivery to leucocytes using various nanomedicines.

6.2 Nanomedicine

By definition, nanomedicine is the medical application of nanotechnology [26, 27]. Nano-vehicles made from different materials with different size, geometry and charge with at least one dimension within the range of 1–100 nm that are used for delivery of imaging agents (for diagnostics) and/or for delivery of therapeutic entities are termed nanomedicine [26, 27] or nano-drug. In addition, a new class of vehicles that can deliver both diagnostics agents and therapeutic agents and can report in real-time the location of a diseased cell/tissue is termed theranostics (diagnosis and therapy), and it is considered as the next-generation approach of current nano-scale delivery platforms. The reader is referred to recent excellent reviews on nanomedicines with various applications, which are beyond the scope of this chapter [28–31]. In this chapter, the term "nanomedicine" will be used for any delivery system at the nanoscale.

6.3 Translation of RNAi into Clinical Practice

Despite the large diversity of available methods for in vitro siRNA delivery briefly described above, there are additional hurdles to translate these methods into a clinical therapeutic. As detailed below, the biggest hurdle facing the translation of RNAi therapeutic potential into the clinic is their delivery.

6.3.1 In Vivo Delivery of siRNAs

Most common routes of administration of siRNAs include the preferred non-invasive per-oral, topical (skin), transmucosal (nasal, pulmonary, buccal, sublingual, vaginal, ocular and rectal) as well as systemic (intravenous) administration.

Local delivery of siRNAs has been demonstrated in various animal models [22, 32–34] and is employed in several ongoing clinical trials. Based on local injections of naked or cationic lipid/polymer-formulated siRNAs, this method of treatment that has shown very promising outcomes is suitable for mucosal diseases or subcutaneous tissues.

For intravenous systemic administration, the delivery approaches need to protect the RNAi molecules from degradation by serum nucleases, avoid clearance from the circulatory system by renal filtration, transport the siRNA molecules across the vascular endothelial barrier, facilitate effective biodistribution and accumulation at the appropriate tissue, and promote efficient uptake and endosomal escape into the cytoplasm of the target cells where they can associated with RISC and guide the cleavage of target mRNA. The reader is referred to an excellent review on the anatomical barriers [35].

Systemic delivery of naked siRNAs has been demonstrated by the hydrodynamic method. This method, whose precise mechanism of action is unknown but may involve pressure-mediated cellular penetration, involves rapid injection of a large volume of siRNA in physiologic solutions (about 10% of the body weight administered within 5–10 s) [36, 37]. Hepatocyte cells in the liver are the main target of this approach. Different studies have been performed with this method, demonstrating functional knock-down of specific genes in the liver of animals [36–39]. Nevertheless, due to volume overload sideeffects, the hydrodynamic method is not relevant for human therapeutic use.

Naked siRNAs could also be utilised for targeting the kidney. When systematically administrated, large amount of naked siRNAs are excreted by the glomerulus (<40 kDa Mw) and reabsorbed in the proximal tubule. The accumulation of free siRNA in the kidney is 40 times higher than in any other organ, an ideal propriety for selective gene therapy. Studies in rat models for renal injury indicated functional silencing of p53, a major pro-apoptotic gene, and renal protection, both in single and multiple injection administration [40]. A product based on these studies, QPI-1002, is being developed by Quark Pharmaceuticals for systemic delivery of p53-siRNA in acute renal injury and delayed graft function.

Accumulation in the kidney of systemic naked siRNA is due to physiological consequence. Otherwise, strategies for systemic delivery of siRNA to alternative sites must rely on transport within nanocarriers or as nano-conjugates such as siRNA–cholestrol conjugates or fusion proteins carrying siRNAs. These carriers should be composed of fully degradable materials (to avoid undesired and probably toxic accumulation of the delivery system components in the body) and should act on specific cells or tissues while avoid damaging others.

Systemic siRNA delivery strategies are divided into two major categories: passive and active cellular delivery. Passive delivery exploits the inherited tendency of nanoparticles to accumulate in organs of the reticuloendothelial system (RES) also known as the mononuclear phagocytic system (MPS). The MPS, part of the immune system, consists of phagocytic cells located in reticular connective tissue, primarily monocytes, dendritic cells and macrophages. These cells accumulate in lymph nodes, the spleen, and liver that phagocytose foreign particles such as viruses, bacteria and parasites recognised by size, shape and charge. Passive targeting siRNAs to the liver can be achieved due to the highly perfused nature of the liver and phagocytic capture by resident phagocytic Kupffer cells located in the liver lining the walls of the sinusoids that form part of the MPS. As an alternative approach, active targeting to liver hepatocytes can be achieved by including receptor specific antibodies, ligands or ligand mimetics into the nanocarrier design.

6.4 Nanoparticle and Nano-Conjugate-Based Delivery Strategies

Nanocarrier-based approaches typically involve the supramolecular assembly of a carrier, which may include one or more targeting moieties that can deliver drugs and/or imaging agents in a cell- or tissue-specific manner (Fig. 6.1). Typical nanocarriers may be constructed from lipids, proteins, carbohydrates or synthetic polymers [27, 41]. Recent advances in the engineering of devices at the nanoscale have led to the development of new platform particle-based technologies. Inorganic nanoparticles may be comprised of gold, silver, silicon, silica or iron oxide (including

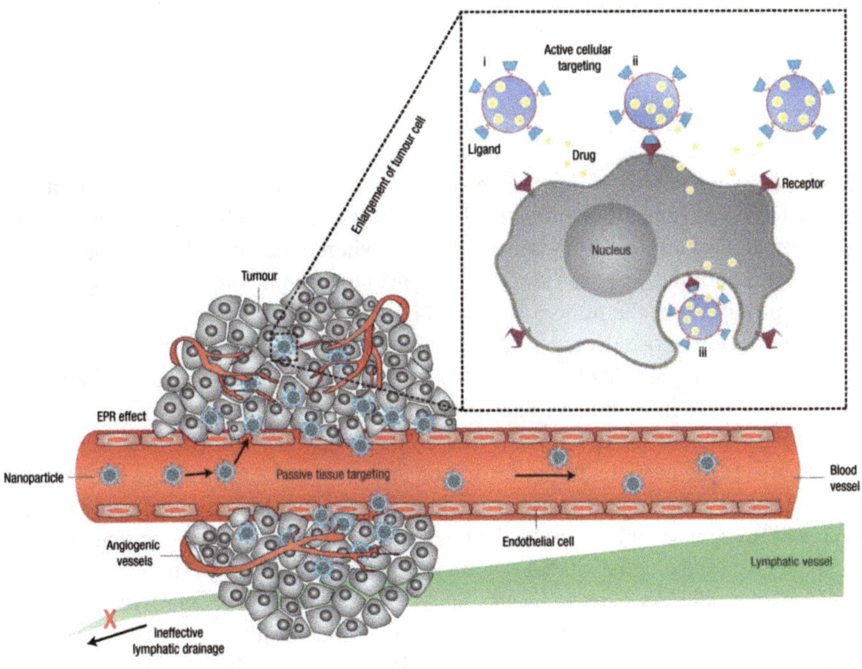

Fig. 6.1 Passive and active tumour targeting. Passive tissue targeting is achieved by extravasation of nanoparticles through increased permeability of the tumour vasculature and ineffective lymphatic drainage (EPR effect). Active cellular targeting (inset) can be achieved by functionalising the surface of nanoparticles with ligands that promote cell-specific recognition and binding. The nanoparticles can (i) release their contents in close proximity to the target cells; (ii) attach to the membrane of the cell and act as an extracellular sustained-release drug depot; or (iii) internalise into the cell. Reprinted with permission from [27] Copyright 2007 Nature Nanotechnology

Nanoshells and Quantum Dots), whilst organic nanoparticles include fullerenes, dendrimers and virus-like structures. These new nanocarriers have been explored for their use in a variety of applications such as drug delivery, imaging, photothermal ablation of tumours, radiation sensitisers, detection of apoptosis and sentinel lymph node mapping [27, 28, 42]. An alternative strategy is construction of nano-conjugates by attachment of moieties such as cholesterol to the siRNA to control the systemic delivery. In choosing an appropriate nanocarrier construct for rapid and effective clinical translation, one should consider the following criteria:

1. A carrier should be made from a material that is biocompatible, well characterised and can be easily functionalised.
2. A nanocarrier should exhibit high differential uptake efficiency by the target cells or tissue.
3. A nanocarrier should be either soluble or colloidal in water and have an extended circulating half-life to increase the likelihood of its effectiveness.
4. A nanocarrier should have a low rate of aggregation and a long shelf-life.

In delivering RNAi payloads, we can subdivide the delivery strategies based on passive and active cellular targeting.

6.4.1 Passive Systemic RNAi Delivery

Passive delivery strategies are the most advanced (from a clinical standpoint) to deliver RNAi payloads systemically. These strategies exploited the enhanced permeability and retention (EPR) effect (Fig. 6.1) where leaky blood vessels and ineffective lymphatic drainage cause accumulation of macromolecules and nanoparticles in close proximity of the tumour [27, 43, 44].

Among the RNAi delivery strategies that utilise the EPR effect are the stable nucleic acid-lipid particles (SNALP). SNALP is a ~100-nm non-targeted liposome with low cationic lipid content that encapsulates siRNAs and is coated with a diffusible polyethylene glycol–lipid (PEG–lipid) conjugate [45, 46]. The PEG-lipid coat stabilises the particle during formation and provides a neutral and hydrophilic exterior that prevents rapid systemic clearance. The lipid bi-layer composed of cationic and fusogenic lipids enables the internalisation of the SNALP and endosomal escape while releasing the siRNA payload. Biodistribution study indicates that most (28%) of the siRNAs carried by the SNALPs accumulated in the liver and only 0.3% in the lungs. Functional study of SNALPs encapsulated ApoB–siRNA has shown significant reduction in ApoB mRNA levels. Despite the presence of cationic lipids known to trigger toxicities [24, 25], studies in mice and non-human primates did not reveal any adverse effects except for liver enzyme release. Based on these results, a clinical trial is now been conducted to test the ability of SNALPs to deliver siRNAs for liver cancer treatment. SNALPs encapsulating siRNA against the polymerase gene of the Zaire strain have been shown to protect guinea pigs from a lethal challenge of the Ebola virus [47]. Other formulations of cationic liposomes, with greater cationic lipid content than the SNALPs, have induced not only effective gene silencing, but also cytokine induction, interferon response and complement activation as well as coagulation cascades and liver toxicity and, thus, cannot be used for clinical evaluation.

Considering the significant toxicities that have been associated with cationic liposomes, neutral charged liposomes are very promising carriers for systemic delivery of siRNAs. 1,2-dioleoyl-sn-glycero-3-phosphatidylcholine (DOPC) non-PEGylated liposomes encapsulating siRNAs against different molecules expressed on melanoma and ovarian cancers inhibited tumour growth in human xenograft models [48, 49]. The accumulation of these liposomes in cancerous tissues is attributed to the EPR effect.

Cationic lipidoid (synthetic lipid-like molecules)-containing liposomes are another siRNA delivery system that has been shown to induce effective gene silencing (80% reduction in murine ApoB and Factor VII mRNAs levels) in the liver. Single intravenous injection of cationic lipidoid-containing liposomes encapsulating ApoB–siRNA resulted in 50% decrease in the protein level 3 days and up to 2 weeks after the treatment. Although no immune response was indicated, increases in the levels of two liver enzymes suggest some degree of liver toxicity [50, 51].

HK peptides are another effective delivery system for siRNAs. This system is based on the addition of histidine into poly-lysine peptides. While lysine is important for binding the siRNAs, histidine stabilises the particles and has an important role in buffering acidic endosomes, thereby leading to endosomal disruption and payload release. Specific ratios and patterns of histidine and lysine have been found to augment the siRNA delivery, while carriers with a higher ratio of histidine to lysine content seemed to be more effective [52]. HK peptides carrying Raf-1-siRNA or human rhomboid family-1-siRNA induced significant silencing of target genes and growth inhibition of tumour xenografts [53, 54].

Atelocollagen is a biomaterial that consists of a low-immunogenic fraction of pepsin-digested type I collagen from calf dermis. Rich in positively charged lysine and hydroxylysine residues it complexes the negatively charged siRNA and interacts with the plasma membrane, hence facilitates incorporation of the siRNA into the cells. Although these particles have not been modified to target tumours, passive targeting due to the EPR effect causes the selective accumulation within the cancerous tissues as shown in several studies with different tumour xenografts [21, 55–57]. Initial studies indicated that atelocollagen particles could be administered safely without induction of cytokines or observed toxicity to the tissues.

6.4.2 Active Cellular Systemic RNAi Delivery

siRNAs conjugated to a targeting moiety or to specific cell penetrating entity (such as cholesterol, specific phospholipids, natural receptor ligands, monoclonal antibodies and their fragments) is a common strategy for active cellular delivery. This class of cell-specific delivery is becoming more popular and feasible due to

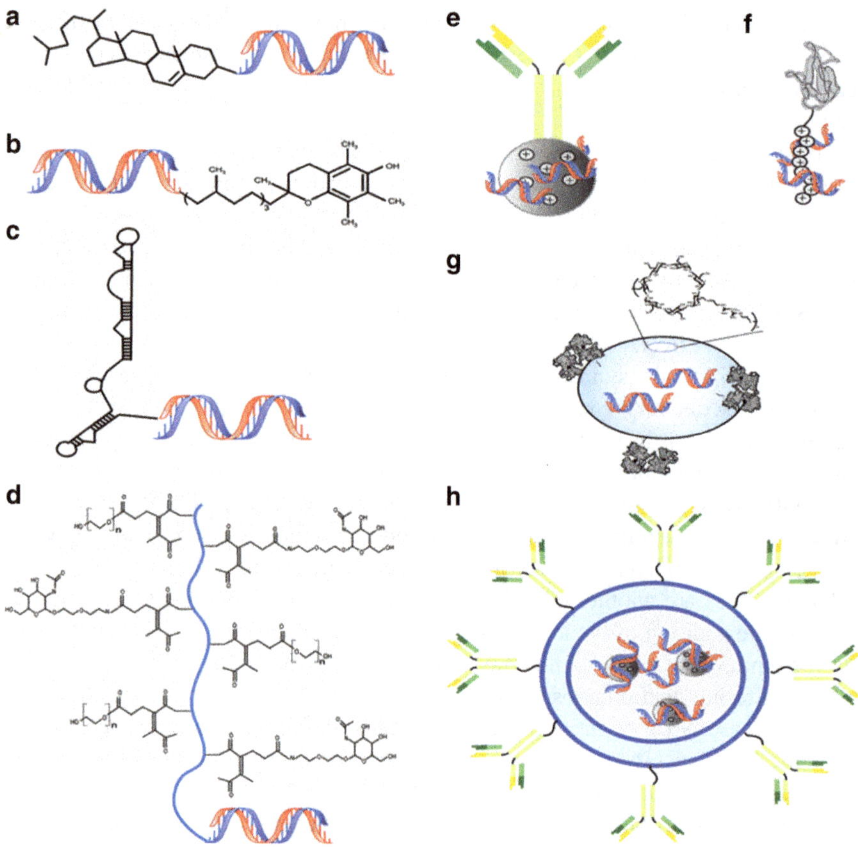

Fig. 6.2 Strategies for cell type-specific siRNA delivery in vivo. A variety of approaches have been used for the cell type-specific delivery of siRNAs, including the direct conjugation of targeting molecules, such as ligands [cholesterol (**a**) and α-tocopherol (**b**)], aptamers (**c**) or polymers [dynamic polyconjugates (**d**)]. In addition, the negatively charged siRNA will spontaneous bind to positively charged proteins or peptides fused to an antibody (**e**) or a ligand [e.g. a portion of the rabies virus glycoprotein (RVG) (**f**)]. siRNAs can be encapsulated into polymeric [e.g. cyclodextrin (**g**) coated with the ligand transferring] or lipid nanoparticles [e.g. immunoliposomes (**h**), lipid nanoparticles coated with an antibody]. Reprinted with permission from [41] Copyright 2011 Gene Therapy

understanding of selective biological molecules and the era of "omics" such as genomics, proteomics and lipidomics [27, 31]. Examples for active cellular targeting include the cholesterol–siRNA conjugate (Fig. 6.2a). The specificity of this delivery system is determined by the lipoprotein to which the cholesterol–siRNAs conjugates are attached in the circulation. When the conjugates bind low-density lipoproteins (LDL), the particles are mainly taken up by the liver due to its LDL-receptors' expression, whereas when binding high-density lipoproteins (HDL), they accumulate in the liver, the gut, the kidney and steroidogenic organs, all of which express scavenger receptor class B, type I (SR-BI) receptors, which bind HDL [58].

Cholesterol–ApoB–siRNA conjugate as well as α-tocopherol [59] (Fig. 6.2b) and lithocholic acid or lauric acid conjugated to ApoB–siRNA [60] reduced serum cholesterol and ApoB mRNA levels in the liver.

Aptamer–siRNA chimeras (Fig. 6.2c) are RNA-based particles for specific delivery of siRNAs. This approach relies only on the fact that structured RNAs are capable of binding a variety of proteins with high affinity and specificity. The chimera includes both a targeting moiety, the aptamer, and an RNA-silencing moiety, the siRNA. The aptamer–siRNA chimeras have demonstrated specific binding and delivery of siRNAs into a xenograft model of prostate cancer. The aptamer portion of the chimeras mediates binding to *Prostate-specific membrane antigen* (PSMA), a cell-surface receptor overexpressed in prostate cancer cells and tumour vascular endothelium, whereas the siRNA reduce the expression of survival genes [61]. This approach eliminates various side effects, hence aptamers and siRNAs exhibit low immunogenicity. Additionally advantages are the possibility to synthesise large quantities at a relatively low cost and the smaller size of aptamers compared with that of antibodies (<15 versus 150 kDa), which promotes better tissue penetration.

A similar strategy of conjugation is the dynamic polyconjugates [62] (Fig. 6.2d). This delivery approach includes membrane-active polymers whose activity is masked until reaching the acidic environment of the endosomes. Due to the employment of *N*-acetylgalactosamine, which binds to the asialoglycoprotein receptor, they target hepatocytes. Like the SNALPs, these particles decreased ApoB mRNA levels in the liver when ApoB–siRNA is used.

Polyethylenimine (PEI) is a cationic polymer used to condense nucleic acids into polyplexes and is endosomolytic due to its buffering capacity at the endosomal pH. PEI polyplexes incorporating siRNAs have demonstrated functional silencing in subcutaneously transplanted tumours in nude mice. Particles composed of RGD (Arg-Gly-Asp) peptide coupled via PEG–PEI allow prolonged circulatory half-life, reduced immunogenicity and active targeting. When complexed with siRNA, RGD–PEG–PEI molecules form a polyplex, with the positively charged RGD–PEG components exposed on its surface. The targeting ability of this particle is based on the overexpression of a_v integrins, which RGD peptides bind in certain cancers and in tumour vasculature [63]. Similar to the previous two examples, cyclodextrin-containing polycation (CDP) particles have been successfully used for siRNAs delivery into murine subcutaneous tumours [64] (Fig. 6.2g). CDP is a polymer with a cyclic oligomeric glucose backbone that complexes with siRNAs to assemble into a colloidal 50–70 nm particle. To achieve targeting, transferrin-coupled PEG is attached to the surface of the particles exploiting the upregulation of transferrin (Tf) receptors in cancers. However, despite considered less toxic than conventional cationic polymers (such as PEI), safety evaluation in non-human primates revealed that intravenous injection of these particles induced elevation in blood urea (that might indicate kidney toxicity), mild increase in liver enzyme levels and a mild increase in IL-6 levels at high concentrations. Multiple injections of the particles induced antibodies to human-Tf. Despite these disadvantages, Tf-coupled CDP containing siRNAs for melanoma cancer treatment has been taken into clinical trials [65].

Antibody-protamine fusion carriers (Fig. 6.2e) are a promising approach for systemic siRNA delivery. Protamines are relatively small (5–8 kDa) and highly basic proteins composed of 55–79% arginine residues [66]. Positively charged protamine interacts with the negatively charged siRNAs to stabilise, neutralise and condense the siRNAs. ErbB2-protamine fusion protein in complex with siRNA significantly inhibited growth of breast cancer cells [67].

Different formulations of targeted cationic liposomes served for selective targeting of hepatic stellate cells, which are the major cell population involved in the formation of scar tissue in response to liver damage such as fibrosis or solid tumours. Stellate cells express receptors for retinol binding protein, which efficiently uptake vitamin A. Taking advantage of this physiological condition, injection of cationic liposomes coupled to vitamin A and complexed with siRNA to a murine key fibrogenesis factor (gp46) into cirrhotic mice silenced the specific gene in mouse liver and resolved fibrosis [68]. 1,2-dioleoyl-3-trimethylammonium-propane (DOTAP) liposomes encapsulating HER2-siRNA and containing endosomolytic histidine-lysine peptides and a single-chain antibody fragment targeting transferrin receptors coupled to the liposomes' surface have been targeted to tumour xenograft and inhibited its growth [69] (Fig. 6.2h). Anisamide-PEG-Liposomes-Polycation-DNA (anisamide-PEG-LPDs) are unilamellar cationic liposomes coated with PEG-linked anisamide (a small-molecule compound that binds to sigma receptors) on their surface, and protamine-condensed mixture of siRNA and a carrier calf thymus DNA in the core. Encapsulating EGFR-siRNA, anisamide-PEG-LPDs injected intravenously into tumour-bearing mice have been shown to increase sensitivity to chemotherapy [70]. These particles induced a significant increase in serum cytokine levels that limits the potential for clinical translation. However, it is important to note that cytokine response is not always deleterious with therapy and there are cases when immune activation could enhance the therapeutic effects [71].

6.4.3 *Targeted Nanomedicines for Leukocytes*

Utilising siRNAs to manipulate gene expression in leucocytes holds great promise for drug discovery, as well as for facilitating the development of new therapeutic platforms for leucocyte-implicated diseases such as inflammation, blood cancers and leucocyte-tropic viral infections. However, due to their resistance to conventional transfection methods and the fact that they are dispersed throughout the body, systemic delivery to leucocytes remains a challenge, although several promising approaches have been documented from 2007.

Kortylewski et al. [72] used siRNA synthetically linked to a CpG oligonucleotide agonist of toll-like receptor (TLR) 9 for targeting myeloid cells and B cells (both are key components of tumour microenvironment) that express this receptor. These particles simultaneously silenced STAT3 by siRNA and activated TLRs responses by their agonists. Consequently, they effectively shifted the tumour microenvironment from pro-oncogenic to anti-oncogenic (by causing activation of tumour-associated immune cells and potent anti-tumour immune responses).

Two studies from the same group presented novel strategies to target leucocytes for treating viral infections. scFvCD7Cys is a single-chain antibody against CD7 (a surface antigen present on the majority of human T cells) that was modified to include a Cys residue for conjugation to a 9 Arg peptide (Fig. 6.2f). This conjugate was used for targeted delivery of CCR5 (a chemokine receptor that functions as a co-receptor for HIV) and Vif/Tat (HIV replication proteins)-siRNA payloads into T cells, and has been demonstrated to suppress HIV infection in humanised mice without inducing toxicity in the target cells [73]. A similar approach for treating dengue virus infected cells employed DC3 (12-mer peptide that targets dendritic cells)-9dR for targeting, with TNF-α (which plays a major role in dengue pathogenesis) or specific highly conserved sequence in the viral envelope. These complexes significantly reduced virus-induced production of TNF-α and succeeded to suppress the viral replication in monocytes-derived dendritic cells and macrophages in vitro. In vivo, treatment of mice with intravenous injection of DC3-9dR-complexes containing TNF-α–siRNA effectively suppressed this cytokine production by dendritic cells [74].

Another approach for targeting leucocytes is based on leucocytes' integrins, which are the largest family of cell adhesion molecules that mediate cell–cell and cell–matrix interactions [75]. Antibody–protamine fusion proteins utilise the lymphocyte function-associated antigen-1 (LFA-1) integrin, a pan leucocyte cell surface marker for selective targeting. The use of LFA-1 for targeting leucocytes is supported by its exclusive expression on leucocytes, its constitutive internalisation and recycling activity and its ability to undergo activation-dependent conformational changes. Using antibody–protamine fusion proteins, RNAi payloads could be directed into leucocytes both in vitro and in vivo. Importantly, neither lymphocytes activation nor interferon response induction was observed. Furthermore, by targeting these fusion proteins to the high affinity conformation of LFA-1 that characterises activated lymphocytes, the ability to induce more selective gene silencing was demonstrated, which unlike most immunosuppressive therapies, could provide a way to overcome the unwanted immune stimulation without global immunosuppressive effects on bystander immune cells. Additionally, due to the prevalence of aberrant affinity modulation of integrins in a variety of leucocyte-implication diseases [76, 77], targeting the high-affinity conformation of LFA-1 seems to be very promising therapeutic strategy [78, 79].

In order to increase payload and achieve more robust targeted gene silencing, a different strategy was generated. Integrin-targeted and stabilised nanoparticles (I-tsNP) that successfully deliver siRNAs into specific leucocytes subset involved in gut inflammation were developed. Using this system, cyclin D1, a regulator protein of the entry into, and the progression throughout the cell cycle, was identified as a potential new therapeutic target for treating inflammation. The I-tsNPs were developed from neutral charged lipids of about 80 nm in diameter liposomes that were loaded with siRNAs pre-condensed with human recombinant protamine. The particles were coated with hyaluronan (HA) for stabilisation during siRNA entrapment and for prolonging the circulation time in vivo. The targeting of the particles to subsets of leucocytes was achieved by attaching a monoclonal antibody against β_7 integrin (which is highly expressed in gut mononuclear leucocytes) to the HA-coated liposomes [80]. Composed of natural biomaterials, these nanoparticles offer a safe platform for siRNAs delivery, avoiding cytokine induction and liver damage. Enabling usage of fairly low siRNA doses (2.5 mg/kg), this system, in addition to advantages such as high payload capacity (~4,000 siRNA molecules per particle) and low off-target effects and toxicities, is economically worthy. This strategy was also used with an LFA-1 antibody decorated particles for the delivery of CCR5-siRNAs to human lymphocytes and monocytes. This system has been shown to protect mice from HIV challenge [81]. LFA-1 I-tsNPs with CCR5-siRNAs did not induce interferon response or TNF-α (inflammatory cytokine) secretion, hence strengthens the potential for clinical relevance.

6.5 Future Perspectives

In summary, although there is no clinically approved siRNA delivery system yet, we are convinced that in the coming years this situation will be changed. Approximately 22 different siRNA/shRNA therapeutics have reached clinical evaluation for the treatment of at least 16 diseases [82]. The results from the ongoing ALN-PCS02 and ALN-TTR01 SNALP-based clinical studies will have a major impact on the RNAi therapeutics era for the next few years. In addition, positive data from the AtuPLEX-based Atu027 trial have the potential to amplify the SNALP-driven approach as well as cyclodextrin-based approach that is currently under clinical evaluation.

Similar to other RNA-based therapeutics, the efficacy of RNAi drugs relies on maximising targeted delivery while minimising off-target toxicity and degradation. We are convinced that the delivery efforts currently explored will be successful. We base this assumption on one of the major advantages of siRNA delivery systems—the flexibility to change either the payloads entrapped inside the nanoparticles (by using different sequences of siRNAs, or with combination to other drugs) or the targeting agent (by replacing the antibody or the ligand decorating the nanoparticle's surface). This opens new avenues for treating a wide variety of diseases as well as adjusting the treatment to the unique molecular abnormalities of a specific patient.

References

1. Sledz CA, Williams BR (2005) RNA interference in biology and disease. Blood 106(3):787–794
2. de Fougerolles A, Vornlocher HP, Maraganore J, Lieberman J (2007) Interfering with disease: a progress report on siRNA-based therapeutics. Nat Rev Drug Discov 6(6):443–453
3. Elbashir SM, Harborth J, Lendeckel W, Yalcin A, Weber K, Tuschl T (2001) Duplexes of 21-nucleotide RNAs mediate RNA interference in cultured mammalian cells. Nature 411(6836):494–498
4. Amarzguioui M, Rossi JJ, Kim D (2005) Approaches for chemically synthesized siRNA and vector-mediated RNAi. FEBS Lett 579(26):5974–5981

5. Bartlett DW, Davis ME (2006) Insights into the kinetics of siRNA-mediated gene silencing from live-cell and live-animal bioluminescent imaging. Nucleic Acids Res 34(1):322–333
6. Robbins M, Judge A, Ambegia E, Choi C, Yaworski E, Palmer L et al (2008) Misinterpreting the therapeutic effects of small interfering RNA caused by immune stimulation. Hum Gene Ther 19(10):991–999
7. Kleinman ME, Yamada K, Takeda A, Chandrasekaran V, Nozaki M, Baffi JZ et al (2008) Sequence- and target-independent angiogenesis suppression by siRNA via TLR3. Nature 452(7187):591–597
8. Ma M, Zhou L, Guo X, Lv Z, Yu Y, Ding C et al (2009) Decreased cofilin1 expression is important for compaction during early mouse embryo development. Biochim Biophys Acta 1793(12):1804–1810
9. Bose S, Leclerc GM, Vasquez-Martinez R, Boockfor FR (2010) Administration of connexin43 siRNA abolishes secretory pulse synchronization in GnRH clonal cell populations. Mol Cell Endocrinol 314(1):75–83
10. Wiese M, Castiglione K, Hensel M, Schleicher U, Bogdan C, Jantsch J (2010) Small interfering RNA (siRNA) delivery into murine bone marrow-derived macrophages cells by electroporation. J Immunol Methods 353(1–2):102–110
11. Honjo K, Takahashi KA, Mazda O, Kishida T, Shinya M, Tokunaga D et al (2010) MDR1a/1b gene silencing enhances drug sensitivity in rat fibroblast-like synoviocytes. J Gene Med 12(2):219–227
12. Donze O, Picard D (2002) RNA interference in mammalian cells using siRNAs synthesized with T7 RNA polymerase. Nucleic Acids Res 30(10):e46
13. Tsubouchi A, Sakakura J, Yagi R, Mazaki Y, Schaefer E, Yano H et al (2002) Localized suppression of RhoA activity by Tyr31/118-phosphorylated paxillin in cell adhesion and migration. J Cell Biol 159(4):673–683
14. Huang YZ, Zang M, Xiong WC, Luo Z, Mei L (2003) Erbin suppresses the MAP kinase pathway. J Biol Chem 278(2):1108–1114
15. Zhang M, Bai CX, Zhang X, Chen J, Mao L, Gao L (2004) Downregulation enhanced green fluorescence protein gene expression by RNA interference in mammalian cells. RNA Biol 1(1):74–77
16. Gosain AK, Machol JA 4th, Gliniak C, Halligan NL (2009) TGF-beta1 RNA interference in mouse primary dura cell culture: downstream effects on TGF receptors, FGF-2, and FGF-R1 mRNA levels. Plast Reconstr Surg 124(5):1466–1473
17. Cheng SQ, Wang WL, Yan W, Li QL, Wang L, Wang WY (2005) Knockdown of survivin gene expression by RNAi induces apoptosis in human hepatocellular carcinoma cell line SMMC-7721. World J Gastroenterol 11(5):756–759
18. Baker BE, Kestler DP, Ichiki AT (2006) Effects of siRNAs in combination with Gleevec on K-562 cell proliferation and Bcr-Abl expression. J Biomed Sci 13(4):499–507
19. Crombez L, Charnet A, Morris MC, Aldrian-Herrada G, Heitz F, Divita G (2007) A non-covalent peptide-based strategy for siRNA delivery. Biochem Soc Trans 35(Pt 1):44–46
20. Muratovska A, Eccles MR (2004) Conjugate for efficient delivery of short interfering RNA (siRNA) into mammalian cells. FEBS Lett 558(1–3):63–68
21. Minakuchi Y, Takeshita F, Kosaka N, Sasaki H, Yamamoto Y, Kouno M et al (2004) Atelocollagen-mediated synthetic small interfering RNA delivery for effective gene silencing in vitro and in vivo. Nucleic Acids Res 32(13):e109
22. Takei Y, Kadomatsu K, Yuzawa Y, Matsuo S, Muramatsu T (2004) A small interfering RNA targeting vascular endothelial growth factor as cancer therapeutics. Cancer Res 64(10):3365–3370
23. Puebla I, Esseghir S, Mortlock A, Brown A, Crisanti A, Low W (2003) A recombinant H1 histone-based system for efficient delivery of nucleic acids. J Biotechnol 105(3):215–226
24. Kedmi R, Ben-Arie N, Peer D (2010) The systemic toxicity of positively charged lipid nanoparticles and the role of Toll-like receptor 4 in immune activation. Biomaterials 31(26): 6867–6875
25. Lv H, Zhang S, Wang B, Cui S, Yan J (2006) Toxicity of cationic lipids and cationic polymers in gene delivery. J Control Release 114(1):100–109

26. Ferrari M (2005) Cancer nanotechnology: opportunities and challenges. Nat Rev Cancer 5(3):161–171
27. Peer D, Karp JM, Hong S, Farokhzad OC, Margalit R, Langer R (2007) Nanocarriers as an emerging platform for cancer therapy. Nat Nanotechnol 2(12):751–760
28. Mizrahy S, Peer D (2012) Polysaccharides as building blocks for nanotherapeutics. Chem Soc Rev 41(7):2623–2640
29. Schroeder A, Heller DA, Winslow MM, Dahlman JE, Pratt GW, Langer R et al (2012) Treating metastatic cancer with nanotechnology. Nat Rev Cancer 12(1):39–50
30. Wang J, Byrne JD, Napier ME, DeSimone JM (2011) More effective nanomedicines through particle design. Small 7(14):1919–1931
31. Moghimi SM, Peer D, Langer R (2011) Reshaping the future of nanopharmaceuticals: ad iudicium. ACS Nano 5(11):8454–8458
32. Bitko V, Musiyenko A, Shulyayeva O, Barik S (2005) Inhibition of respiratory viruses by nasally administered siRNA. Nat Med 11(1):50–55
33. Palliser D, Chowdhury D, Wang QY, Lee SJ, Bronson RT, Knipe DM et al (2006) An siRNA-based microbicide protects mice from lethal herpes simplex virus 2 infection. Nature 439(7072):89–94
34. Kim B, Tang Q, Biswas PS, Xu J, Schiffelers RM, Xie FY et al (2004) Inhibition of ocular angiogenesis by siRNA targeting vascular endothelial growth factor pathway genes: therapeutic strategy for herpetic stromal keratitis. Am J Pathol 165(6):2177–2185
35. Juliano R, Bauman J, Kang H, Ming X (2009) Biological barriers to therapy with antisense and siRNA oligonucleotides. Mol Pharm 6(3):686–695
36. Zender L, Kubicka S (2007) Suppression of apoptosis in the liver by systemic and local delivery of small-interfering RNAs. Methods Mol Biol 361:217–226
37. De Souza AT, Dai X, Spencer AG, Reppen T, Menzie A, Roesch PL et al (2006) Transcriptional and phenotypic comparisons of Ppara knockout and siRNA knockdown mice. Nucleic Acids Res 34(16):4486–4494
38. Saito Y, Kon S, Fujiwara Y, Nakayama Y, Kurotaki D, Fukuda N et al (2007) Osteopontin small interfering RNA protects mice from fulminant hepatitis. Hum Gene Ther 18(12): 1205–1214
39. Lewis DL, Wolff JA (2007) Systemic siRNA delivery via hydrodynamic intravascular injection. Adv Drug Deliv Rev 59(2–3):115–123
40. Molitoris BA, Dagher PC, Sandoval RM, Campos SB, Ashush H, Fridman E et al (2009) siRNA targeted to p53 attenuates ischemic and cisplatin-induced acute kidney injury. J Am Soc Nephrol 20(8):1754–1764
41. Peer D, Lieberman J (2011) Special delivery: targeted therapy with small RNAs. Gene Ther 18(12):1127–1133
42. Mizrahy S, Raz SR, Hasgaard M, Liu H, Soffer-Tsur N, Cohen K et al (2011) Hyaluronan-coated nanoparticles: the influence of the molecular weight on CD44-hyaluronan interactions and on the immune response. J Control Release 156(2):231–238
43. Matsumura Y, Maeda H (1986) A new concept for macromolecular therapeutics in cancer chemotherapy: mechanism of tumoritropic accumulation of proteins and the antitumor agent smancs. Cancer Res 46(12 Pt 1):6387–6392
44. Maeda H (2001) The enhanced permeability and retention (EPR) effect in tumor vasculature: the key role of tumor-selective macromolecular drug targeting. Adv Enzyme Regul 41: 189–207
45. Zimmermann TS, Lee AC, Akinc A, Bramlage B, Bumcrot D, Fedoruk MN et al (2006) RNAi-mediated gene silencing in non-human primates. Nature 441(7089):111–114
46. Morrissey DV, Lockridge JA, Shaw L, Blanchard K, Jensen K, Breen W et al (2005) Potent and persistent in vivo anti-HBV activity of chemically modified siRNAs. Nat Biotechnol 23(8):1002–1007
47. Geisbert TW, Hensley LE, Kagan E, Yu EZ, Geisbert JB, Daddario-DiCaprio K et al (2006) Postexposure protection of guinea pigs against a lethal ebola virus challenge is conferred by RNA interference. J Infect Dis 193(12):1650–1657

48. Villares GJ, Zigler M, Wang H, Melnikova VO, Wu H, Friedman R et al (2008) Targeting melanoma growth and metastasis with systemic delivery of liposome-incorporated protease-activated receptor-1 small interfering RNA. Cancer Res 68(21):9078–9086
49. Landen CN Jr, Chavez-Reyes A, Bucana C, Schmandt R, Deavers MT, Lopez-Berestein G et al (2005) Therapeutic EphA2 gene targeting in vivo using neutral liposomal small interfering RNA delivery. Cancer Res 65(15):6910–6918
50. Anderson DG, Akinc A, Hossain N, Langer R (2005) Structure/property studies of polymeric gene delivery using a library of poly(beta-amino esters). Mol Ther 11(3):426–434
51. Akinc A, Zumbuehl A, Goldberg M, Leshchiner ES, Busini V, Hossain N et al (2008) A combinatorial library of lipid-like materials for delivery of RNAi therapeutics. Nat Biotechnol 26(5):561–569
52. Leng Q, Goldgeier L, Zhu J, Cambell P, Ambulos N, Mixson AJ (2007) Histidine-lysine peptides as carriers of nucleic acids. Drug News Perspect 20(2):77–86
53. Leng Q, Scaria P, Lu P, Woodle MC, Mixson AJ (2008) Systemic delivery of HK Raf-1 siRNA polyplexes inhibits MDA-MB-435 xenografts. Cancer Gene Ther 15(8):485–495
54. Yan Z, Zou H, Tian F, Grandis JR, Mixson AJ, Lu PY et al (2008) Human rhomboid family-1 gene silencing causes apoptosis or autophagy to epithelial cancer cells and inhibits xenograft tumor growth. Mol Cancer Ther 7(6):1355–1364
55. Honma K, Iwao-Koizumi K, Takeshita F, Yamamoto Y, Yoshida T, Nishio K et al (2008) RPN2 gene confers docetaxel resistance in breast cancer. Nat Med 14(9):939–948
56. Fujii T, Saito M, Iwasaki E, Ochiya T, Takei Y, Hayashi S et al (2006) Intratumor injection of small interfering RNA-targeting human papillomavirus 18 E6 and E7 successfully inhibits the growth of cervical cancer. Int J Oncol 29(3):541–548
57. Mu P, Nagahara S, Makita N, Tarumi Y, Kadomatsu K, Takei Y (2009) Systemic delivery of siRNA specific to tumor mediated by atelocollagen: combined therapy using siRNA targeting Bcl-xL and cisplatin against prostate cancer. Int J Cancer 125(12):2978–2990
58. Wolfrum C, Shi S, Jayaprakash KN, Jayaraman M, Wang G, Pandey RK et al (2007) Mechanisms and optimization of in vivo delivery of lipophilic siRNAs. Nat Biotechnol 25(10):1149–1157
59. Nishina K, Unno T, Uno Y, Kubodera T, Kanouchi T, Mizusawa H et al (2008) Efficient in vivo delivery of siRNA to the liver by conjugation of alpha-tocopherol. Mol Ther 16(4):734–740
60. Lorenz C, Hadwiger P, John M, Vornlocher HP, Unverzagt C (2004) Steroid and lipid conjugates of siRNAs to enhance cellular uptake and gene silencing in liver cells. Bioorg Med Chem Lett 14(19):4975–4977
61. McNamara JO 2nd, Andrechek ER, Wang Y, Viles KD, Rempel RE, Gilboa E et al (2006) Cell type-specific delivery of siRNAs with aptamer-siRNA chimeras. Nat Biotechnol 24(8):1005–1015
62. Rozema DB, Lewis DL, Wakefield DH, Wong SC, Klein JJ, Roesch PL et al (2007) Dynamic PolyConjugates for targeted in vivo delivery of siRNA to hepatocytes. Proc Natl Acad Sci USA 104(32):12982–12987
63. Schiffelers RM, Ansari A, Xu J, Zhou Q, Tang Q, Storm G et al (2004) Cancer siRNA therapy by tumor selective delivery with ligand-targeted sterically stabilized nanoparticle. Nucleic Acids Res 32(19):e149
64. Hu-Lieskovan S, Heidel JD, Bartlett DW, Davis ME, Triche TJ (2005) Sequence-specific knockdown of EWS-FLI1 by targeted, nonviral delivery of small interfering RNA inhibits tumor growth in a murine model of metastatic Ewing's sarcoma. Cancer Res 65(19):8984–8992
65. Davis ME (2009) The first targeted delivery of siRNA in humans via a self-assembling, cyclodextrin polymer-based nanoparticle: from concept to clinic. Mol Pharm 6(3):659–668
66. Andrabi SM (2007) Mammalian sperm chromatin structure and assessment of DNA fragmentation. J Assist Reprod Genet 24(12):561–569
67. Song E, Zhu P, Lee SK, Chowdhury D, Kussman S, Dykxhoorn DM et al (2005) Antibody mediated in vivo delivery of small interfering RNAs via cell-surface receptors. Nat Biotechnol 23(6):709–717

68. Sato Y, Murase K, Kato J, Kobune M, Sato T, Kawano Y et al (2008) Resolution of liver cirrhosis using vitamin A-coupled liposomes to deliver siRNA against a collagen-specific chaperone. Nat Biotechnol 26(4):431–442
69. Pirollo KF, Rait A, Zhou Q, Hwang SH, Dagata JA, Zon G et al (2007) Materializing the potential of small interfering RNA via a tumor-targeting nanodelivery system. Cancer Res 67(7):2938–2943
70. Li SD, Chen YC, Hackett MJ, Huang L (2008) Tumor-targeted delivery of siRNA by self-assembled nanoparticles. Mol Ther 16(1):163–169
71. Goldsmith M, Mizrahy S, Peer D (2011) Grand challenges in modulating the immune response with RNAi nanomedicines. Nanomedicine (Lond) 6(10):1771–1785
72. Kortylewski M, Swiderski P, Herrmann A, Wang L, Kowolik C, Kujawski M et al (2009) In vivo delivery of siRNA to immune cells by conjugation to a TLR9 agonist enhances antitumor immune responses. Nat Biotechnol 27(10):925–932
73. Kumar P, Ban HS, Kim SS, Wu H, Pearson T, Greiner DL et al (2008) T cell-specific siRNA delivery suppresses HIV-1 infection in humanized mice. Cell 134(4):577–586
74. Subramanya S, Kim SS, Abraham S, Yao J, Kumar M, Kumar P et al (2010) Targeted delivery of siRNA to human dendritic cells to suppress Dengue viral infection and associated proinflammatory cytokine production. J Virol 84(5):2490–2501
75. Shimaoka M, Takagi J, Springer TA (2002) Conformational regulation of integrin structure and function. Annu Rev Biophys Biomol Struct 31:485–516
76. Cauli A, Yanni G, Pitzalis C, Challacombe S, Panayi GS (1995) Cytokine and adhesion molecule expression in the minor salivary glands of patients with Sjogren's syndrome and chronic sialoadenitis. Ann Rheum Dis 54(3):209–215
77. Tanaka Y, Mine S, Figdor CG, Wake A, Hirano H, Tsukada J et al (1998) Constitutive chemokine production results in activation of leukocyte function-associated antigen-1 on adult T-cell leukemia cells. Blood 91(10):3909–3919
78. Peer D, Zhu P, Carman CV, Lieberman J, Shimaoka M (2007) Selective gene silencing in activated leukocytes by targeting siRNAs to the integrin lymphocyte function-associated antigen-1. Proc Natl Acad Sci USA 104(10):4095–4100
79. Elfakess R, Peer D (2010) Overcoming RNAi transduction in leukocytes using targeted and stabilized nanoparticles. IDrugs 13(9):626–631
80. Peer D, Park EJ, Morishita Y, Carman CV, Shimaoka M (2008) Systemic leukocyte-directed siRNA delivery revealing cyclin D1 as an anti-inflammatory target. Science 319(5863): 627–630
81. Kim SS, Peer D, Kumar P, Subramanya S, Wu H, Asthana D et al (2010) RNAi-mediated CCR5 silencing by LFA-1-targeted nanoparticles prevents HIV infection in BLT mice. Mol Ther 18(2):370–376
82. Burnett JC, Rossi JJ (2012) RNA-based therapeutics: current progress and future prospects. Chem Biol 19(1):60–71

Chapter 7
Lipidoids: A Combinatorial Approach to siRNA Delivery

Michael Goldberg

Abstract The safe and effective delivery of siRNA remains the principal challenge to the realization of its clinical potential. The present collection of delivery materials and their diversity remains limited, in part due to their slow, multistep syntheses. This chapter will describe a class of lipid-like delivery molecules, termed "lipidoids," as carriers for RNAi therapeutics. Specifically, the chapter will address the rationale underlying the combinatorial approach; the synthetic chemical methods employed; the screening assay utilized; the structure–activity relationships determined; the formulation considerations learned; several applications of the platform; and the evolution of the strategy to generate next-generation libraries.

7.1 Introduction

RNA interference (RNAi) is a conserved process by which double-stranded RNA molecules direct the sequence-specific degradation of complementary mRNA targets [1]. The utility of this powerful phenomenon in the dissection of gene function through genome-wide screening experiments was rapidly realized [2–4]. This application allowed for the large-scale investigation of mammalian genetics, enabling direct analysis of human cells rather than relying on the use of traditional model organisms such as *Caenorhabditis elegans*, *Saccharomyces cerevisiae*, and *Drosophila melanogaster*. Though it was immediately apparent that this powerful application of gene silencing could be complemented by the knockdown of disease-causing genes in patients [5], the delivery of siRNAs into cells in animals was a profound obstacle to the realization of this potential. The ability to generate systems that could deliver these molecules in a safe and efficacious manner was highly desirable.

M. Goldberg (✉)
Department of Cancer Immunology and AIDS, Dana-Farber Cancer Institute, Harvard Medical School, 450 Brookline Avenue, Boston, MA 02215, USA
e-mail: michael_goldberg1@dfci.harvard.edu

K.A. Howard (ed.), *RNA Interference from Biology to Therapeutics*, Advances in Delivery Science and Technology, DOI 10.1007/978-1-4614-4744-3_7,

The translation from in vitro experiments to proof-of-concept in vivo studies was rapid. The report of successful silencing in vivo in 2002 [6] served as an exciting advance in the field, and the first demonstration of RNAi-mediated disease inhibition in an animal model was reported only 1 year later [7]. In 2004, the first example of therapeutic silencing of an endogenous gene following systemic administration of cholesterol-modified siRNAs was described [8]. More advanced systems would soon follow, and the utility of RNAi as a platform technology was validated in non-human primates (NHP) within 2 years [9]. Importantly, some of the safety concerns attributed to shRNA-mediated gene silencing [10] were not found to be relevant to siRNA-mediated gene silencing. siRNA enters the RNAi pathway downstream of nuclear export and Dicer processing, so it does not interrupt the endogenous microRNA pathway [11]. Saturation of Ago2 loading is possible in vitro but is unlikely to be of concern in vivo, where efficient delivery remains a challenge. Clearly, the ability to deliver siRNAs in vivo would enable a fundamentally new way of treating disease, catalytically preventing protein production rather than stoichiometrically inactivating aberrant proteins following their translation, as traditional pharmacological methods do.

There are, however, several significant barriers to entry. siRNAs are large, polyanionic molecules that do not readily cross the hydrophobic, tightly packed cell membrane, so carriers must be used. These carriers must protect the drug payload from degradation, filtration, and phagocytosis in the bloodstream; facilitate transport across the vascular endothelial barrier; allow for diffusion through the extracellular matrix; and promote cellular uptake as well as cargo release into the cytosol [12]. While extremely efficient mediators of DNA delivery, viruses do not deliver synthetic siRNA and can result in insertional mutagenesis when delivering shRNA, whose expression levels are difficult to control. For these reasons, synthetic delivery systems have been developed. Polycationic polymers and lipids are often used because they can electrostatically bind and condense the polyphosphate backbone of nucleic acids to form nanometer-sized particles that enhance stability and cellular uptake. A novel class of lipid-like compounds, termed "lipidoids," was developed in order to expand the scale and diversity of systemic siRNA delivery systems. This chapter will focus on the synthesis, screening, and application of these molecules.

7.2 Rational Versus Combinatorial Approaches

Though lipid-based formulations for systemic siRNA delivery represent one of the most promising near-term opportunities for the development of RNAi therapeutics [13], the present collection of delivery materials and their diversity remain limited. The central obstacle to the expansion of scale is the slow, multistep syntheses required to create cationic lipids. The result of this laborious and time-consuming approach is a low-throughput, iterative process.

When performed rigorously, this directed approach can be useful for interrogating the effects of minor differences on functional properties toward the elucidation of focused structure–activity relationships. For example, an analogous series of cationic lipids containing an increasing number of double bonds per alkyl chain was synthesized to evaluate the correlation between lipid saturation, fusogenicity, and efficiency of intracellular nucleic acid delivery [14]. It was shown that increasing saturation decreases fusogenicity, resulting in lower gene silencing efficiency despite being more readily internalized by cells. The findings are, however, restricted to the lipid backbone analyzed because the importance of a functional group for one lipid should not be extrapolated to other lipids.

Such rational design is very useful in the optimization of effective lead candidate lipids that have been shown to be effective in vivo; however, because there are few examples of such validated lipids, this strategy is of limited utility. One notable example was the use of medicinal chemistry to improve the efficiency of an efficacious ionizable cationic lipid based on its proposed mechanism of action of endosomal membrane disruption [15]. The endeavor was highly focused. Maintaining a constant hydrocarbon chain—based on the results of the aforementioned study [14]—the two other components of the cationic lipid, the linker and headgroup, were systematically varied.

Though the sample size was modest in this analysis, trends were observed. Lipids containing alkoxy linkers were consistently more effective than those containing ester, thioester, or carbamate linkers. While changing the structure of the amine head group did not greatly influence activity, increasing the distance between the dimethylamino group and the dioxolane linker by a single methylene group yielded a surprisingly large improvement in efficacy. This finding highlights the limitation of the rational approach, which does not allow for the exploration of broad chemical space: a substantial element of luck is required in selecting the modifications because it is not obvious which minor changes will confer a major impact. Furthermore, the yield of these reactions is often poor, and the throughput is low, so the development of alternative chemistries would be attractive.

7.3 Lipidoids: Synthetic Scheme

The Michael addition reaction allows for the conjugate addition of an amine to acrylates or acrylamides, enabling the simple and rapid synthesis of large degradable or nondegradable combinatorial libraries, respectively [16]. This approach has yielded thousands of structurally diverse lipid-like materials, which have been dubbed "lipidoids," from the Greek for "lipid-like" [17]. The members of the library chemically resemble lipids and thus protect and enhance the uptake of their siRNA payload. An inspection of this novel class of materials reveals that these unique compounds manifest as a hybrid structure between cationic lipids and first-generation dendrimers [18] (see Fig. 7.1). Notably, unlike traditional cationic lipids, which are generally comprised of a discrete cationic head group and a hydrophobic tail, lipidoids are cationic owing to the presence of reversibly protonatable backbone amines. Additionally, whereas natural lipids typically possess two tails, lipidoids contain up

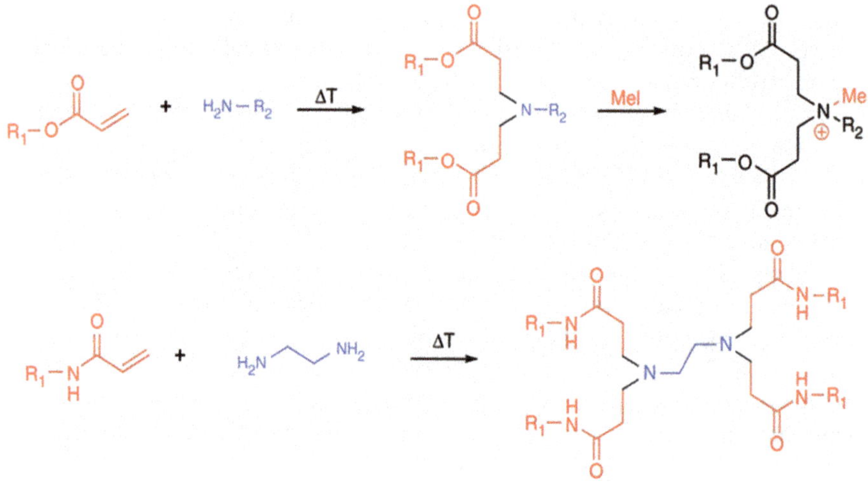

Fig. 7.1 Synthesis occurs through the conjugate addition of amines to an acrylate or acrylamide. Depending on the number of addition sites in the amino monomer, lipidoids can be formed with anywhere from 1 to 7 tails. Amino groups in the lipidoid can be quaternized by treatment with methyl iodide. For ease of nomenclature, lipidoids are named as follows: (amine number)(acrylate or acrylamide name)-(number of tails)("+" if quaternized) (reproduced with permission from [18])

to seven tails emanating from the amine backbone. Another interesting chemical property that renders lipidoids a distinct class of biomaterials is the inversion of its ester linkage with respect to the aliphatic chain when compared to natural lipids such as triglycerides. Lipidoids can be quarternized to impart permanent charge through the addition of alkylating agents such as methyl iodide.

Whereas standard chemistries used to synthesize cationic lipids involve several onerous steps that include protection, deprotection, solvent exchanges, and serial purification [19], this innovative route enables high-throughput, parallel synthesis in a single step. To synthesize a lipidoid, one need only add an amine, an acrylate or acrylamide, and a stir bar. The reaction mixture is then placed in a 90°C oven for one or six days, respectively. There is tremendous potential for high throughput because the starting reagents are commercially available and there are no side reactions. As a consequence of this simplicity, the number of lipofection materials synthesized and studied to date was increased by two orders of magnitude.

The first-generation lipidoid library was generated from starting reagents that provided significant chemical diversity [18] (see Fig. 7.2). The functional groups included linear and cyclic ethers and diamines; diols; primary, secondary, and tertiary amines; piperidines; pyrrolidines; and imidazoles. These materials were selected because they imparted good transfection ability upon poly(beta amino) esters, a class of degradable transfection agents that was previously developed to deliver plasmid DNA [16].

Amines are used because they are thought to promote endosomal escape by acting as "proton sponges" [20, 21]. The ability to escape the endosome is of

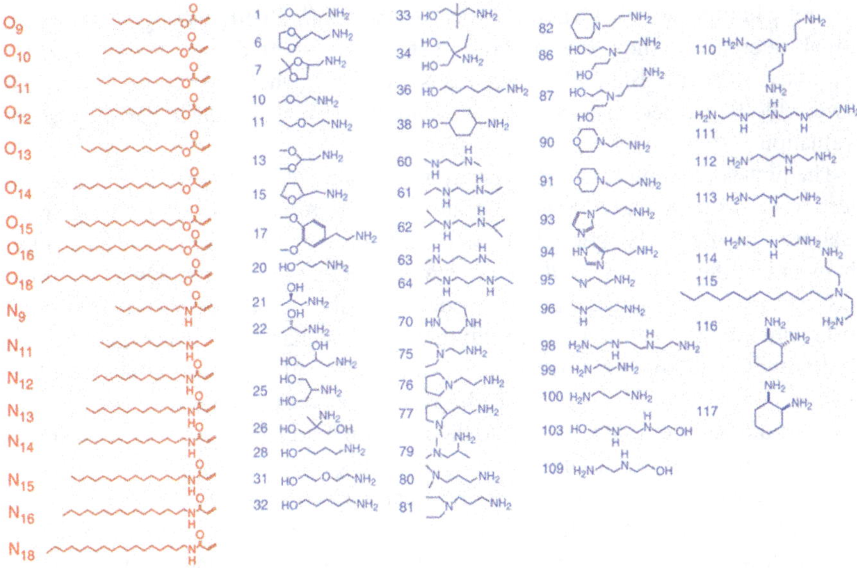

Fig. 7.2 Alkyl-acrylate, alkyl-acrylamide, and amino molecules were used to synthesize a combinatorial library of lipidoids (reproduced with permission from [18])

paramount importance because endocytosis is the main process by which cells uptake nanoparticles, and nucleic acids are membrane impermeable [22, 23]. Briefly, amines possessing the appropriate pK_a can accept protons that are pumped into the endosome as it acidifies to become a catabolic lysosome [24], preventing degradation of the compartment's contents. The net positive charge on the quarternary amines creates an electrochemical gradient, leading to the influx of negatively charged chloride ions. The high salt concentration inside the endosome produces a strong osmotic gradient, leading to the inflow of water molecules, which eventually rupture the endosomal membrane and release the drug cargo into the cytosol.

7.4 Structure–Activity Relationships

The ability of the lipidoids to deliver siRNA to mammalian cells was first investigated in vitro [18]. A HeLa cell line was established to express stably both firely (*Photinus pyralis*) and *Renilla* (*Renilla reniformis*) luciferases. These bioluminescent enzymes have unique substrates and can, thus, be differentiated by performing sequential assays using DualGlo™ (Promega). Cells were plated into 96-well plates and allowed to attach overnight in growth medium. Lipidoids were dissolved in 25 mM sodium acetate (pH 5) and mixed with siRNA targeting firefly luciferase at multiple lipidoid-to-siRNA weight-to-weight ratios. These mixtures were incubated for 20 min to allow for complex formation prior to addition to the cells. After 24 h, the firefly luciferase signal was normalized to the *Renilla* luciferase signal. If neither signal decreased, then the lipidoid tested was ineffective; if both signals decreased, then the lipidoid tested was toxic. Only when the ratio of firefly-to-*Renilla* luciferase decreased was the lipidoid of interest considered for subsequent evaluation.

The first-generation library comprised more than 700 members. Several parameters were systematically varied to enable the determination of structure–activity relationships among the lipidoids. First, the influence of the linker between the alkyl chain and amine was evaluated. It was determined that the more stable amide bond was preferred to the degradable ester bond. Second, it was demonstrated that the postsynthetic introduction of a constitutive positive charge to the acrylate-derived lipidoids by quaternization of the amine did not greatly improve efficacy. While these insights were extremely valuable from a general perspective, more specific information was required to elucidate the functionalities that conferred good transfection ability among acrylamide-derived lipidoids.

This information was extracted from a heat map [18] [(see Fig. 7.3), which is a useful tool for the interrogation of combinatorial libraries. The alkyl chain length series ranged from C10 to C18, and the functional groups surrounding the backbone amines were varied. It was observed that the top-performing materials were composed of multiple primary or secondary amines separated by ethyl or propyl groups (e.g., monomers 61–64, 95–103) and featured more than two short (<C12) tails. These findings were unexpected and would likely never have been discovered using the traditional rational approach, as standard cationic lipids generally have a lone amine head and two longer (C18) tails [25].

To test the hypothesis that critical common features had been elucidated, a second-generation library of 500 lipidoids was synthesized. Screening of this library confirmed that a convergence of structure had indeed been identified. Whereas only 3% of the first-generation library was able to confer knockdown comparable to or greater than the positive control transfection reagent Lipofectamine2000™, more than 50% of those examined from the second-generation library were able to achieve this feat. Because multiple tails can be conjugated to oligoamines, the second-generation library was synthesized using varied reaction stoichiometries to generate lipidoids with a diverse number of tails. It was found that retention of one unreacted secondary amine—that is, having one tail less than complete substitution—was most favorable. Notably, many of the effective lipidoids were structurally different from both conventional lipids and cationic polymers. This set substantially expanded the collection and chemical diversity of materials known to facilitate siRNA delivery into cells.

To confirm that lipidoids could transfect multiple cell types, the top-performing materials were tested in the human hepatocellular carcinoma cell line HepG2 and primary bone marrow-derived murine macrophages at multiple doses in vitro. Interestingly, the lipidoids displayed variable efficacy in the different cell types. Silencing was consistently observed to be dose dependent, indicating that the dynamic range under investigation was nonsaturating. While the overall silencing at low lipidoid–siRNA complex concentrations in HeLa and HepG2 cells lines is less

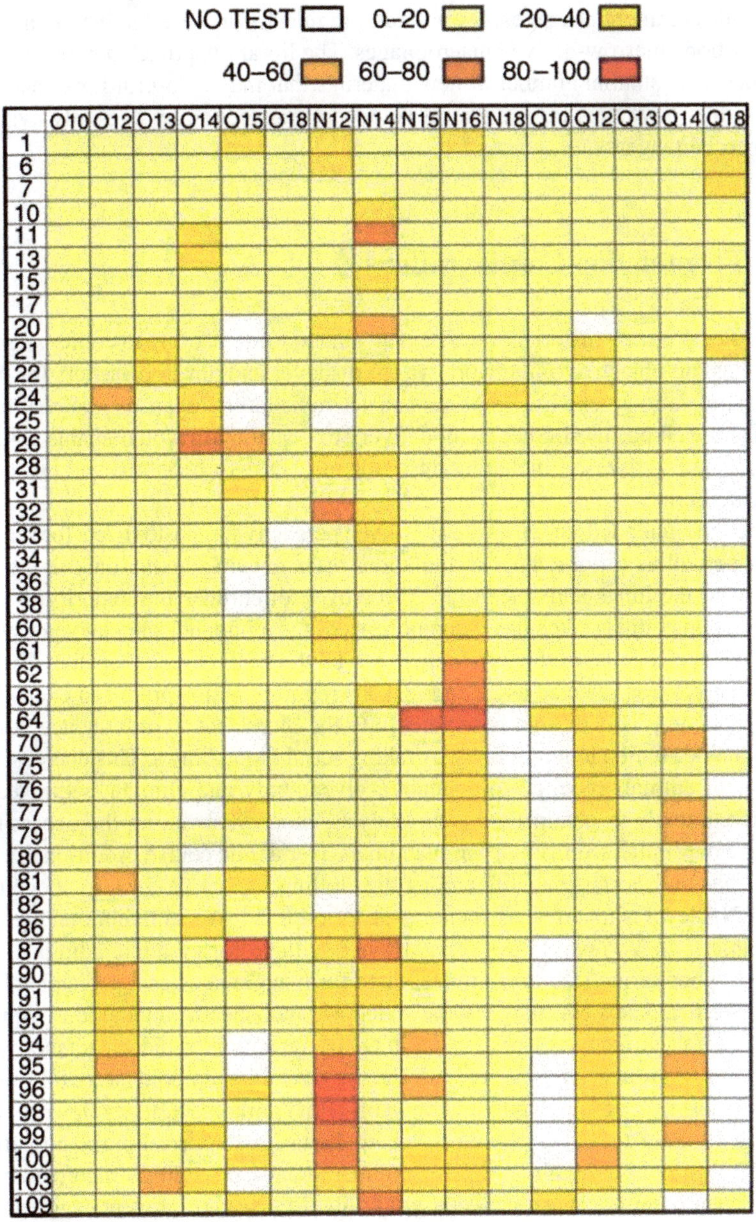

Fig. 7.3 HeLa cells expressing both firefly and *Renilla* luciferase were treated with firefly luciferase targeting siRNA–lipidoid complexes. The average percent reduction in firefly luciferase activity after treatment with siRNA–lipidoid complexes at a 5:1 (wt/wt) ratio in quadruplicate is shown. For ease of analysis, data are grouped as follows: no test, 0–20% knockdown, 20–40%, 40–60%, 60–80%, 80–100% (reproduced with permission from [18])

than with commercial reagents, lipidoids provide for more effective silencing in primary bone marrow-derived macrophages. The library approach herein described provides a substantial number of new materials that may be useful for transfection of cells that have thus far proven refractory to transfection by currently available commercial reagents.

7.5 Formulation Considerations

The top-performing materials were next evaluated for the ability to confer silencing in vivo as the objective of the work was to discover and develop carriers for therapeutic applications. The liver represents a desirable target organ, as it is well fenestrated and well perfused. Successful delivery to hepatocytes would enable RNAi to address sundry diseases, including dyslipidemia, fibrosis, hepatitis, and hepatocellular carcinoma.

Since the intended target tissue was liver parenchyma, customized formulation was important as cationic lipid formulations often accumulate not only in the liver but also in the lungs and spleen [26]. Moreover, oftentimes the majority of liver-associated material is taken up by components of the mononuclear phagocyte system (MPS) (Kupffer cells) of the liver rather than by the parenchyma (hepatocytes).

Factor VII was selected as an endogenous reporter as it is produced exclusively by hepatocytes. This protein is secreted into the blood. As a consequence, knockdown can be studied longitudinally by taking serial blood draws. This decreases the number of animals required to conduct a given study and eliminates the need to sacrifice animals to obtain tissue for analysis, thereby obviating the necessity to compare data collected from different animals over a time course. Additionally, factor VII has the shortest half-life among the clotting factors (2–5 h), so silencing at the mRNA level is manifest as silencing at the protein level with minimal lag.

Simple ionic complexes are generally not suitable for systemic administration owing to poor serum stability, a tendency to form aggregates, and poor tolerability [27]. Specifically, excess charge is associated with nonspecific interactions with biological surfaces, increased protein binding, opsonization, rapid clearance by the MPS, hemolysis, and cytotoxicity [28]. For these reasons, lipidoids were formulated with cholesterol and polyethylene glycol (PEG)-lipid. These excipients were selected because they had previously been shown to stabilize lipid nanoparticles [29]. Cholesterol is naturally found in lipid membranes, providing structure to bilayers by occupying the space between lipid tails. PEGylation prevents aggregation, decreases uptake by the MPS, and increases circulation time. Inclusion of PEG-lipid in the formulation was critical, as it resulted in the formation of small, nonaggregated particles that were accessible to hepatocytes via 100–150 nm-sized endothelial fenestrae [30]. Literature values for molar ratios of cholesterol (30–50%) and PEG-lipid (5–10%) were used as starting points. The lipid composition was systematically optimized to achieve a robust formulation with high in vivo activity.

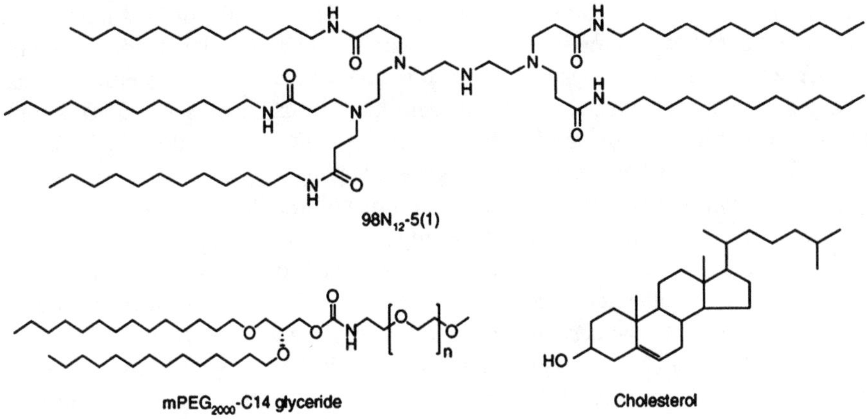

Fig. 7.4 The structures of $98N_{12}$-5(1), the five-tailed isomer of triethylenetetramine–laurylamino-propionate with a free internal amine, cholesterol, and $mPEG_{2000}$-C14 glyceride (reproduced with permission from [31])

In addition to formulation composition, the effects of siRNA loading, particle PEGylation, and particle size on in vivo delivery efficacy were investigated [31]. Loading is an important consideration because, for a given dose of siRNA, a higher siRNA loading translates to a lower dose of administered lipid. Decreasing the dose of lipid in vivo could decrease toxicity; however, it could also decrease efficacy. It was determined that maximal siRNA loading occurs at a lipid:siRNA of ~7.5:1 (wt:wt). A comparison of efficacy and tolerability at multiple lipid:siRNA loading ratios confirmed that this minimum amount of lipid required to entrap the entire siRNA dose was optimal.

Small changes in the anchor chain length of PEG-lipids were observed to affect efficacy significantly. Pharmacokinetics are strongly influenced by the extent of particle PEGylation and the rate at which deshielding occurs [32]. The lipid portion of PEG-lipids is incorporated into lipid particles hydrophobically rather than covalently, so this deshielding rate can be controlled by altering the length of the alkyl chain lipid anchor [33]. To determine the optimal length of the hydrophobic lipid tail, a glycerol backbone was derivatized with one PEG chain (MW 2000) and two alkyl chains of various lengths, ranging from C10 to C16. It was found that C14 lipid anchor length maximized the combination of desirable efficacy and tolerability. Next, the effect of particle size on the performance of this formulation was examined. Particles were extruded through filters with pore sizes ranging from 50 to 150 nm, and it was confirmed that smaller particles increase efficacy.

The final optimized formulation (see Fig. 7.4) had a lipid composition of lipidoid $98N_{12}$-5(1):cholesterol:PEG-lipid = 42:48:10 (mol:mol:mol), total lipid:siRNA = ~7.5:1 (wt:wt), C14 alkyl chain length on the PEG-lipid (MW 2000), and a mean particle size of ~50–60 nm. The surface charge was measured to be ~ neutral (~+3 mV). The particles effectively protected the entrapped siRNA, increasing its half-life in serum from ~15 min to >24 h in vitro. The formulation was monitored for 5 months following storage at 4, 25, or 37°C, and particle size distribution and siRNA entrapment were unchanged over this time course. No loss in efficacy was detected for a batch of particles that had been stored for 20 months at 4°C. This lead formulation efficiently accumulated in the liver (>90% injected dose in 1 h) and induced fully reversible, long-duration gene silencing—up to 4 weeks—without loss of activity following repeated administrations of modified siRNA [31].

The pharmacodynamics of this systemically administered siRNA formulation were assessed using a mouse that expresses luciferase ubiquitously [34]. Since the entire animal emits bioluminescence in the presence of substrate, the mice must be euthanized and organs must be harvested prior to analysis. Nonetheless, this mouse strain is a useful tool for the rapid screening and optimization of siRNA formulations, particularly for tissue-targeted delivery systems. Importantly, it was confirmed that the formulation is relevant to multiple hepatocyte-expressed target genes, as apolipoprotein B (ApoB) was also effectively silenced [11]. Furthermore, the observed effects were dose dependent and specific compared to mismatch control siRNA [18].

The specificity of siRNA can be lost when it induces a general innate immune response, which sometimes occurs when unmodified siRNAs are used [35]. siRNAs can interact with toll-like receptors (TLRs) in a structure- and sequence-dependent manner [36]. The inclusion of chemically modified nucleotides, such as of 2′-fluoro pyrimidines and 2′-*O*-methyl purines, not only increase nuclease resistance [37] but also abrogate the interferon response [38], suppressing interaction with TLR7 and TLR8 [39]. In some instances, such as antiviral applications, the combination of specific target gene silencing (mediated by siRNA) and general immunostimulation (mediated by so-called immunostimulatory RNA, or isRNA) can confer synergistic outcomes [40].

Interestingly, it was found that not only is the structure of the siRNA critical to the modulation of innate immunity, but also the lipidoid formulation is important [41]. The formulation of the lipidoid $98N_{12}$-5(1) was optimized to maximize the immunostimulatory capacity of unmodified siRNA in a mouse model of influenza. Specifically, lyophilization of the particles resulted in increased particle size, which is believed to have led to increased uptake by interferon-producing resident plasmacytoid dendritic cells (cf., hepatocytes) in the liver. This formulation method enhanced the antiviral effects of the siRNA payload, providing efficient prophylactic inhibition of influenza A in mouse lungs. Significantly, lipidoids can also be applied to afford highly specific gene knockdown in the absence of an innate immune response, and this has been demonstrated in several disease models.

7.6 Applications

Lipidoids were first used to address targets relevant to hypercholesterolemia. siRNA was delivered efficiently to the liver to silence proprotein convertase subtilisin/kexin type 9 (PCSK9) [42] and ApoB [18] in non-human primates (NHP). The silencing of PCSK9 increases low-density lipoprotein (LDL) receptor protein levels in liver and decreases plasma LDL cholesterol (LDLc). A single dose of lipidoid-formulated siRNA targeting PCSK9 in NHP resulted in a rapid and durable—lasting 3 weeks—reduction of plasma PCSK9, ApoB, and LDLc, without measurable effects on either high-density lipoprotein cholesterol (HDLc) or triglycerides. The silencing of ApoB similarly reduced LDLc in a specific and persistent manner. While these approaches might be useful for the treatment of hypercholesterolemia, the utility of siRNA in addressing this and other disease types requires a safe platform technology.

The lipidoid platform was used to verify the safety of siRNA as a therapeutic modality [11]. shRNAs had previously been shown to impart morbidity and mortality in mice owing to saturation of the endogenous microRNA processing pathway [10]. It was thus important to confirm that synthetic siRNAs, which enter the pathway downstream of miRNA biogenesis, do not induce deleterious consequences. Systemic administration of lipidoid-formulated siRNA resulted in robust target knockdown in mouse liver without any demonstrable effect on miRNA levels or activity, as determined by Northern blotting and qRT-PCR. These findings were also confirmed following multiple administrations in hamsters using a nuclease protection assay. The results of this study were significant because they established that siRNA-mediated gene silencing could be achieved either acutely or chronically without altering cellular miRNA biogenesis or function [11].

These studies confirmed the safety and utility of siRNA in addressing the pathophysiology of an isolated organism. Lipidoids were next applied to affect the biology of an organism in the context of its environment, influencing its interaction with a pathogen. Specifically, lipidoids were shown to decrease the pathogenicity of invading parasites by facilitating the knockdown of host proteins involved in the inflammatory response [43]. Mice treated with lipidoid-formulated siRNA targeting heme oxygenase 1 (HO-1) were protected from the development of blood-stage malaria infection after infection with *Plasmodium berghei* sporozoites. In contrast, all mice treated with control siRNA developed patent parasitemia. In addition to being suggestive of a therapeutic application in the mitigation of infectious diseases, these data validated that the induction of HO-1 in the liver is required for the efficient establishment of malarial infection, representing an important biological insight.

While lipidoids are very effective mediators of siRNA delivery to the liver, they are also adept at transfecting other cell types, including peritoneal macrophages [18] and ovarian cancer cells following intraperitoneal (i.p.) administration [44, 45]. Injection into the i.p. space minimizes systemic exposure and clearance of the drug and serves to concentrate the treatment at the intended site of action. Indeed, it has been shown that i.p. administration provides improved outcomes for ovarian cancer patients treated with the standard-of-care chemotherapeutic regimen relative to intravenous administration [46].

Multiple lipidoids were shown to be effective in this context. $98N_{12}$-5(1) lipidoids loaded with siRNA targeting the tight junction protein Claudin-3 was delivered to the i.p. cavity of mice bearing either autochthonous ovarian tumors or tumors derived from mouse ovarian surface epithelial cells, resulting in decreased tumor growth [44]. The concomitant reduction of ascites development suggested that the treatment suppressed metastasis. $100N_{11}$-3 lipidoids loaded with siRNA targeting the DNA repair enzyme Parp1 was used to generate a synthetic lethal phenotype in the context of Brca1 deficiency [45], which is a hallmark of the familial forms of ovarian and breast cancer [47]. The treatment of mice harboring disseminated tumors derived from genetically defined Brca1-deficient cells resulted in survival extension, with a marked increase in apoptotic cells relative to treatment with control siRNA or to tumors derived from Brca1 wild-type cells.

Lipidoids can also transfect lung epithelia upon intranasal administration. Lipidoid-formulated siRNA targeting the respiratory syncytial virus (RSV) provided greater than two log reduction in viral plaques relative to saline and mismatch siRNA controls in a mouse model of RSV [18]. Together, these data demonstrate that lipidoids are relevant to nonsystemic applications of RNAi and can deliver siRNA to nonhepatic cell types.

Notably, lipidoids can also be used to deliver other classes of nucleic acid therapeutics. Specifically, lipidoids were used to transfect hepatocytes with single-stranded oligoribonucleotides targeting miRNAs (anti-miRs) in vivo [18]. The successful delivery of anti-miR122 resulted in silencing of this miRNA and concomitant derepression of its target genes. The lipidoid formulation was found to confer greater repression than a 16-fold higher dose of cholesterol-conjugated oligonucleotide, highlighting the efficiency of its delivery capacity. The successful delivery of anti-miRs suggests that lipidoids can be used to deliver nucleic acid therapeutics that not only silence genes but also upregulate them, enabling the manipulation of targeted pathways at multiple nodes. Clearly, the ability of lipidoids to deliver multiple payloads to multiple cell types in vitro and in vivo represents a robust platform.

7.7 Evolution of a Platform

Several approaches have been adopted to improve the lipidoid system. Because lipidoids possess multiple tails, it is possible to fix the structure of some of the tails while varying the structure of others. This approach also allows for the determination of functional group effects in a combinatorial manner [48]. The top amine backbones from the first-generation library, amine 98 (triethylenetetramine) and amine 100 (diaminopropane), were reacted with N_{12} to yield precursors with $n-2$ tails. A final tail was added in a combinatorial manner with acrylate- or acrylamide-derived tails that contained hydroxyl, carbamate, ether, or amine functional groups as well as variations in alkyl chain length and branching. Such heterofunctional materials bear differential capacities for hydrogen bonding, hydrophobic interactions, and protonation states. The results of the study supported previous findings that suggested that the most-active lipidoids contain three or four secondary or tertiary amines [18, 49]. Lipidoids featuring final tails containing ethylene or propylene glycol units were observed to confer silencing in vivo, though none of the members of this new library was as effective as the parent molecules that contained the N_{12} aliphatic final tail. While valuable for extending the structure–activity relationships from the previous library, this strategy requires knowledge of an effective starting point material for optimization and involves the undesirable use of protection/deprotection chemistry.

A second approach was used to extend the utility of the first-generation lipidoid library. Rather than modify existing core structures, scientists combined suboptimal-performing lipidoids to generate synergistic combinations [50]. Using a modest starting library of 36 compounds, 630 binary pairs were generated and screened at six weight fractions (0, 0.2, 0.4, 0.6, 0.8, 1.0), resulting in a total of 3,780 unique formulations for a single siRNA concentration (20 nM) and lipidoid:siRNA (5:1) weight ratio. A synergy value was defined for each formulation to quantify improvements in gene silencing relative to the expected additive effects. It was confirmed mechanistically that successful binary combinations could leverage the complementary abilities of the individual components. Specifically, it was observed that one lipidoid promoted cellular uptake, while the second lipidoid promoted endosomal escape [50]. These in vitro data were supported by successful silencing experiments in vivo. Interestingly, the lead combination influenced not only knockdown but also biodistribution, enabling uptake by the liver and spleen, which had not been observed using nanoparticles comprising either of the parent materials alone. Notably, this combination approach is relevant to all classes of delivery materials. The efficacy of lipidoids may thus be further enhanced through combination with other carrier types, such as polymers or siRNA conjugates.

While interesting, these two examples represent iterative extensions of the original library; the application of creative chemistry can be used to generate disruptive advances. A novel library was created through the efficient ring-opening of epoxides by amine substrates [49]. The resultant nondegradable amino alcohols contained the key structural elements that had been previously defined—protonatable amines and hydrophobic aliphatic tails—and showed vastly superior activity, presumably owing to the more stable linkage. The reaction ran to completion after 3 days, which is intermediate relative to the 1 day for acrylates and 7 days for acrylamides. A library of 126 lipid-like members was generated by parallel synthesis, and structure–activity relationships were again evaluated. It was found that C14 was optimal. Amine 113 (*N*-methyl-2,2′-diaminodiethylamine) was common to the two top-performing materials in vitro. This amine, as well as the customly synthesized amine 200, is very closely related to the original optimal backbone, amine 98, further supporting the convergence of structure identified. Amine 200 was observed to confer the greatest silencing ability in vivo, affording virtually complete silencing at two orders of magnitude lower dose than other delivery systems.

This improved potency confers several advantages. First, it translates to less siRNA and less formulation material being administered to patients, thereby increasing tolerability. Second, it allows for the use of a larger dose than minimally required efficacious dose, which was shown to increase the duration of silencing. Third, it enables the administration of RNAi cocktails. The combination of siRNAs targeting multiple genes would be useful for treating infectious diseases or multifactorial diseases, such as cancer, in which multiple genes or pathways have been implicated. To this end, it was shown that five hepatic genes could be simultaneously knocked down following a single injection [49]. The new lead material, C12–200, was also used to deliver siRNA targeting the therapeutically relevant gene transthyretin in NHP and was shown to be the most potent carrier examined in primates to date [49].

Thus, by varying the chemical structure of the final tail, combining existing lipidoids in binary pairs to identify synergistic relationships, and leveraging a related synthetic scheme, the utility of the original lipidoid library was extended. These studies yielded structure–activity relationships, insights into mechanism of action, and materials exhibiting enhanced potency.

7.8 Conclusion

Chemical methods were developed to allow the rapid synthesis of large libraries of structurally diverse lipidoids. The members of the libraries were screened for the ability to transfect mammalian cells in vitro, and the top-performing materials were formulated for evaluation in vivo. The lead compound from the first-generation library, $98N_{12}$-5(1), was tested in four species, including NHP, and was shown to facilitate durable, potent, and specific silencing of therapeutically relevant endogenous gene transcripts. Importantly, the knockdown resulted in physiologically relevant outcomes in disease models of hypercholesterolemia, infectious disease, and cancer. Lipidoids can be administered locally—to the lung or intraperitoneal cavity—or systemically. The formulation was optimized to maximize uptake by hepatocytes and was demonstrated to be safe. Notably, inspection of the chemical functional groups common to the top-performing lipidoids from the first-generation library informed the synthesis of future-generation libraries, which exhibited a dramatic increase in the percentage of effective members, including members with greatly enhanced potency. This combinatorial synthetic approach led to a convergence of structure and represents an important strategy to expand the scale and diversity of siRNA delivery reagents. Such expansion enhances the likelihood of identifying compounds with improved efficacy, resulting in decreased dose as well as increased tolerability and silencing, facilitating the realization of the potential of RNAi therapeutics in the clinic.

7.9 Future Perspectives

The realization of the tremendous potential of RNAi therapeutics depends on the ability to deliver siRNA successfully into cells in vivo. The single-step synthetic methodologies employed allow for the simple, parallel generation of large combinatorial libraries. The development of lipidoids has expanded the scale and diversity of available delivery options. These materials confer efficient delivery to liver [49], lung [18], and peritoneal macrophages [18]. Lipidoids were also used to treat multiple models of ovarian cancer [44, 45]. It is likely that consideration of the following issues will yet improve the lipidoid platform as well as other delivery systems.

The application of additional single-step chemistries to the generation of large, chemically diverse libraries is likely to lead to the discovery of novel delivery materials. Examples of such synthetic methods include "click" chemistry [51] and the Staudinger ligation [52], though they both require the azide functionality, which limits the breadth of commercially available reagents. Creative chemists are sure to play a large role in the progress of the field.

The most significant challenge in the delivery space remains the effective and selective targeting of specific cells in vivo. Such targeting can decrease dose as well as side effects. Positively charged nanoparticles often accumulate in the liver, spleen, and kidneys. To reach distal sites, the particles must avoid clearance by these natural filtration sites. Such systemically targeted siRNA delivery has been achieved through the use of aptamer–siRNA chimeras, which have been applied to address mouse models of prostate cancer [53] and HIV [54, 55]. These molecules are not expected to escape endosomes efficiently, though they seem to be very efficacious notwithstanding. A second successful approach, which has been validated in patients, involves nanoparticle surface decoration with a targeting ligand [56]. This strategy, in which PEG is employed to enhance circulation time and amines are incorporated into the carrier to enhance endosomal escape, is likely to serve as the best means to target lipidoids as well. Additionally, it seems that some lipidoids inherently confer cell type-specific delivery in the absence of targeting ligands. The ability to minimize the number of components in a formulation is desirable because it decreases variability between particles, cost, and likelihood of immunogenicity (to targeting moieties such as peptides or antibodies). Further studies are required to elucidate the mechanism underlying the preference of certain lipidoid structures for certain cells.

Notably, targeted siRNA delivery is different from general targeted drug delivery (monoclonal antibodies or targeted nanoparticles containing small molecules) because receptor binding, while necessary, is not sufficient; siRNA delivery requires internalization of the payload. Interestingly, the modeling of transferrin-mediated tumor targeting suggested that the efficacy of these targeted nanoparticles was owing to increased cellular uptake by tumor cells rather than to increased overall tumor localization relative to nontargeted particles [57]. Thus, the optimization of internalization may be the critical step in the development of highly effective, nanoparticle-based targeted RNAi therapeutics.

Advances in understanding the mechanism of particle uptake and subsequent intracellular transport are thus expected to lead to significant progress in this field. The bridging of the chemistry and engineering aspects of drug delivery with the biological and medical aspects of target selection will lead to synergies in the realm of therapeutics [58]. Receptor type, density, heterogeneity, shedding, and internalization should all be considered, along with ligand type, density, size, and number. Particle physicochemical properties are derived from the combination of multiple constituents, so the contributions of composition, size, charge, shape, and surface properties to function should also be regarded.

While viruses have evolved over millions of years to deliver nucleic acids into cells efficiently, nonviral vectors have been developed over the course of the past 20 years. The amount of progress that has been made in this short period of time is astounding, through the complementary principles of rational design and combinatorial approaches. Both methods afford the ability to learn about delivery and uptake empirically, albeit at different scales of study. Understanding the mechanisms of uptake and intracellular trafficking and determining the critical physical properties that facilitate these steps—with clarification from the analysis of structure–activity relationships—will hopefully inform the judicious design of next-generation vectors that display maximal efficacy and minimal toxicity [17].

References

1. Fire A et al (1998) Potent and specific genetic interference by double-stranded RNA in Caenorhabditis elegans. Nature 391(6669):806–811
2. Berns K et al (2004) A large-scale RNAi screen in human cells identifies new components of the p53 pathway. Nature 428(6981):431–437
3. Kittler R et al (2004) An endoribonuclease-prepared siRNA screen in human cells identifies genes essential for cell division. Nature 432(7020):1036–1040
4. Paddison PJ et al (2004) A resource for large-scale RNA-interference-based screens in mammals. Nature 428(6981):427–431
5. Bumcrot D et al (2006) RNAi therapeutics: a potential new class of pharmaceutical drugs. Nat Chem Biol 2(12):711–719
6. McCaffrey AP et al (2002) Gene expression: RNA interference in adult mice. Nature 418(6893):38–39
7. McCaffrey AP et al (2003) Inhibition of hepatitis B virus in mice by RNA interference. Nat Biotechnol 21(6):639–644
8. Soutschek J et al (2004) Therapeutic silencing of an endogenous gene by systemic administration of modified siRNAs. Nature 432(7014):173–178
9. Zimmermann TS et al (2006) RNAi-mediated gene silencing in non-human primates. Nature 441(7089):111–114
10. Grimm D et al (2006) Fatality in mice due to oversaturation of cellular microRNA/short hairpin RNA pathways. Nature 441(7092):537–541
11. John M et al (2007) Effective RNAi-mediated gene silencing without interruption of the endogenous microRNA pathway. Nature 449(7163):745–747
12. Whitehead KA, Langer R, Anderson DG (2009) Knocking down barriers: advances in siRNA delivery. Nat Rev Drug Discov 8(2):129–138
13. Schroeder A et al (2010) Lipid-based nanotherapeutics for siRNA delivery. J Intern Med 267(1):9–21
14. Heyes J et al (2005) Cationic lipid saturation influences intracellular delivery of encapsulated nucleic acids. J Control Release 107(2):276–287
15. Semple SC et al (2010) Rational design of cationic lipids for siRNA delivery. Nat Biotechnol 28(2):172–176
16. Lynn DM et al (2001) Accelerated discovery of synthetic transfection vectors: parallel synthesis and screening of a degradable polymer library. J Am Chem Soc 123(33):8155–8156
17. Goldberg M (2008) Screening, synthesis, and applications of "lipidoids": a novel class of molecules developed for the delivery of RNAi therapeutics, in biological chemistry. Massachusetts Institute of Technology, Cambridge
18. Akinc A et al (2008) A combinatorial library of lipid-like materials for delivery of RNAi therapeutics. Nat Biotechnol 26(5):561–569
19. Behr JP et al (1989) Efficient gene transfer into mammalian primary endocrine cells with lipopolyamine-coated DNA. Proc Natl Acad Sci USA 86(18):6982–6986

20. Boussif O et al (1995) A versatile vector for gene and oligonucleotide transfer into cells in culture and in vivo: polyethylenimine. Proc Natl Acad Sci USA 92(16):7297–7301
21. Sonawane ND, Szoka FC, Verkman AS (2003) Chloride accumulation and swelling in endosomes enhances DNA transfer by polyamine-DNA polyplexes. J Biol Chem 278(45): 44826–44831
22. Medina-Kauwe LK, Xie J, Hamm-Alvarez S (2005) Intracellular trafficking of nonviral vectors. Gene Ther 12(24):1734–1751
23. Wolff JA, Rozema DB (2007) Breaking the bonds: non-viral vectors become chemically dynamic. Mol Ther 16(1):8–15
24. Wattiaux R et al (2000) Endosomes, lysosomes: their implication in gene transfer. Adv Drug Deliv Rev 41(2):201–208
25. Miller AD (1998) Cationic liposomes for gene therapy. Angew Chem Int Ed 37(13–14): 1768–1785
26. Maurer N et al (1999) Lipid-based systems for the intracellular delivery of genetic drugs. Mol Membr Biol 16(1):129–140
27. Shmueli RB, Anderson DG, Green JJ (2010) Electrostatic surface modifications to improve gene delivery. Expert Opin Drug Deliv 7(4):535–550
28. Kedmi R, Ben-Arie N, Peer D (2010) The systemic toxicity of positively charged lipid nanoparticles and the role of toll-like receptor 4 in immune activation. Biomaterials 31(26): 6867–6875
29. Wheeler JJ et al (1999) Stabilized plasmid-lipid particles: construction and characterization. Gene Ther 6(2):271–281
30. Wisse E et al (2008) The size of endothelial fenestrae in human liver sinusoids: implications for hepatocyte-directed gene transfer. Gene Ther 15(17):1193–1199
31. Akinc A et al (2009) Development of lipidoid-siRNA formulations for systemic delivery to the liver. Mol Ther 17(5):872–879
32. Adlakha-Hutcheon G et al (1999) Controlled destabilization of a liposomal drug delivery system enhances mitoxantrone antitumor activity. Nat Biotechnol 17(8):775–779
33. Webb MS et al (1998) Comparison of different hydrophobic anchors conjugated to poly(ethylene glycol): effects on the pharmacokinetics of liposomal vincristine. Biochim Biophys Acta 1372(2):272–282
34. Svensson RU et al (2008) Assessing siRNA pharmacodynamics in a luciferase-expressing mouse. Mol Ther 16(12):1995–2001
35. Sioud M (2008) Does the understanding of immune activation by RNA predict the design of safe siRNAs? Front Biosci 13:4379–4392
36. Eberle F et al (2008) Modifications in small interfering RNA that separate immunostimulation from RNA interference. J Immunol 180(5):3229–3237
37. Zhang HY et al (2006) RNA interference with chemically modified siRNA. Curr Top Med Chem 6(9):893–900
38. Morrissey DV et al (2005) Potent and persistent in vivo anti-HBV activity of chemically modified siRNAs. Nat Biotechnol 23(8):1002–1007
39. Judge AD et al (2006) Design of noninflammatory synthetic siRNA mediating potent gene silencing in vivo. Mol Ther 13(3):494–505
40. Schlee M, Hornung V, Hartmann G (2006) siRNA and isRNA: two edges of one sword. Mol Ther 14(4):463–470
41. Nguyen DN et al (2009) Drug delivery-mediated control of RNA immunostimulation. Mol Ther 17(9):1555–1562
42. Frank-Kamenetsky M et al (2008) Therapeutic RNAi targeting PCSK9 acutely lowers plasma cholesterol in rodents and LDL cholesterol in nonhuman primates. Proc Natl Acad Sci USA 105(33):11915–11920
43. Epiphanio S et al (2008) Heme oxygenase-1 is an anti-inflammatory host factor that promotes murine plasmodium liver infection. Cell Host Microbe 3(5):331–338
44. Huang Y-H et al (2009) Claudin-3 gene silencing with siRNA suppresses ovarian tumor growth and metastasis. Proc Natl Acad Sci USA 106(9):3426–3430

45. Goldberg MS et al (2011) Nanoparticle-mediated delivery of siRNA targeting Parp1 extends survival of mice bearing tumors derived from Brca1-deficient ovarian cancer cells. Proc Natl Acad Sci USA 108(2):745–750
46. Armstrong DK et al (2006) Intraperitoneal cisplatin and paclitaxel in ovarian cancer. N Engl J Med 354(1):34–43
47. O'Donovan PJ, Livingston DM (2010) BRCA1 and BRCA2: breast/ovarian cancer susceptibility gene products and participants in DNA double-strand break repair. Carcinogenesis 31(6):961–967
48. Mahon KP et al (2010) Combinatorial approach to determine functional group effects on lipidoid-mediated siRNA delivery. Bioconjug Chem 21(8):1448–1454
49. Love KT et al (2010) Lipid-like materials for low-dose, in vivo gene silencing. Proc Natl Acad Sci USA 107(5):1864–1869
50. Whitehead KA, et al. (2011) Synergistic silencing: combinations of lipid-like materials for efficacious siRNA delivery. Mol Ther
51. Kolb HC, Finn MG, Sharpless KB (2001) Click chemistry: diverse chemical function from a few good reactions. Angew Chem Int Ed 40(11):2004–2021
52. Saxon E, Armstrong JI, Bertozzi CR (2000) A "traceless" staudinger ligation for the chemoselective synthesis of amide bonds. Org Lett 2(14):2141–2143
53. Dassie JP et al (2009) Systemic administration of optimized aptamer-siRNA chimeras promotes regression of PSMA-expressing tumors. Nat Biotechnol 27(9):839–846
54. Neff CP et al (2011) An aptamer-siRNA chimera suppresses HIV-1 viral loads and protects from helper CD4+ T cell decline in humanized mice. Sci Transl Med 3(66):66ra6
55. Wheeler LA et al (2011) Inhibition of HIV transmission in human cervicovaginal explants and humanized mice using CD4 aptamer-siRNA chimeras. J Clin Invest 121(6):2401–2412
56. Davis ME et al (2010) Evidence of RNAi in humans from systemically administered siRNA via targeted nanoparticles. Nature 464(7291):1067–1070
57. Bartlett DW et al (2007) Impact of tumor-specific targeting on the biodistribution and efficacy of siRNA nanoparticles measured by multimodality in vivo imaging. Proc Natl Acad Sci USA 104(39):15549–15554
58. Sharp PA, Langer R (2011) Promoting convergence in biomedical science. Science 333(6042):527

Chapter 8
Polymeric Micelles for siRNA Delivery

Frederico Pittella and Kazunori Kataoka

Abstract Polymeric micelles as nanoassemblies incorporating siRNA have been receiving much attention due to their high capacity to accomplish several functions owing to the flexibility in designing functional block of copolymers by synthetic polymer chemistry. This chapter presents the basis for formation and examples of designs of polymeric micelles for siRNA delivery. The barriers found to reach the cytoplasm of diseased cells are presented, while the chemical solutions available for the preparation of polymers used in the assembly of smart polymeric micelles are described. Further, this chapter introduces the interactions that conduct the micellization process as well as examples of rational strategies for the formation of micelles using forces that facilitate this process. Advantageous features of polymeric micelles are summarized, and the future directions in this field are highlighted. This chapter provides basis for a better understanding of polymeric micelles design and the feasibility of these vehicles in translating RNAi biology into RNAi therapeutics.

8.1 Introduction

The recent combinatory application of supramolecular nanotherapeutics self-assembled from bioactive compounds, such as small interfering RNA (siRNA), and engineered polymers with multifunctionality holds great promise to the development of new protocols for the treatment of diseases. Pioneering research on new reactions for polymerization as well as characterization methods to obtain well-defined polymers integrated with multifunctionality arise from synthetic polymer chemistry born

F. Pittella • K. Kataoka (✉)
Division of Clinical Biotechnology, Center for Disease Biology and Integrative Medicine, Graduate School of Medicine, The University of Tokyo,
7-3-1 Hongo, Bunkyo-ku, Tokyo 113-0033, Japan
e-mail: kataoka@bmw.t.u-tokyo.ac.jp

K.A. Howard (ed.), *RNA Interference from Biology to Therapeutics*,
Advances in Delivery Science and Technology, DOI 10.1007/978-1-4614-4744-3_8,

in the twentieth century. Among other characteristics, the flexibility of synthetic polymer chemistry allows the synthesis of polymers with different solvent affinities and tailored molecular weight, as well as the addition of biomimetic features and bioresponsive elements to a construct [1].

Blocks of two or more repeating monomers in the same polymer chain can result in molecules with regions that have opposite affinities for a solvent. These amphiphilic block copolymers were found to have the ability to form micelles in selective solvents by organizing themselves according to the solvent affinity. Polymeric micelles are spherical supramolecular nanoassemblies with the size range of several tens of nanometers and are characterized by a core-shell architecture in which the inner compartment can be used as nanocarriers for versatile compounds [2, 3]. The concept of nanoassembly arrangement of amphiphilic block copolymers through the formation of a hydrophilic corona surrounding a water-incompatible core can be extended to include macromolecular association through electrostatic interaction [3]. Supramolecular assembly is obtained by utilizing the opposite charges of negative oligonucleotides with positive polycations to form polyion complexes through electrostatic interactions [3–6].

Although viral vectors are natural vehicles for nucleic acids, the inherent immunogenic characteristic and safety concerns have substantially impeded their translation into approved pharmaceuticals [7]. As an alternative, nonviral vectors for nucleic acid encapsulation that mimic viral size and structure can be produced by the application of nanotechnology. Among the nonviral nanocarrier systems, the polymeric nanocarriers offer great biological stability and versatility in design due to copolymerization and inclusion of chemical components such as targeting ligands during the polymer synthesis to achieve surface functionalization [8].

A fundamental pathway in eukaryotic cells by which short sequences of RNA can promote the cleavage of a complementary endogenous messenger RNA (mRNA) transcript is called RNA interference (RNAi). Although researchers had been using exogenous single-stranded antisense oligonucleotides in cell experiments to cleave specific mRNA and silence genes, Mello and Fire [9] described the process of double-stranded RNA-mediated gene silencing, while Elbashir and colleagues proved the principle in mammalian cells [10]. These studies provide the basis for the potential application of RNAi-based therapeutic intervention to control gene expression at the posttranscriptional level. Clinical translation of RNAi, however, requires nontoxic and biocompatible delivery systems. In this sense, the knowledge obtained from the development of plasmid DNA (pDNA) nanocarriers as gene delivery vehicles can be applied to RNAi-based therapeutics due to similar polyanionic and macromolecular characteristics.

This chapter gives an overview of the rational design of polymeric micelles for siRNA delivery directed toward therapeutic application. The design of block copolymers for the assembly into multifunctional vehicles takes into account requirements needed to overcome extracellular and intracellular delivery barriers. Strategies for the formation of stable polymeric micelles are discussed, and smart polymers to overcome barriers are presented.

8.2 Barriers to Synthetic siRNA Therapeutics

Small interfering RNA has a well-defined short structure of double-stranded RNA of 19–25 nucleotides in length with 2 nucleotide overhangs at either 3′ ends [11]. The synthetic siRNA molecule can induce mRNA cleavage by sequence specificity once it is delivered to the cell cytoplasm. Synthetic siRNA is particularly attractive for clinical applications as they enter the RNAi pathway in a later phase that avoids the Dicer step and are less likely to interfere with gene regulation by endogenous imperfectly paired noncoding hairpin RNA structures (microRNAs) that are naturally transcribed by the genome [12]. In addition, they can be chemically modified to improve stability, reduce immunogenicity, and off-target effects. The RNAi effect induced by exogenous siRNA in vertebrates, however, is transient that lasts for a number of days [13–15], which requires continuous doses for maintenance of a therapeutic effect.

The delivery of siRNA into the target cell is the key to the translation of this new technology into a therapeutic. The polyanionic and macromolecular (~13 kDa) characteristics restrict its uptake across cellular membranes, while nonmodified siRNA is unstable within the blood circulation due to degradation by serum nucleases and the rapid clearance by renal excretion [14, 15]. Hence, a nanocarrier system is required to provide systemic stability and prolonged circulation within the bloodstream as well as facilitating cellular uptake. Thus, nanoparticle-based systems have been developed to improve the therapeutic effectiveness of siRNA. But even with protection of the siRNA by incorporation within nanocarrier systems, there are several extracellular and intracellular barriers for successful in vivo siRNA delivery using nanocarriers (Table 8.1). The nanocarrier should (1) be stable and exhibit prolonged circulation in order to reach the diseased site, (2) extravasate from the bloodstream into the diseased tissue, (3) show stability in the extracellular matrix, (4) enter the diseased cells, (5) escape from the endosomal compartment to avoid degradation in the lysosome, and (6) disassemble in the cytoplasm to allow interaction of free siRNA molecules with the RNAi machinery.

Table 8.1 Main barriers for the systemic delivery of siRNA within nanocarriers

Level	Challenges
Circulation	Interaction with biomacromolecules and clearance by mononuclear phagocyte system (MPS)
Biodistribution	Nonspecific accumulation
	Unwanted systemic effects
Toxicology	Nanocarrier toxicity
	Immune response to RNAi
	Oversaturation of RISC
Tissue permeability	Endothelium penetration (extravasation)
Extracellular	Extracellular stability and diffusion
Internalization	Cellular uptake (endocytosis)
Intracellular	Escape from endosomal compartment
	Dissociation from the nanocarrier and trafficking

Systemic administration of a nanocarrier system offers great potential due to dissemination within the bloodstream that allows access to a wide variety of tissues and cells. Therefore, the first challenge of a system is to be stable in the blood and protect siRNA from nuclease degradation. In addition, other nonspecific interactions with serum proteins can induce the nanocarrier dissociation or aggregation. The interaction with biomacromolecules can induce embolization in microvessels or capture by the mononuclear phagocyte system (MPS) mainly in the liver and spleen. The size of the nanocarrier reduces renal filtration that is needed to prolong circulation within the bloodstream. Undesirable biodistribution of nanocarriers may lead to nonspecific accumulation in healthy tissues and to unwanted systemic effects.

The possible toxicity of a nanocarrier system needs to be addressed before clinical experiments. Complete in vitro assessment allied to further preclinical evaluation of the biodegradability of components is necessary to avoid toxic effects in humans. Correspondingly, the evaluation of immune stimulation after administration of siRNA is an important matter. This stimulation is mediated by immune cells, via the Toll-like receptor pathway and can activate high levels of inflammatory cytokines such as interleukin-6 (IL-6) and interferon (IFN) [16–18]. Another toxicology concern is the oversaturation of RISC in the target cell. Studies suggest that the exogenous introduction of siRNA may result in the competition with endogenous miRNAs for RISC [19, 20]. The repression of miRNA-regulated genes leads to their re-expression, and this can ultimately disturb the normal tissue physiology.

The extravasation from the bloodstream to a desired tissue may occur passively or actively. The passive accumulation is generally explained by the microvascular hyperpermeability to circulating macromolecules, whereas the active accumulation is facilitated by targeting ligands that associate with specific cell-surface receptors (refer to Sect. 8.3).

The extracellular medium is also a challenge to the system. The microenvironment of the tissue presents differences in the pH, enzymes, or ions that can damage the nanocarrier causing dissociation before cellular entry. Moreover, for an extended RNAi effect in the tissue, the particle should diffuse in the extracellular matrix to reach cells located distant from the blood vessels. The siRNA nanocarrier normally enters cells by the process of endocytosis that facilitates entry into the endocytic pathway. Following cellular uptake, complexes are mainly transported to endosomes with decreasing pH. The inability to escape this compartment will lead to the degradation of the nanocarrier within the lysosomes. Thus, escaping the endosomes and selective disassembly in cytoplasm are the ultimate challenges to be surpassed.

The success of nanoparticle-delivered RNAi therapeutics will depend on the ability of the nanocarrier to overcome these biological barriers over a specific period, selectively target the diseased tissue, degrade predictably, be well tolerated and provide a high therapeutic index [21]. Thus, the therapeutic application of RNAi relies on the strategy to safely and efficiently deliver the siRNA to the diseased tissue. The in vivo application requires the design of formulations that are capable to perform several functions in order to overcome these hurdles and reach the cytoplasm of a diseased cell. As cancer is a potential target disease for RNAi-based therapeutics, this chapter will focus on nanocarrier delivery of siRNA to cancer cells.

8.3 Passive and Active Targeting in Cancer

8.3.1 *Passive Targeting*

Rapidly dividing cancerous cells require a continuous supply of nutrients to maintain proliferation. Cancerous cells, therefore, secrete growth factors that induce angiogenesis and result in a fast and disordered neovascularization around the tumor area. The new blood capillaries present loose interendothelial junctions that allow enhanced permeability to extravascular tissue. Through this leaky vasculature of the tumor, some substances including polymeric micelles are able to extravasate from the blood vessels to the tumor tissue. Moreover, the development of a lymphatic system is insufficient in tumor tissue, resulting in poor drainage of macromolecular substances from the tissue. This preferential macromolecular accumulation in solid tumor tissue is known as enhanced permeability and retention (EPR) effect [22, 23]. The EPR effect found by Matsumura and Maeda has become the guideline for passive targeting in cancer.

For an effective passive accumulation by the EPR effect, the nanocarriers should be small enough to permeate through the gaps in the endothelium from the blood compartment into solid tumors. Secondly, nanocarriers should have long circulation properties. Theoretically, long-circulating nanocarriers have an increased chance to find these gaps between endothelial cells and extravasate to tumor tissue. The long-circulation property is, in fact, a consequence of its stability in the bloodstream, avoiding glomerular excretion by the kidney and nonspecific complement activation and opsonization that could lead to uptake by the MPS in the liver, spleen, and lung. As the renal filtration cutoff is 50 kDa (or 5–6 nm), rapid removal and excretion of particles smaller than this cutoff is expected [24]. On the other hand, larger particles can be recognized and removed by the MPS, resulting in short half-life in the blood [25].

8.3.2 *PEGylated Micelles*

A common strategy to improve the blood circulation and the pharmacokinetic properties of a nanocarrier to successfully passive target solid tumors by EPR is PEGylation. PEGylation is the addition of a poly(ethylene glycol) (PEG) shell to the nanoparticles to utilize its protective properties and produce stealth particles. PEG has a general structure of $HO–(CH_2CH_2O)_n–CH_2CH_2–OH$, encompassing a flexible polyether backbone, with ether oxygen molecules forming hydrogen bonds with water molecules in solution. The protective effect of PEG is due to the formation of a dense, hydrated layer of long flexible chains on the surface of the colloidal particle that reduces nonspecific interactions with plasma proteins [26]. Importantly, an increase of PEG molecular weight >2,000 Da improves the blood circulation half-life of the PEGylated particles, probably due to the increased chain flexibility of higher MW PEG polymers [27]. Thus, PEGylation provides nanocarriers with

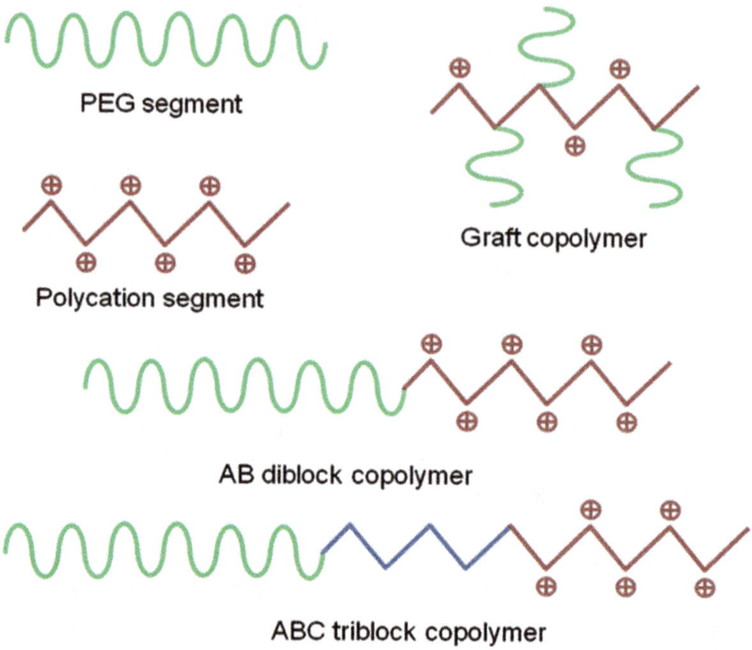

Fig. 8.1 Structures of block copolymers comprising PEG and polycations segments

improved solubility and lower surface charge and prevents aggregation. The outer PEG hydrophilic shell layer increases the system stability in serum, prolongs its circulation time, and reduces polymer toxicity. As a result of this, PEG has become an essential component in the design of block copolymers and micelles for siRNA delivery in vivo.

For the preparation of PEG-polycation block copolymers, PEG is utilized as macroinitiator for further polymerization of a cationic segment [5, 28]. Usually, these two segments can be coupled together, resulting in AB diblock or even ABC triblock architecture when a neutral B block is included with different functionality (Fig. 8.1). In contrast to AB or ABC block copolymer, the PEG chains in graft copolymers are attached to the main chain of cationic segments which may hinder the electrostatic interaction of polycationic segment to siRNA and interfere with micellization. PEG used in the block copolymer synthesis usually has a functional α or/and ω end groups such as amino groups [5, 28–30].

8.3.3 Active Targeting

When a receptor-mediated endocytosis mechanism is employed, it is possible to deliver the siRNA to a particular cell population in the target region. In this case, a targeting moiety is introduced into the polymer for a specific affinity towards a cellular receptor. The incorporation of targeting ligands into the micelle shell provides

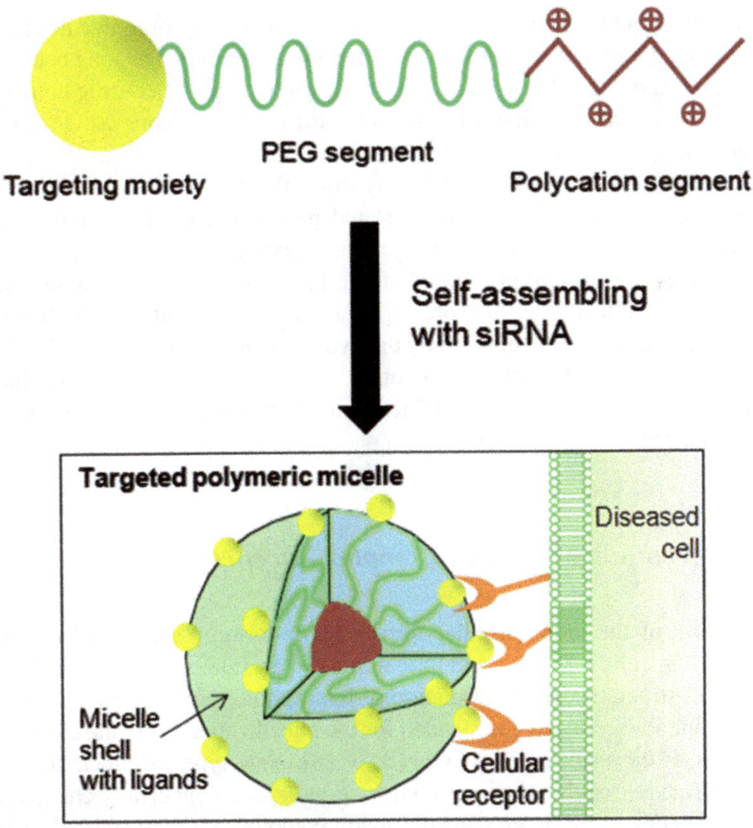

Fig. 8.2 Incorporation of targeting moiety into AB block copolymer for the preparation of surface functionalized micelle to active target diseased cells

control of biodistribution and site-specific cellular uptake, in a process referred to as active targeting (Fig. 8.2). Due to increased stability in serum provided by PEG, the strategy used by researchers is the functionalization of the distal end of PEG chain to introduce the targeting moieties. The moieties studied for active targeting include lactose (to bind asialoglycoprotein receptors in hepatocytes) [31, 32], cyclic RGD peptide (to bind $\alpha_v\beta_3$ integrin receptors expressed in several cells including cancerous) [33, 34], transferrin [35, 36], hyaluronic acid [37], and various others [38, 39].

8.4 Formation of Polymeric Micelles

The unique characteristic of amphiphilic block copolymers to form supramolecular structures of different size and shape due to distinct solvent affinity facilitates the formation of polymeric micelles. Core-shell architecture is often observed when the hydrophobic segment is segregated from the aqueous phase to form the core which is surrounded by a palisade of the hydrophilic segments. This structure provides the use of micelles as therapeutic nanocarriers due to the variety of molecules that can be incorporated into the core. In fact, the scientific community has long been studying the application of polymers as carrier of drugs (e.g., polymer–drug conjugates) [40–42] as well as forming micelles [43–46] with the core as a separated nanocontainer.

Micellization through self-assembly is often the result of a delicate balance between intermolecular forces [3]. As stated previously, the polymeric micelles have the hydrophobic interactions as the main driving force for the micellization [47] when they are composed of amphiphilic block copolymers and assembled in aqueous media. In this case, the core segregation from aqueous milieu directs the formation of micelles and is induced by the hydrophobic interactions [2, 47, 48] and proceeds through the combination of other intermolecular forces which include metal complexation [49, 50], organic/inorganic interactions [51–54], and hydrogen bonding [30, 55].

8.4.1 Formation of Polyion Complex Micelles

The variation of the intermolecular force of core-forming segments enables the regulation of micelle formation and its stability. In fact, the concept of micellization of core-shell structures by the formation of hydrophilic shell surrounding a segregated hydrophobic core can be extended for the micellization using the electrostatic interactions as the main driving force [3, 6, 28]. New types of block copolymer micelles formed through electrostatic interaction between oppositely charged molecules in aqueous medium are named polyion complex (PIC) micelles. The oppositely charged molecules include, among others, synthetic polymers of various architectures and biopolymers (proteins, enzymes, and oligonucleotides).

In this sense, the homopolymer as the polycation and negatively charged oligonucleotides precipitate to form water-insoluble organic salts. However, in order to provide dispersibility to the aggregates, additional segments to the homopolymer are necessary for the assembly into micelles. Hence, the design of block copolymers comprising a block of neutral hydrophilic polymer such as poly(ethylene glycol) (PEG) and a block of polyion is often used to the micellization process to achieve colloidal stability. The hydrophilic segments will form a shell surrounding the polyionic core and provide the dispersibility through the stealth property.

The micellization in this case arises from the association of a pair of polyelectrolytes with opposite charge which induces the release of the counterions from each polyelectrolyte as free ions to increase the translational entropy, while the electrostatic attraction of opposite charges occurs due to Coulomb force [28, 56–58]. With the increase in residual energy at the interface between PIC and an aqueous phase, the size of the complexes should consequently increase to decrease the relative surface area, allowing a reduction of the interfacial free energy [47]. This balance between interfacial energy and conformational entropy of the polymer strands determines the thermodynamically stable size of the PIC micelles [3].

The formation of polyion complex micelles (PIC micelles) was first verified by our group for the pair of oppositely charged block copolymers PEG-poly(L-lysine) and PEG-poly(α,β-aspartic acid) [28]. These first PIC micelles presented size around 30 nm and polydispersity of <0.1. Obviously, the core of the PIC micelle can serve as a reservoir for molecules, once it forms a separated phase from the outer space. In a similar association fashion, supramolecular assembly is easily obtained when using opposite charges of oligonucleotides and polycations: the negatively charged oligonucleotides bind the positively charged polycation to form a polyion complex through electrostatic interactions [3–5]. In addition, negatively charged oligonucleotides such as double-stranded DNA and siRNA share many common characteristics; thus, siRNA micelle design can benefit from the ideas and knowledge previously developed for DNA delivery [59].

8.5 Polymeric Micelles for siRNA Delivery

A new in vivo vector for oligonucleotides formed through the self-association of block copolymers, and antisense DNA was reported soon after the concept of PIC micelles [6]. As a result of the use of PEG-poly(L-lysine) (PEG-PLL) as the oligonucleotide counterpart, spherical PIC micelles were formed entrapping the antisense DNA in the core (Fig. 8.3). Similarly, plasmid DNA (pDNA) was loaded into

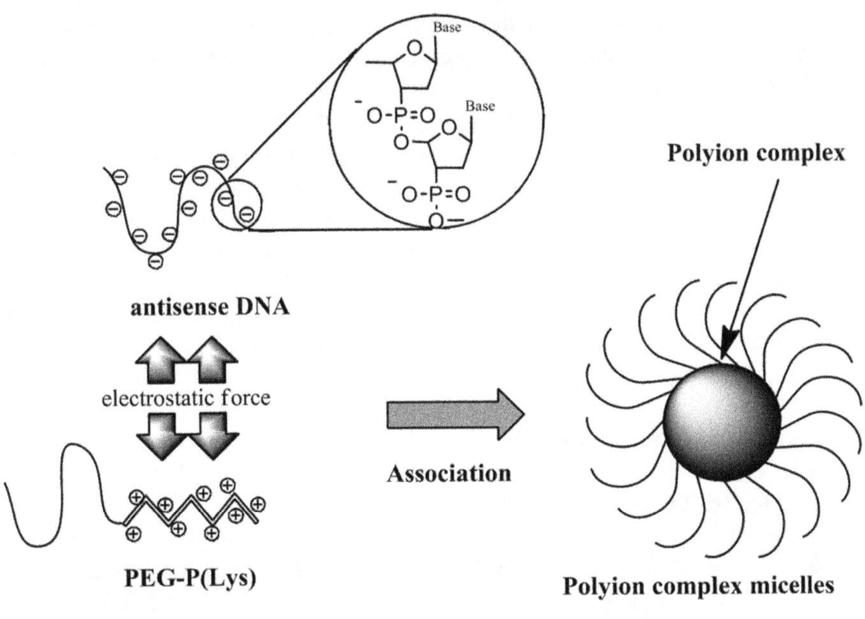

Fig. 8.3 Formation of PIC micelles between PEG-PLL and antisense DNA. Adapted with permission from [6]. Copyright 1996 American Chemical Society

PIC micelles formed with PEG-PLL, and the regulated release of free DNA from the polymeric micelle was reported to be triggered through replacement by the counterpolyanion [5].

Similarly to the micellar assembly of the PIC micelles with DNA, block copolymers consisting of PEG-polycation segments lead to the self-assembly of a structure with a core-shell architecture, when mixed with siRNA. The first passive targeting PIC micelle prepared for siRNA delivery comprised of a PEG-block catiomer carrying diamine side chain with distinctive pK_a [60]. PEG-poly(3-[(3-aminopropyl) amino] propyl aspartamide (PEG-DPT) was synthesized by a side chain aminolysis reaction of PEG-poly(β-benzyl L-aspartate) block copolymer (PEG-PBLA) with dipropylene triamine (DPT). The resulted block copolymer could efficiently incorporate the siRNA to form micelles and showed luciferase gene knockdown.

A siRNA nanocarrier system for active targeting was successfully reported by Schiffelers and colleagues [33]. In this research, the authors utilized a copolymer comprising PEI as the core-forming agent, PEG for steric stabilization, and peptide ligand containing a disulfide-stabilized Arg-Gly-Asp (RGD) motif at the distal end of PEG to provide tumor selectivity due to its ability to target integrins expressed on activated endothelial cells in tumor vasculature. The self-assembly of the RGD–PEG–PEI conjugates with siRNA was found to form sterically stabilized "layered" PIC micelles with exposed RGD ligands. This work suggested that the introduction of the targeting ligand resulted in a selective uptake of the polyplexes in the tumor vasculature after intravenous administration.

Besides the electrostatic force, alternative approaches to provide higher stability for the PIC micelles are being studied. This includes the use of additional forces such as hydrophobic interactions through the incorporation of hydrophobic groups for facilitated core formation, the chemical conjugation of siRNA to PEG, and inorganic/organic interactions.

In an attempt to improve the stability of polymeric micelles in the blood, Kim et al. prepared siRNA covalently bond to PEG segment via disulfide linkage [61]. The micelles were assembled with the addition of a core-forming polycation polyethylenimine (PEI) to produce a more stable core-shell structure. The hydrophilic PEG segment covalently attached to the siRNA forms a corona-type surface layer, while the core is formed by charge neutralization between siRNA and cationic agents. This system formed micelles with a hydrodynamic diameter of ~99 nm and was evaluated for in vivo delivery of antiangiogenic siRNA to female nude mice bearing subcutaneous prostate tumors. Intravenously injected PEG-siRNA/PEI micelles were two times more effective than siRNA/PEI complexes by the reduction of angiogenic vascular endothelium growth factor (VEGF) protein in the tumor. Interestingly, micelles accumulated passively in the tumor by the EPR effect, most likely taking advantage of the protective PEG corona.

An alternative approach to design micelles for siRNA delivery based on the conjugate PEG-siRNA was reported by Oishi et al. [31]. Remarkably, enhanced gene silencing was achieved by PIC micelles composed of lactosylated-PEG-siRNA conjugate bearing acid-labile β-thiopropionate linkage and poly(L-lysine) (Fig. 8.4). PIC micelles were formed by the interaction of the polycation poly(L-lysine) and

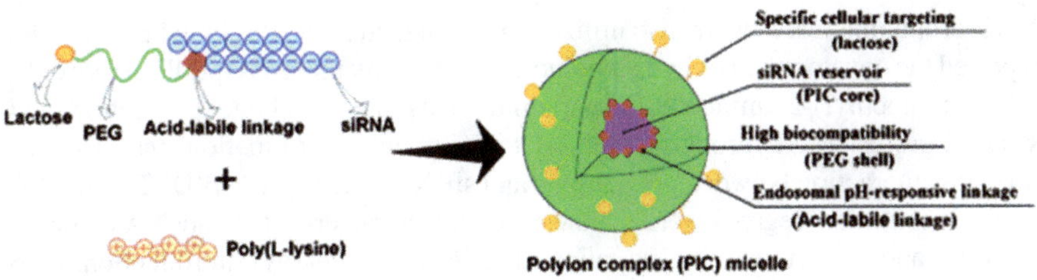

Fig. 8.4 Formation of PIC micelles designed for receptor mediated endocytosis. Reprinted with permission from [31]. Copyright 2005 American Chemical Society

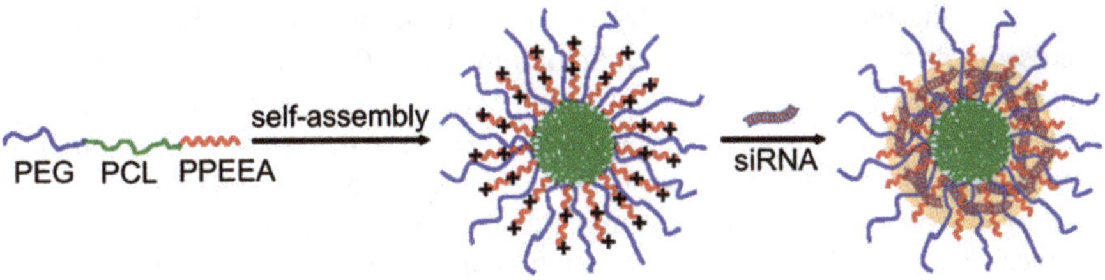

Fig. 8.5 Schematic drawing of self-assembled cationic micellar nanoparticles and loading of siRNA. Reprinted from [62], Copyright 2008, with permission from Elsevier

the anionic siRNA with a corona of hydrophilic PEG. The clustered lactose moieties at the end of PEG enabled active uptake into hepatoma cells by a receptor-mediated mechanism. Furthermore, the efficient transport of free siRNA into the cytoplasm was achieved through PEG detachment from the conjugates in response to the decreased pH (~5.5) in the endosomal/lysosomal compartment that triggered cleavage of the β-thiopropionate linkage.

Hydrophobicity is known to drive phase separation in aqueous environment. Consequently, the use of hydrophobic interactions as an additional force to assist PIC micelle formation is an interesting strategy to improve stability. In this context, the group of Wang presented an elegant design for the preparation of polymeric micelles to encapsulate siRNA [62, 63]. They prepared an amphiphilic triblock copolymer consisting of a monomethoxy PEG (mPEG) block, a poly(ε-caprolactone) (PCL) block and a cationic poly(2-aminoethyl ethylene phosphate) (PPEEA) block (mPEG-PCL-PPEEA). The rationale behind the choice is the hydrophobic interactions between PCL segments which induces the micellar core formation and stabilizes the nanoparticles, while the positively charged PPEEA block serves as siRNA binding site. Tri-layered micellar nanoparticles were formed by self-assembly in aqueous solution (Fig. 8.5). The siRNA loaded in these polymeric micelles was efficiently delivered and resulted in green fluorescent protein (GFP) gene silencing in HEK293 cells in the presence of serum [62], while reduced subcutaneous tumor growth was observed by knocking down acid ceramidase in vivo [63].

Hydrophobic forces were also utilized to stabilize the micelles by Kim et al. who reported the development of hydrophobic polycations by using PAsp(DET) (poly{*N*-[*N*-(2-aminoethyl)-2-aminoethyl]aspartamide}) as the backbone polycation and stearoyl groups as a hydrophobic moiety to facilitate core formation [64]. To optimize the interaction between polycations and siRNA, stearoyl PAsp(DET) with different substitution degrees were synthesized and characterized for siRNA complex stability and RNAi activity in cultured cells. The stearoyl introduction onto PAsp(DET) side chains led to the complex stabilization via hydrophobic interaction and micelles presented low cytotoxicity. Additionally, the PAsp(DET) polycations backbone contributed to the excellent endosomal escape of siRNA complexes, resulting in an improved RNAi activity in vitro. The in vivo application of micelles comprising stearoyl PEG-SS-PAsp(DET) complexed with siRNA presented longer stability in the blood as verified by intravital real-time confocal laser scanning microscopy [65]. The extended circulation time of PEGylated micelles that were hydrophobically stabilized allowed effective passive accumulation in solid tumors. The significant in vivo gene silencing through the use of this smart siRNA complex after systemic administration strongly indicates successful siRNA delivery.

Inorganic/organic interactions can facilitate the formation of nanoparticles and the encapsulation of siRNA. The precipitation of inorganic salts apparently has the electrostatic force as the main driving force. When mixed in an aqueous environment, inorganic compounds precipitate to form insoluble particles, which soon aggregate with each other. In this case, physical barriers surrounding the salt are needed to avoid further aggregation and crystal growth. The use of additives to provide colloidal stability through steric repulsion of the precipitates includes block copolymers and other molecules.

Interestingly, negatively charged nucleic acids are entrapped in inorganic nanoparticles during wet precipitation. Inorganic/organic hybrid nanoparticles carrying siRNA hold great promise for clinical application since they are easily prepared, water dispersible, and stable in biological environments. In addition, due to physiological stability, hybridizing inorganic nanoparticles with organic compounds such as polymers as the physical barrier to aggregation offers a potential alternative for effective nanocarriers [66–68]. This hypothesis was realized through the use of PEG-polyanion as the physical barrier to calcium phosphate nanoparticles aggregation in the report of the first calcium phosphate hybrid system [51]. Kakizawa et al. obtained stable PEGylated hybrid nanoparticles as a low cytotoxic nanocarrier of antisense DNA.

The first application of hybrid nanoparticles for siRNA delivery that involved calcium phosphate (CaP) nanoparticles coated with poly(ethylene glycol)-block-poly(aspartic acid) (PEG-P(Asp)) shows promising results [69]. Through a simple mixing of separate solutions containing calcium, phosphate, siRNA, and PEG-P(Asp), nanoparticles were formed having a size smaller than 100 nm. While the loading capacity of siRNA reached almost 100% under optimal conditions, the particle size could be regulated to some extent by adjusting the PEG-P(Asp) concentration. Eventually, the intracellular environment with appreciably lowered calcium ion concentration compared to the exterior allowed the release of the incorporated

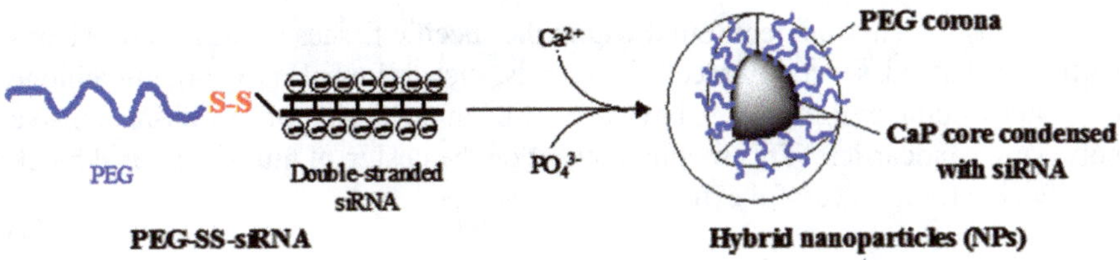

Fig. 8.6 Formation of inorganic/organic hybrid nanoparticles by the use of PEG-SS-siRNA and calcium phosphate. Reprinted with permission from [52]. Copyright 2009 Wiley

siRNA in a controlled manner. Furthermore, using a similar strategy to form polymeric nanoparticles, the same authors employed the polyanion PEG-block-poly(methacrylic acid) (PEG-PMA). The strategy was to incorporate the molecular units facilitating endosomal escape directly into the block copolymer structure triggered by PMA conformational transition at pH 4–6 which approximately corresponds to the endosomal pH [70]. Finally, the use of PEG-polyanion was essential to the formation of stable hybrid nanoparticles that showed an appreciable silencing of the reporter gene in vitro.

Interestingly, other reports have proposed the use of different strategies to obtain stable hybrid nanoparticles through inorganic/organic interactions. Zhang et al. hypothesized that the integration of siRNA into block copolymers via a cleavable disulfide bond to form PEG-ss-siRNA conjugate could be useful to form the calcium phosphate hybrid nanoparticles and obtain a high loading efficiency (Fig. 8.6) [52]. Theoretically, an increased efficacy of siRNA entrapment is a trade-off relationship in the regulated crystal growth since siRNA and PEG-*block*-polyanion compete for binding with the positive charges on the CaP crystal [71]. Indeed, using PEG and siRNA bound via disulfide, the authors obtained nanoparticles of size between 90 and 120 nm with very narrow particle size distribution (polydispersity index <0.1). Nanoparticles were confirmed to be spherical and exhibited well-defined core-shell architecture, with up to 86% siRNA incorporation efficiency.

8.6 Designing Polymeric Nanocarriers for Cytoplasmic Delivery

A great advantage of using synthetic polymers in nanocarrier systems is the possibility to include blocks or segments with distinct functions to overcome a specific delivery barrier. Hence, it is of great importance to understand the pathways and environments in which the nanocarriers undergo, and the physical, chemical or biological changes in the microenvironment that are related to the disease to be treated or tissue to be reached.

Many block copolymers with smart functions, such as chemical [72–74] or physical [75] stimuli-sensitivity, and the targetability to specific tissues [38, 39] have already been designed for drug or gene delivery. These functions utilize the stimuli provided by the environment found within the specific tissues or intracellular compartments. Indeed, several intracellular signals, such as low pH [73, 74], glutathione [76], and specific enzymes [77], have been used in the design of stimuli-responsive polymeric nanocarriers. This section focuses on the design of multifunctional block copolymers for a successful cytoplasmic delivery.

8.6.1 Stimuli-Responsive Systems

A variety of biochemical environments encountered by the micelles during the path to the diseased tissue can be utilized to promote physicochemical changes in polymeric nanocarriers [78, 79]. For instance, the pH in the bloodstream is ~7.4 while ~6.5 in the early endosomes and 5.5 in the late endosomes, rising to 7.2 in the cytoplasm [80]. These pH changes allow the integration of an endosomal escape function to the polymer, and thus to the micelle, by utilizing polymers possessing acidic pH sensitivity.

Surprisingly, simple polycations such as PEI have presented significant endosomal escape due to low pK_a amines that undergo protonation in an acidic environment. The mechanism through which low pK_a amines promote endosomal escape is the "proton sponge effect," in which the protonation of amines in the endosome stimulates the influx of protons and chloride ions into this compartment, increasing the osmotic pressure that results in the disruption of the compartment [81]. Although the "proton sponge effect" is the theory most accepted in this case, its effectiveness is under debate [82]. Concerns regarding PEI cytotoxicity have restricted its clinical use; therefore, new biodegradable PEI derivatives for siRNA delivery are being designed to enhance transfection efficacy using substitution of degradable moieties and copolymers [83, 84]. Interestingly, Park and colleagues developed a siRNA vector comprising PEI grafted to hyaluronic acid (HA) via a disulfide linkage (PEI-SS-HA). Described as a nontoxic vector, specific cellular uptake by the HA receptor through endocytosis is observed. In addition, intratumoral injection of this complex incorporating VEGF siRNA significantly suppressed subcutaneous tumor growth through the anti-angiogenic effect with decreased VEGF mRNA and protein [37].

A comprehensive study done by our group [80, 85–87] has shown that aligning the aminoethylene structure (the main component of PEI) in the side chain of a polymer strand can significantly decrease the polycation cytotoxicity by generating pH sensitivity. As observed in PAsp(DET) and PAsp(TEP) (PAsp(TEP): poly[*N*-(*N*-{*N*-[*N*-(2-aminoethyl)-2-aminoethyl]-2-aminoethyl}-2-aminoethyl)aspartamide]), a monoprotonated structure is present at pH 7.4 while the drop to pH 5.5 allows the formation of a diprotonated structure in the 1,2-diaminoethane unit, which is essential for the endosome-selective membrane destabilization effect. This selective protonation of the polymer in the endosomes allow efficient but less toxic endosomal escape of the micelles (Fig. 8.7).

Stimuli-responsive polymers not only allow enhanced endosomal escape, but also provide efficient siRNA release from the micelles in the cytoplasm. A review

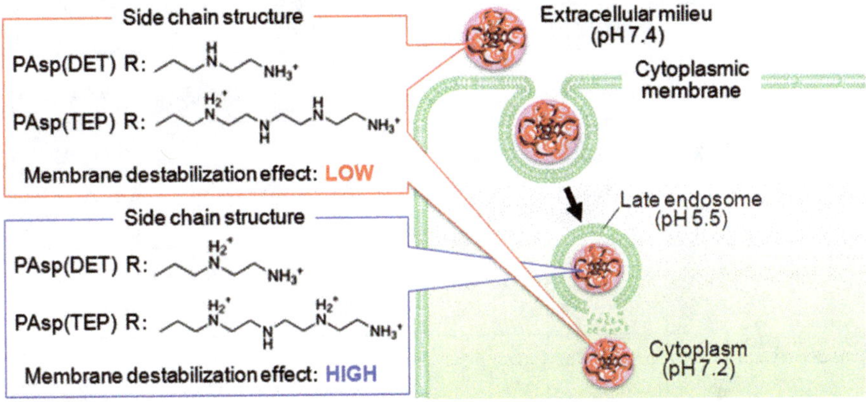

Fig. 8.7 Protonated structures of PAsp(DET) and PAsp(TEP), according to pH changes in the cellular microenvironment. Reproduced from [80] with permission from The Royal Society of Chemistry

by Kwon has presented stimuli-sensitive chemistry that can be included in the design of polymeric micelles for siRNA delivery [78]. Among the structures, acid-cleavable linkers and acid-protonating groups were identified (Fig. 8.8). The synthesis of block copolymers containing stimuli-sensitive groups enables the structure to (1) undergo conformational changes from a hydrophilic structure to a membrane-destabilizing form at the low pH in the endosomes or (2) facilitate disassembly of the nanocarrier resulting in efficient delivery of free siRNA [88].

Some of these chemical groups can be linked to a block copolymer to protect the endosome-disrupting component, converting the charge of (or neutralizing) the primary amines. That is the case of a system designed by our group, in which the cationic charges of the dual-functional polycation PEG-PAsp(DET) (the stealth and endosomal-disrupting functions provided by PEG and PAsp(DET), respectively), was converted to a net negative charge with the addition of *cis*-aconitic acid forming PEG-PAsp(DET-Aco) [poly(ethylene glycol)-*b*-poly{*N*-[*N*-(*N*-cis-aconityl-2-aminoethyl)-2-aminoethyl]aspartamide}]. Namely, charge-conversional polymer (CCP), the newly formed polyanion comprising *cis*-aconitylamide enabled an effective integration of the dual-functional block copolymer to calcium phosphate nanoparticles carrying siRNA through the two carboxylates of the *cis*-aconityl moiety [53]. In the rational hypothesis, the obtained hybrid nanoparticles are stable in pH 7.4, but at the endosomal pH of 5.5, cleavage of the *cis*-aconitylamide will take place to reproduce the endosomal destabilizing polycation PAsp(DET) following to the disassembly of the particles and siRNA release into the cytoplasm (Fig. 8.9).

Confocal microscopy revealed a rapid endosomal escape of the labeled siRNA (Fig. 8.10), indicating that the siRNA was released from the nanoparticles in the cytoplasm and was free to produce the RNAi effect, in agreement with the initial hypothesis for the design of the system. When the red pixels of cy5-siRNA are co-localized with the green pixels of lysotracker, the resultant color is yellow, and it is an indicative that the nanocarrier is still inside the endosomes. When the red pixels

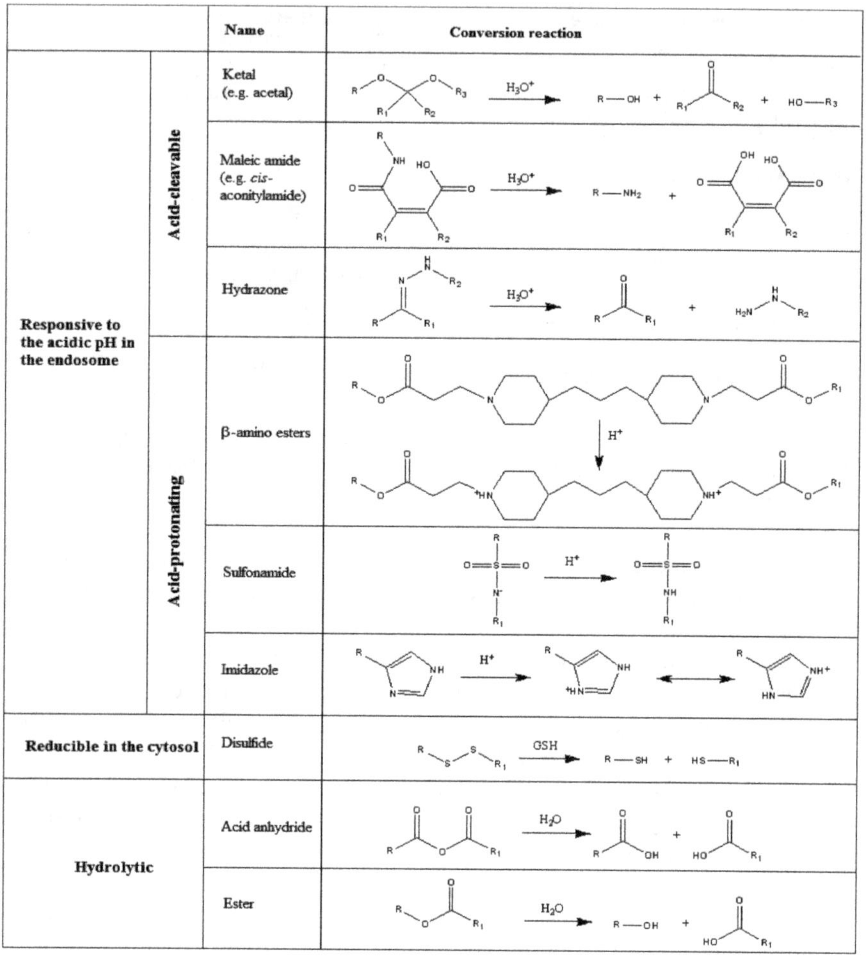

Fig. 8.8 Stimuli-responsive groups for the design of smart polymeric micelles. Adapted with permission from [78]. Copyright 2011 American Chemical Society

are not co-localized with green, the siRNA is already released into the cytoplasm. As a control, a non-charge-conversional polymer was used (PEG-PAsp(DET-Car)). A special feature of this calcium phosphate hybrid system allowed the slow disassembly of calcium crystals in a low ionic compartment, which resulted in the release of siRNA after 24 h in noncharge conversional polymers as well [53], although not as efficiently as the CCP containing system.

Among the pH-sensitive groups, the disulfide linkage is well characterized as a chemical bond that is cleaved specifically in the cytoplasm. Once it is reduced by glutathione (GSH), whose intracellular concentration is much higher than the extracellular, it allows the disassembly of the particles carrying siRNA that are stabilized

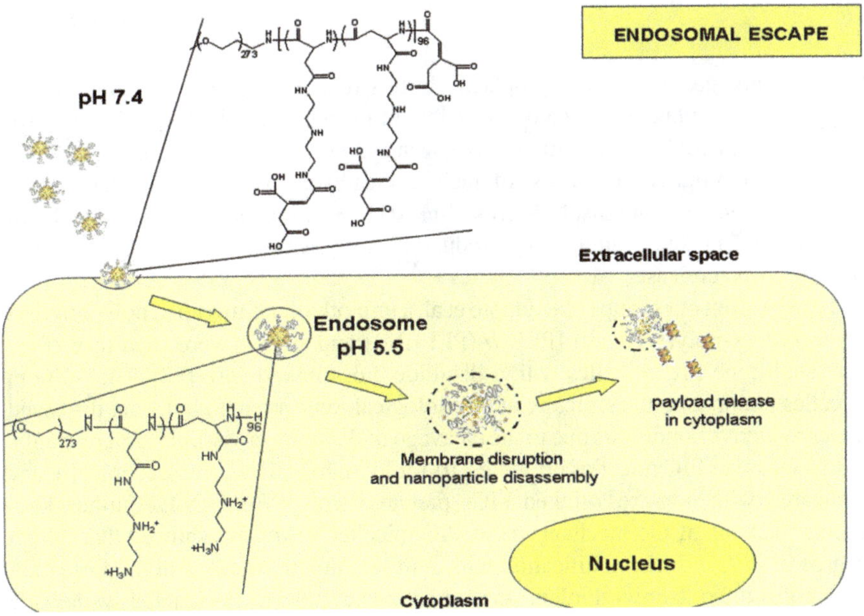

Fig. 8.9 The hypothesis for endosomal escape and disassembly of hybrid nanoparticles comprising of a charge-conversional polymer (CCP). The polyanion containing *cis*-aconitylamide of this system is rapidly cleaved when the pH is reduced to 5.5 to exhibit the membrane destabilizing agent PAsp(DET)

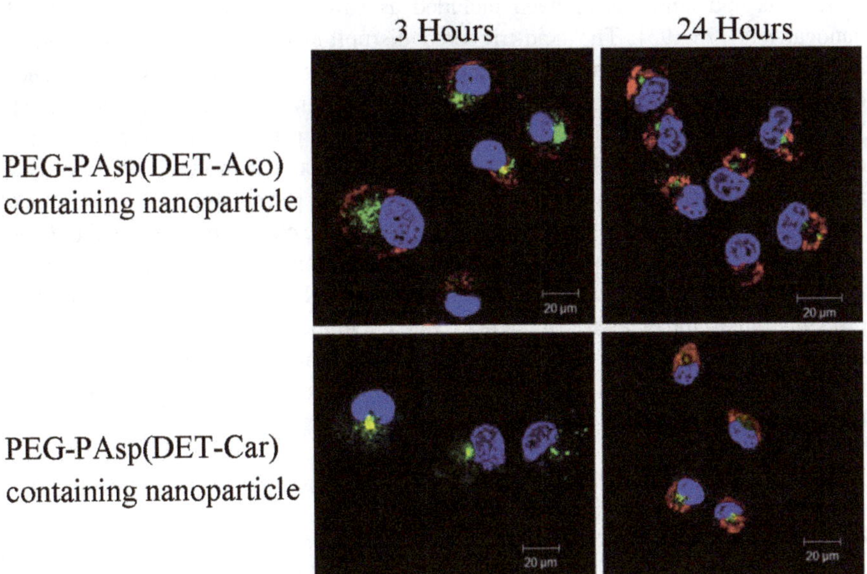

Fig. 8.10 Confocal microscopy images of pancreatic cancer cells (PanC-1) at 3 and 24 h after addition of hybrid particles comprising pH-sensitive PEG-PAsp(DET-Aco) or the non-pH-sensitive control PEG-PAsp(DET-Car). *Blue*=Hoescht at 710 nm; *Green*=Lysotracker Green at 488 nm; *Red*=Cy5-labeled siRNA at 633 nm. Reprinted from [53], Copyright 2011, with permission from Elsevier

by these linkages within the cytoplasm. In this regard, the primary amino groups of PLL are advantageous once it is possible to introduce thiol groups [76, 89, 90]. A variety of novel PLL derivatives have been synthesized and investigated [91–93] in an effort to improve the release of nucleic acids by utilizing these linkages for (1) producing micelles with disulfide cross-linked core for higher stability in the bloodstream and (2) take advantage of its reducible characteristics for a controlled intracellular siRNA release.

In work from our group (Matsumoto et al.), iminothiolane-modified poly(ethylene glycol)-*block*-poly(L-lysine) [PEG-*b*-(PLL-IM)] and siRNA were used to prepare core-shell-type PIC micelles with a disulfide cross-linked core [93]. The ~60 nm micelles maintained the structure at physiological ionic strength but were disrupted under reductive conditions due to the cleavage of disulfide cross-links. Interestingly, redox-sensitive PIC micelles presented 100-fold higher siRNA transfection efficacy compared with non-cross-linked PICs prepared from PEG-*b*-PLL. Furthermore, deeper analysis of the mechanisms of PIC micelles formation showed that a high degree of PEG-*b*-PLL modification with 2-iminothiolane resulted in the formation of *N*-substituted 2-iminothiolane structures in the majority of reacted lysine side chains [94]. Although micelles formed with PEG-*b*-PLL(N2IM-IM) and siRNA showed high stability in vitro and in the bloodstream, there was a loss of sensitivity to disulfide reducing conditions which resulted in lower RNAi effect on the cellular level. Thus, reversible micelle stability is critical to achieve high gene silencing and a balanced ratio in the reducible disulfide crosslink is necessary.

Acetals and ketals have been included as stimuli-responsive components of nanocarriers [78, 95]. The acid-triggered disruption of the interaction between siRNA and polymer can be achieved by incorporating the polymers with branches carrying internal acetal/ketal linkages into the polymeric siRNA micelles [78]. Kwon's group has synthesized linear PEI with grafted primary amines through acid degradable ketal linkage [96]. The siRNA is complexed with the grafted primary amine of the ketalized PEI to form PIC micelles under physiological pH conditions. The intracellular release of siRNA is accelerated by the cleavage of the ketal linkage when the nanocarrier is subjected to acidic conditions. Confocal laser scanning microscopy observation confirmed the release of siRNA from the ketalized PEI, probably due to the cleavage of the ketal linkage in the low pH of endosomes/lysosomes. Overall, it is important to note that hydrolysis of acetals generates aldehydes that can react with amines of proteins in the cell and result in toxicity, while hydrolysis of ketals will produce ketones.

8.7 Future Perspectives

The effective therapeutic application of RNAi is highly dependent on the development of nanocarriers for siRNA protection and site-specific delivery. Improved delivery will lead to increased RNAi efficiency and the decrease in any possible adverse effects. Successful in vivo delivery requires nanocarriers capable of performing multiple functions at an appropriate timing during the delivery process. In this regard, a great advantage for synthetic polymers is the possibility of preparing different structures and blocks of polymers capable to perform specific functions. These block copolymers have been utilized to form nanoscale polymeric micelles for siRNA delivery based on previous knowledge obtained for drug and gene delivery systems.

The rational design of novel stimuli-responsive block copolymers for the formation of new polymeric micelles has been shown to be a promising strategy toward effective RNAi. Future directions on polymer synthesis will couple biological information on tissue microenvironment changes and the available library of responsive chemical groups and moieties for the preparation of novel stimuli-responsive block copolymers. Biodegradable functional polymers as well as biocompatible materials may play an important role for the approval in toxicological tests. New options of hydrophilic polymers are also desirable as alternatives for PEG, such as poly(*N*-[2-hydroxypropyl]methacrylamide). Furthermore, although non-targeting system are exploiting the leaky vasculature of tissue for passive accumulation by the EPR effect, targeting systems holds the promise to sharply regulate the biodistribution by active targeting in vivo. Finally, polymeric micelles composed of multifunctional polymers rationally designed to overcome extracellular and intracellular barriers offer the potential to be translated into therapeutics.

The possibility to incorporate siRNA and imaging agents in the same nanocarrier provides new paths for future research. This exciting approach is the design of theranostics, the union of therapeutics and diagnostics to be realized in the future. The strategy will require convenient methods to prepare multifunctional versatile nanocarriers with tunable structure and properties, and synthetic polymer chemistry may play an important role to achieve this.

The feasibility for the clinical translation of polymeric micelles will depend on several factors which include the efficient performance in preclinical tests allied to its safety. A complete preclinical evaluation of nanocarriers can provide robust validation for consistent introduction into clinical tests. Rigorous toxicological assessment and effective pharmacokinetics and biodistribution data are essential.

Much effort has been made to translate RNAi biology into RNAi therapeutics. The following years promise to be crucial for the demonstration of efficacy and safety of polymeric micelles to enable the clinical use of RNAi. With only few years since initial characterization of RNAi, there is still a completely open field for new ideas, approaches, targets, diseases, and nanocarriers systems.

References

1. Duncan R (2003) The dawning era of polymer therapeutics. Nat Rev Drug Discov 2(5):347–360
2. Kataoka K, Kwon GS, Yokoyama M, Okano T, Sakurai Y (1993) Block-copolymer micelles as vehicles for drug delivery. J Control Release 24:119–132

3. Kataoka K, Harada A, Nagasaki Y (2001) Block copolymer micelles for drug delivery: design, characterization and biological significance. Adv Drug Deliv Rev 47:113–131
4. Kabanov AV, Kabanov VA (1998) Interpolyelectrolyte and block ionomer complexes for gene delivery: physicochemical aspects. Adv Drug Deliv Rev 30:49–60
5. Katayose D, Kataoka K (1997) Water-soluble polyion complex associates of DNA and poly(ethylene glycol)–poly(L-lysine) block copolymer. Bioconjug Chem 8:702–707
6. Kataoka K, Togawa H, Harada A, Yasugi K, Matsumoto T, Katayose S (1996) Spontaneous formation of polyion complex micelles with narrow distribution from antisense oligonucleotide and cationic block copolymer in physiological saline. Macromolecules 29(26):8556–8557
7. Grisham J (2000) Inquiries into gene therapy widens. Nat Biotechnol 18:254–255
8. Howard KA, Kjems J (2007) Polycation-based nanoparticle delivery for improved RNA interference therapeutics. Expert Opin Biol Ther 7(12):1811–1822
9. Fire A, Xu S, Montgomery MK, Kostas SA, Driver SE, Mello CC (1998) Potent and specific genetic interference by double-stranded RNA in Caenorhabditis elegans. Nature 391:806–811
10. Elbashir SM, Harborth J, Lendeckel W, Yalcin A, Weber K, Tuschl T (2001) Duplexes of 21-nucleotide RNAs mediate RNA interference in cultured mammalian cells. Nature 411:494–498
11. Hannon GJ (2002) RNA interference. Nature 418:244–251
12. Nellen W, Lichtenstein C (1993) What makes an mRNA anti-sense-itive? Trends Biochem Sci 18:419–423
13. Tavernarakis N, Wang SL, Dorovkov M, Ryazanov A, Driscoll M (2000) Heritable and inducible genetic interference by double-stranded RNA encoded by transgenes. Nat Genet 24:180–183
14. Whitehead KA, Langer R, Anderson DG (2009) Knocking down barriers: advances in siRNA delivery. Nat Rev Drug Discov 8(2):129–138
15. Bartlett DW, Davis ME (2006) Insights into the kinetics of iRNA-mediated gene silencing from live-cell and live-animal bioluminescent imaging. Nucleic Acids Res 34:322–333
16. Alexopoulou L, Holt AC, Medzhitov R, Flavell RA (2001) Recognition of double-stranded RNA and activation of NF-κB by Toll-like receptor 3. Nature 413:732–738
17. Robbins M, Judge A, MacLachlan I (2009) siRNA and innate immunity. Oligonucleotides 19:89–102
18. Samuel-Abraham S, Leonard JN (2010) Staying on message: design principles for controlling nonspecific responses to siRNA. FEBS J 277:4828–4836
19. Khan AA, Betel D, Miller ML, Sander C, Leslie CS, Marks DS (2009) Transfection of small RNAs globally perturbs gene regulation by endogenous microRNAs. Nat Biotechnol 27(6):549–555
20. Koller E, Propp S, Murray H, Lima W, Bhat B, Prakash TP, Allerson CR, Swayze EE, Marcusson EG, Dean NM (2006) Competition for RISC binding predicts in vitro potency of siRNA. Nucleic Acids Res 34(16):4467–4476
21. Shim MS, Kwon YJ (2010) Efficient and targeted delivery of siRNA in vivo. FEBS J 277(23):4814–4827
22. Matsumura Y, Maeda H (1986) A new concept for macromolecular therapeutics in cancer chemotherapy: mechanism of tumoritropic accumulation of proteins and the antitumor agent smancs. Cancer Res 46:6387–6392
23. Maeda H, Wu J, Sawa T, Matsumura Y, Hori K (2000) Tumor vascular permeability and the EPR effect in macromolecular therapeutics: a review. J Control Release 65:271–284
24. Choi HS, Liu W, Misra P, Tanaka E, Zimmer JP, Itty Ipe B, Bawendi MG, Frangioni JV (2007) Renal clearance of quantum dots. Nat Biotechnol 25:1165–1170
25. Alexis F, Pridgen E, Molnar LK, Farokhzad OC (2008) Factors affecting the clearance and biodistribution of polymeric nanoparticles. Mol Pharm 5:505–515
26. Joralemon MJ, McRae S, Emrick T (2010) PEGylated polymers for medicine: from conjugation to self-assembled systems. Chem Commun (Camb) 46(9):1377–1393

27. Gref R, Luck M, Quellec P, Marchand M, Dellacherie E, Harnisch S, Blunk T, Muller RH (2000) 'Stealth' corona-core nanoparticles surface modified by polyethylene glycol (PEG): influences of the corona (PEG chain length and surface density) and of the core composition on phagocytic uptake and plasma protein adsorption. Colloids Surf B Biointerfaces 18:301–313
28. Harada A, Kataoka K (1995) Formation of polyion complex micelles in aqueous milieu from a pair of oppositely-charged block copolymers with poly(ethylene glycol) segments. Macromolecules 28:5294–5299
29. Yokoyama M, Inoue S, Kataoka K, Yui N, Okano T, Sakurai Y (1989) Molecular design for missile drug-synthesis of adriamycin conjugated with immunoglobulin-G using poly(ethylene glycol)-block-poly (aspartic acid) as intermediate carrier. Makromol Chem 190:2041–2054
30. Kataoka K, Ishihara A, Harada A, Miyazaki H (1998) Effect of the secondary structure of poly(L-lysine) segments on the micellization in aqueous milieu of poly (ethylene glycol)–poly(L-lysine) block copolymer partially substituted with a hydrocinnamoyl group at the Nepsilon-position. Macromolecules 31:6071–6076
31. Oishi M, Nagasaki Y, Itaka K, Nishiyama N, Kataoka K (2005) Lactosylated poly(ethylene glycol)-siRNA conjugate through acid-labile beta-thiopropionate linkage to construct pH-sensitive polyion complex micelles achieving enhanced gene silencing in hepatoma cells. J Am Chem Soc 127(6):1624–1625
32. Kataoka K, Harada A, Wakebayashi D, Nagasaki Y (1999) Polyion complex micelles with reactive aldehyde groups on their surface from plasmid DNA and end functionalized charged block copolymers. Macromolecules 32:6892–6894
33. Schiffelers RM, Ansari A, Xu J, Zhou Q, Tang Q, Storm G, Molema G, Lu PY, Scaria PV, Woodle MC (2004) Cancer siRNA therapy by tumor selective delivery with ligand-targeted sterically stabilized nanoparticle. Nucleic Acids Res 32(19):e149
34. Oba M, Fukushima S, Kanayama N, Aoyagi K, Nishiyama N, Koyama H, Kataoka K (2007) Cyclic RGD peptide-conjugated polyplex micelles as targetable gene delivery system directed to cells possessing $\alpha_v\beta_3$ and $\alpha_v\beta_5$ integrins. Bioconjug Chem 18:1415–1423
35. Davis ME (2009) The first targeted delivery of siRNA in humans via a self-assembling, cyclodextrin polymer-based nanoparticle: from concept to clinic. Mol Pharm 6(3):659–668
36. Davis ME, Zuckerman JE, Choi CH, Seligson D, Tolcher A, Alabi CA, Yen Y, Heidel JD, Ribas A (2010) Evidence of RNAi in humans from systemically administered siRNA via targeted nanoparticles. Nature 464(7291):1067–1070
37. Park K, Lee MY, Kim KS, Hahn SK (2010) Target specific tumor treatment by VEGF siRNA complexed with reducible polyethyleneimine-hyaluronic acid conjugate. Biomaterials 31:5258–5265
38. Nagasaki Y, Yasugi K, Yamamoto Y, Harada A, Kataoka K (2001) Sugar-installed block copolymer micelles: their preparation and specific interaction with lectin molecules. Biomacromolecules 2:1067–1070
39. Bae Y, Jang W-D, Nishiyama N, Fukushima S, Kataoka K (2005) Multifunctional polymeric micelles with folate-mediated cancer cell targeting and pH-triggered drug releasing properties for active intracellular drug delivery. Mol Biosyst 1:242–250
40. Ringsdorf H (1975) Structure and properties of pharmacologically active polymers. J Polym Sci Symp 51:135–153
41. Lloyd JB, Duncan R, Kopecek J (1986) Synthetic polymers as carriers for chemotherapeutic agents. Biochem Soc Trans 14(2):391–392
42. Yokoyama M, Inoue S, Kataoka K, Yui N, Sakurai Y (1987) Preparation of adriamycin-conjugated poly(ethylene glycol)-poly(aspartic acid) block copolymer. Makromol Chem 8:431–435
43. Plestil J, Baldrian J (1973) Small-angle X-ray scattering of the block copolymer polystyrene/polybutadiene/polystyrene in ethyl methyl ketone. Makromol Chem 174:183–191
44. Kabanov AV, Chekhonin VP, Alakhov VY, Batrakova EV, Lebedev AS, Melik-Nubarov NS, Arzhakov SA, Levashov AV, Morozov GV, Severin ES, Kabanov VA (1989) The neuroleptic activity of haloperidol increases after its solubilization in surfactant micelles. FEBS Lett 258:343–345

45. Yokoyama M, Miyauchi M, Yamada N, Okano T, Sakurai Y, Kataoka K, Inoue S (1990) Polymer micelles as novel drug carrier: adriamycin-conjugated poly(ethylene glycol)–poly(aspartic acid) block copolymer. J Control Release 11:269–278
46. Yokoyama M, Miyauchi M, Yamada N, Okano T, Sakurai Y, Kataoka K, Inoue S (1990) Characterization and anticancer activity of the micelle forming polymeric anti-cancer drug adriamycin-conjugated poly(ethylene glycol)–poly(aspartic acid) block copolymer. Cancer Res 50:1693–1700
47. Harada A, Kataoka K (2006) Supramolecular assemblies of block copolymers in aqueous media as nanocontainers relevant to biological applications. Prog Polym Sci 31(11):949–982
48. Bader HR, Ringsdorf H, Schmidt B (1984) Water soluble polymers in medicine. Angew Makromol Chem 123(124):457–485
49. Yokoyama M, Okano T, Sakurai Y, Suwa S, Kataoka K (1996) Introduction of cisplatin into polymeric micelle. J Control Release 39:351–356
50. Nishiyama N, Yokoyama M, Aoyagi T, Okano T, Sakurai Y, Kataoka K (1999) Preparation and characterization of self-assembled polymer–metal complex micelle from cis-dichlorodiammineplatinum(II) and poly (ethylene glycol)–poly(α, β-aspartic acid) block copolymer in an aqueous medium. Langmuir 15:377–383
51. Kakizawa Y, Kataoka K (2002) Block copolymer self-assembly into monodispersive nanoparticles with hybrid core of antisense DNA and calcium phosphate. Langmuir 18:4539–4543
52. Zhang M, Ishii A, Nishiyama N, Matsumoto S, Ishii T, Yamasaki Y, Kataoka K (2009) PEGylated calcium phosphate nanocomposites as smart environment-sensitive carriers for siRNA delivery. Adv Mater 21(34):3520–3525
53. Pittella F, Zhang M, Lee Y, Kim HJ, Tockary T, Osada K, Ishii T, Miyata K, Nishiyama N, Kataoka K (2011) Enhanced endosomal escape of siRNA-incorporating hybrid nanoparticles from calcium phosphate and PEG-block charge-conversional polymer for efficient gene knockdown with negligible cytotoxicity. Biomaterials 32:3106–3114
54. Giger EV, Puigmartí-Luis J, Schlatter R, Castagner B, Dittrich PS, Leroux JC (2011) Gene delivery with bisphosphonate-stabilized calcium phosphate nanoparticles. J Control Release 150(1):87–93
55. Venkatesan P, Cheng Y, Kahne D (1994) Hydrogen bonding in micelle formation. J Am Chem Soc 116(15):6955–6956
56. Kabanov AV, Bronich TK, Kabanov VA, Yu K, Eisenberg A (1996) Soluble stoichiometric complexes from poly(N-ethyl-4-vinylpyridinium) cations and poly(ethylene-oxide)-block-polymethacrylate anions. Macromolecules 29:6797–6802
57. Harada A, Kataoka K (1999) Chain length recognition: core-shell supramolecular assembly from oppositely charged block copolymers. Science 283:65–67
58. Hofs B, Voets IK, de Keizer A, Stuart MAC (2006) Comparison of complex coacervate core micelles from two diblock copolymers or a single diblock copolymer with a polyelectrolyte. Phys Chem Chem Phys 8(36):4242–4251
59. Gary DJ, Puri N, Won Y-Y (2007) Polymer-based siRNA delivery: perspectives on the fundamental and phenomenological distinctions from polymer-based DNA delivery. J Control Release 121:64–73
60. Itaka K, Kanayama N, Nishiyama N, Jang WD, Yamasaki Y, Nakamura K, Kawaguchi H, Kataoka K (2004) Supramolecular nanocarrier of siRNA from PEG-based block catiomer carrying diamine side chain with distinctive pK(a) directed to enhance intracellular gene silencing. J Am Chem Soc 126(42):13612–13613
61. Kim SH, Jeong JH, Lee SH, Kim SW, Park TG (2008) Local and systemic delivery of VEGF siRNA using polyelectrolyte complex micelles for effective treatment of cancer. J Control Release 129:107–116
62. Sun T, Du J, Yan L, Mao H, Wang J (2008) Self-assembled biodegradable micellar nanoparticles of amphiphilic and cationic block copolymer for siRNA delivery. Biomaterials 29: 4348–4355
63. Mao C, Du J, Sun T, Yao Y, Zhang P, Song E, Wang J (2011) A biodegradable amphiphilic and cationic triblock copolymer for the delivery of siRNA targeting the acid ceramidase gene for cancer therapy. Biomaterials 32:3124–3133

64. Kim HJ, Ishii A, Miyata K, Lee Y, Wu S, Oba M, Nishiyama N, Kataoka K (2010) Introduction of stearoyl moieties into a biocompatible cationic polyaspartamide derivative, PAsp(DET), with endosomal escaping function for enhanced siRNA-mediated gene knockdown. J Control Release 145:141–148
65. Kim HJ, Oba M, Pittella F, Nomoto T, Cabral H, Matsumoto Y, Miyata K, Nishiyama N, Kataoka K (2012) PEG-detachable cationic polyaspartamide derivatives bearing stearoyl moieties for systemic siRNA delivery toward subcutaneous BxPC3 pancreatic tumor. J Drug Target 20:33–42
66. Chowdhury EH (2008) Self-assembly of DNA and cell-adhesive proteins onto pH-sensitive inorganic crystals for precise and efficient transgene delivery. Curr Pharm Des 14(22): 2212–2228
67. Sokolova VV, Radtke I, Heumann R, Epple M (2006) Effective transfection of cells with multi-shell calcium phosphate DNA nanoparticles. Biomaterials 27:3147–3153
68. Kakizawa Y, Miyata K, Furukawa S, Kataoka K (2004) Size-controlled formation of a calcium phosphate-based organic-inorganic hybrid vector for gene delivery using poly(ethylene glycol)-block-poly(aspartic acid). Adv Mater 16(8):699–702
69. Kakizawa Y, Furukawa S, Kataoka K (2004) Block copolymer-coated calcium phosphate nanoparticles sensing intracellular environment for oligodeoxynucleotide and siRNA delivery. J Control Release 97(2):345–356
70. Kakizawa Y, Furukawa S, Ishii A, Kataoka K (2006) Organic–inorganic hybrid-nanocarrier of siRNA constructing through the self-assembly of calcium phosphate and PEG-based block aniomer. J Control Release 111:368–370
71. Zhang M, Kataoka K (2009) Nano-structured composites based on calciumphosphate for cellular delivery of therapeutic and diagnostic agents. Nano Today 4:508–517
72. Bae Y, Fukushima S, Harada A, Kataoka K (2003) Design of environment-sensitive supramolecular assemblies for intracellular drug delivery: polymeric micelles that are responsive to intracellular pH change. Angew Chem Int Ed 42:4640–4643
73. Bae Y, Nishiyama N, Fukushima S, Koyama H, Matsumura Y, Kataoka K (2005) Preparation and biological characterization of polymeric micelle drug carriers with intracellular pH-triggered drug release property: tumor permeability, controlled subcellular drug distribution, and enhanced in vivo antitumor efficacy. Bioconjug Chem 16:122–130
74. Lee Y, Miyata K, Oba M, Ishii T, Fukushima S, Han M, Koyama H, Nishiyama N, Kataoka K (2008) Charge conversion ternary polyplex with endosomes disruption moiety: a technique for efficient and safe gene delivery. Angew Chem Int Ed Engl 120:5241–5244
75. Kohori F, Sakai K, Aoyagi T, Yokoyama M, Yamato M, Sakurai Y, Okano T (1990) Control of adriamycin cytotoxic activity using thermally responsive polymeric micelles composed of poly(N-isopropylacrylamide-co-N, Ndimethylacrylamide)-b-poly(D, L-lactide). Colloids Surf B Biointerfaces 16:195–205
76. Miyata K, Kakizawa Y, Nishiyama N, Harada A, Yamasaki Y, Koyama H, Kataoka K (2004) Block catiomer polyplexes with regulated densities of charge and disulfide cross-linking directed to enhance gene expression. J Am Chem Soc 126:2355–2361
77. Katayama Y, Sonoda T, Maeda M (2001) A polymer micelle responding to the protein kinase A signal. Macromolecules 34:8569–8573
78. Kwon YJ (2012) Before and after endosomal escape: roles of stimuli-converting siRNA/polymer interactions in determining gene silencing efficiency. Acc Chem Res 45(7):1077–1088
79. Tamura A, Nagasaki Y (2010) Smart siRNA delivery systems based on polymeric nanoassemblies and nanoparticles. Nanomedicine 5(7):1089–1102
80. Miyata K, Nishiyama N, Kataoka K (2012) Rational design of smart supramolecular assemblies for gene delivery: chemical challenges in the creation of artificial viruses. Chem Soc Rev 41:2562–2574
81. Boussif O, Lezoualc'h F, Zanta MA, Mergny MD, Scherman D, Demeneix B, Behr JP (1995) A versatile vector for gene and oligonucleotide transfer into cells in culture and in vivo: polyethylenimine. Proc Natl Acad Sci USA 92:7297–7301

82. Funhoff AM, van Nostrum CF, Koning GA, Schuurmans-Nieuwenbroek NME, Crommelin DJA, Hennink WE (2004) Endosomal escape of polymeric gene delivery complexes is not always enhanced by polymers buffering at low pH. Biomacromolecules 5:32–39
83. Meyer M, Philipp A, Oskuee R, Schmidt C, Wagner E (2008) Breathing life into polycations: functionalization with pH-responsive endosomolytic peptides and polyethylene glycol enables siRNA delivery. J Am Chem Soc 130:3272–3273
84. Philipp A, Zhao X, Tarcha P, Wagner E, Zintchenko A (2009) Hydrophobically modified oligoethylenimines as highly efficient transfection agents for siRNA delivery. Bioconjug Chem 20:2055–2061
85. Miyata K, Oba M, Nakanishi M, Fukushima S, Yamasaki Y, Koyama H, Nishiyama N, Kataoka K (2008) Polyplexes from poly(aspartamide) bearing 1,2-diaminoethane side chains induce pH-selective, endosomal membrane destabilization with amplified transfection and negligible cytotoxicity. J Am Chem Soc 130(48):16287–16294
86. Uchida H, Miyata K, Oba M, Ishii T, Suma T, Itaka K, Nishiyama N, Kataoka K (2011) Odd-even effect of repeating aminoethylene units in the side chain of N-substituted polyaspartamides on gene transfection profiles. J Am Chem Soc 133(39):15524–15532
87. Suma T, Miyata K, Ishii T, Uchida S, Uchida H, Itaka K, Nishiyama N, Kataoka K (2012) Enhanced stability and gene silencing ability of siRNA-loaded polyion complexes formulated from polyaspartamide derivatives with a repetitive array of amino groups in the side chain. Biomaterials 33(9):2770–2779
88. Convertine AJ, Diab C, Prieve M, Paschal A, Hoffman AS, Johnson PH, Stayton PS (2010) pH-responsive polymeric micelle carriers for siRNA drugs. Biomacromolecules 11: 2904–2911
89. Kakizawa Y, Harada A, Kataoka K (1999) Environment-sensitive stabilization of core-shell structured polyion complex micelle by reversible cross-linking of the core through disulfide bond. J Am Chem Soc 121(48):11247–11248
90. Kakizawa Y, Harada A, Kataoka K (2001) Glutathione-sensitive stabilization of block copolymer micelles composed of antisense DNA and thiolated poly(ethylene glycol)-block-poly(L-lysine): a potential carrier for systemic delivery of antisense DNA. Biomacromolecules 2:491–497
91. Inoue Y, Kurihara R, Tsuchida A, Hasegawa M, Nagashima T, Mori T, Niidome T, Katayama Y, Okitsu O (2008) Efficient delivery of siRNA using dendritic poly(L-lysine) for loss-of-function analysis. J Control Release 126:59–66
92. Watanabe K, Harada-Shiba M, Suzuki A, Gokuden R, Kurihara R, Sugao Y, Mori T, Katayama Y, Niidome T (2009) In vivo siRNA delivery with dendritic poly(L-lysine) for the treatment of hypercholesterolemia. Mol Biosyst 5:1306–1310
93. Matsumoto S, Christie RJ, Nishiyama N, Miyata K, Ishii A, Oba M, Koyama H, Yamasaki Y, Kataoka K (2009) Environment-responsive block copolymer micelles with a disulfide cross-linked core for enhanced siRNA delivery. Biomacromolecules 10(1):119–127
94. Christie RJ, Miyata K, Matsumoto Y, Nomoto T, Menasco D, Lai TC, Pennisi M, Osada K, Fukushima S, Nishiyama N, Yamasaki Y, Kataoka K (2011) Effect of polymer structure on micelles formed between siRNA and cationic block copolymer comprising thiols and amidines. Biomacromolecules 12:3174–3185
95. Gillies ER, Goodwin AP, Frechet JMJ (2004) Acetals as pH-sensitive linkages for drug delivery. Bioconjug Chem 15:1254–1263
96. Shim MS, Kwon YJ (2009) Acid-responsive linear polyethylenimine for efficient, specific, and biocompatible siRNA delivery. Bioconjug Chem 20(3):488–499

Chapter 9
RNAi Therapeutic Delivery by Exosomes

Samira Lakhal, Samir El Andaloussi, Aisling J. O'Loughlin, Jinghuan Li, and Matthew M.J. Wood

Abstract Since the discovery of short interfering RNAs (siRNAs), their potential as a therapeutic platform has been widely recognized. However, clinical translation has been stalled by inefficient delivery in vivo. While some success has been achieved with cationic lipids and lipid-like materials for therapeutic RNAi delivery to liver, delivery across the blood–brain barrier (BBB) to the central nervous system for the treatment of neurological disorders such as Parkinson's, Alzheimer's, and Huntington's disease remains a challenge. To address the problem of inefficient delivery across the BBB, our laboratory exploited one of nature's mechanisms for intercellular communication, named exosomes. They are a class of membrane vesicles derived from the endolysosomal compartment implicated in cell–cell communication by shuttling various proteins, lipids, and RNAs between cells. We have developed a method to target exosomes with brain-specific peptides and subsequently load them with siRNA for targeted delivery to brain. This chapter aims at providing an insight into membrane vesicle-mediated RNA delivery and how these vectors can be utilized for RNAi therapy.

9.1 Introduction

RNAi has revolutionized molecular biology by allowing silencing of virtually any gene. A key mediator of RNAi is the 19–21 nucleotide duplex RNA (siRNA) that associate with the RNA-inducing silencing complex (RISC) in the cytoplasm and direct the highly specific cleavage of complementary mRNAs [1, 2]. Since the initial discovery in 1998 by Fire and Mello [3], siRNA have been exploited both as a

---

Samira Lakhal and Samir El Andaloussi have equally contributed.

S. Lakhal • S.E. Andaloussi • A.J. O'Loughlin • J. Li • M.M.J. Wood (✉)
Department of Physiology, Anatomy and Genetics, University of Oxford, Oxford, UK
e-mail: matthew.wood@dpag.ox.ac.uk

K.A. Howard (ed.), *RNA Interference from Biology to Therapeutics*,
Advances in Delivery Science and Technology, DOI 10.1007/978-1-4614-4744-3_9,

basic research tool to investigate gene function and as potential therapeutics to silence disease-causing genes (reviewed in [4]). In contrast to conventional single-stranded antisense oligonucleotides (ASOs), siRNAs are considered significantly more potent since only few siRNA effectors are needed to promote complete RNAi in a given cell. Furthermore, improved sequence design and incorporation of chemically modified nucleotides at selected positions within siRNA duplexes have dramatically improved the potency further. However, the large size and hydrophilic nature of siRNAs make them essentially impermeable for cells. For that reason, various delivery vectors, some of which were initially developed for plasmids and ASOs, have been used to improve the bioavailability of siRNAs. The majority of these are based on polycations or cationic lipids that form nanocomplexes with siRNAs that are subsequently taken up by cells through endocytosis [4–6]. In addition to cationic liposomes, a class of peptide-based vectors referred to as cell-penetrating peptides (CPPs) have been utilized for nanoparticle formation with siRNA with subsequent RNAi induction in various cell types in vitro as well as in vivo following systemic delivery [7–9].

Different vectors have been developed that hold clinical potential, in particular, for RNAi-based treatment of liver-related malignancies [6, 10]; however, only a very small number of mechanistic-based treatments against CNS diseases exist in the clinic, compared to diseases of other biological tissues. Major advances have been made in the last decade in understanding the molecular mechanisms, genetic basis, and pathogenesis underlying the pathology of major neurological disorders, including Huntington's disease, Parkinson's disease, Alzheimer's disease, spinocerebellar ataxias, stroke, and brain cancers. However, this advanced understanding of disease and identification of therapeutic targets in the context of CNS pathology have not resulted in successful development of CNS-targeting therapies, due primarily to the impermeability of the BBB to most therapeutic agents. There is an urgent need, therefore, for the development of novel approaches for efficient and safe delivery of macromolecular cargoes across the BBB. One plausible strategy would be to exploit one of nature's own cell-to-cell information transmitters such as membrane vesicles, in particular exosomes.

This chapter provides an overview of existing methods for delivery of therapeutics to brain, with focus on siRNA, and the challenges associated with delivery across the BBB. Furthermore, membrane vesicles and their role in exchange of genetic information between different cells will be discussed with emphasis on endogenous exosome-mediated mRNA and microRNA (miRNA) transport and how this system could be exploited for therapeutic RNAi delivery to the brain.

9.2 Drug Delivery to Brain

9.2.1 The Blood–Brain Barrier

The BBB is a tightly regulated biological barrier composed of several cell types that cooperatively act to separate the blood from the brain parenchyma. The neurovascular unit consists of endothelial cells, basement membrane, pericytes, astrocytes, and microglial cells [11]. The brain endothelial cells are the central cells in the BBB, and these are tightly interconnected through complexes of tight and adherence junctions that form an impermeable barrier for almost all substances, even small hydrophilic molecules. Furthermore, these cells display remarkably low levels of transcytosis, the process of vesicular trafficking of macromolecules across cells, yet express specific receptors for transport of vital nutrients for the brain parenchyma [12]. In addition, endothelial cells of the BBB express high levels of drug efflux proteins that protect the brain from toxic substances [13]. Pericytes, which line the entire surface of the BBB, were thought to only provide structural support to the vasculature but have recently also been identified as key regulators of vesicular trafficking in neighboring endothelial cells [14, 15]. The two glial cell types have distinct functions in the BBB; astrocytes display extensions, called end feet, which cover the entire cerebral vasculature and are thought to control endothelial junctions as well as regulate water homeostasis at the BBB [16], whereas microglial cells are the resident immune cells of the CNS [17]. Finally, the basement membrane is the extracellular matrix that connects the above-mentioned cells and provides structural support. In conclusion, the BBB is a tightly regulated biological barrier that protects the brain from potentially harmful substances while allowing highly regulated uptake of vital nutrients. In light of this, it is not surprising that only few drugs are active in the brain following systemic delivery, and it has actually been estimated that 98% of all small molecules do not cross the BBB.

9.2.2 Drug Delivery to Brain

As aforementioned, there is no paracellular pathway of solute exchange between blood and brain, and there is minimal endocytosis, and consequently no significant transcellular pathway for free solute exchange. Basically, there are only two known mechanisms by which molecules in the blood can get access to brain interstitial fluid (1) lipid-mediated transport of lipophilic substances below 400 Da and (2) catalyzed transport, which includes carrier-mediated transport for small molecules such as glucose and amino acids and receptor-mediated transport (RMT) for large molecules such as insulin and transferrin [11].

Several approaches have been attempted for the delivery of macromolecular drugs across the BBB. These include local invasive delivery by direct injection or infusion (e.g., nerve growth factor) [18], the olfactory route (e.g., insulin) [19], and nonselective osmotic or biochemical opening of the BBB [20, 21]. These routes of delivery have all been hindered by several setbacks, including low efficacy of transport across the BBB and serious limiting safety issues associated with neurosurgical intervention or with nonselective permeation of the BBB.

Only a few successful systems have been reported for noninvasive systemic delivery to brain. All of these utilize the RMT system for targeted delivery. Exploiting RMT for brain delivery of macromolecules has been extensively studied by the group of Pardridge [22, 23]. This has been accomplished by utilizing chimeric monoclonal antibodies (mAb), specific for insulin or transferrin receptors. By recombinantly expressing various neuroprotective proteins, or segments of proteins, as fusion proteins to the carboxyl terminus of the heavy chain of these mAbs, it has been shown that the brain uptake increases up to 50- to 100-fold. This "Trojan horse" concept has been exploited for the delivery of various macromolecular protein compounds including erythropoietin, TNF-α receptor decoys, aβ-amyloid antibodies, and GDNF to confer neuroprotection in various neuronal disorders including Parkinson's disease, Alzheimer's disease, and stroke [24–27]. Albeit a relatively efficient strategy for BBB delivery of therapeutic proteins, this technology has major limitations in several ways. First, this delivery system relies on cumbersome expression and purification of fusion proteins that are needed in high quantities in order to achieve activity, and, second, these large fusion proteins are novel antigens and hence are potentially immunogenic upon repeated systemic administration.

9.2.3 RNAi Delivery Across the BBB

To our knowledge, there are only two synthetic nanoparticle-based systems reported to convey nucleic acids across the BBB to date. One exploits the previously described Trojan horse strategy by coating synthetic, nucleic acid-loaded liposomes with BBB-homing antibodies [23], and the second system makes use of a chimeric peptide comprising a polyarginine sequence for siRNA condensation and a rabies virus glycoprotein (RVG) peptide sequence for BBB targeting [28]. RVG is a 29-mer peptide derived from the rabies virus glycoprotein that specifically binds to acetylcholine receptors expressed on neuroendothelial cells as well as neuronal cells and allows transvascular delivery to the nervous system [29]. Using the RVG-R9 chimeric peptide complexed with siRNA, the authors observed a strong RNAi response in neuronal cells in vitro, and following systemic injection into mice, siRNAs localized to neuronal cells that resulted in specific gene silencing within the brain. Furthermore, intravenous treatment with RVG-R9-bound antiviral siRNA mediated robust protection against fatal viral encephalitis in mice [28].

Despite providing an efficient means of reaching the CNS, these systems are hindered by the same major drawbacks as other polycation and liposome-based nanoparticle vehicles. The fact that they are entirely synthetic makes them potentially immunogenic and kidney-, liver-, and lung-related toxicities have been reported for these and related delivery modalities [30, 31]. Thus, a major scientific challenge is to find enabling delivery technologies with a greatly enhanced ability to target and traverse the BBB following systemic administration in a safe, nontoxic manner and without triggering organ toxicity or immunological response upon repeated administration. The exploitation of natural biological nanoparticles such as exosomes could fulfill these criteria and provide a paradigm shift in technology for BBB penetration and targeted macromolecular drug delivery to brain.

9.3 The Origin and Biology of Membrane Vesicles

Transmission of information between cells in the body occurs through several mechanisms. Communication can occur through (1) secreted growth factors, cytokines, or other small molecules including nucleotides and bioactive lipids; (2) cell-to-cell adhesions mediated by specialized adhesion molecules; and (3) tunneling nanotubes that interconnect cells. In addition, cells can communicate through membrane transfer by the secretion of membrane vesicles, a mechanism that for years was largely overlooked. Historically, when viewed under a microscope, small membrane-bound vesicles were thought to be a result of cell damage or death. Recently, release of these vesicles has been increasingly recognized as a novel, universally conserved mode of intercellular communication with a role in many physiological and pathological processes. Such vesicles increase the complexity of cell signaling by transmitting important information via the proteins, lipids, and nucleic acids they contain, which can alter signaling in any recipient cell [32, 33].

Currently, this heterogeneous group of intercellular vesicle messengers is divided into subgroups based on their biogenesis, biophysical properties, and functions, although they share some common characteristics of an internal microenvironment protected from the extracellular milieu by a lipid bilayer and retain the same membrane topology as the cell of origin. A precise definition and nomenclature of the diverse group of membrane vesicles is lacking as current understanding is limited by the lack of supporting technologies necessary for their accurate isolation and characterization. This review will use the term membrane vesicles to refer collectively to these membrane-derived particles with a role in intercellular communication. Depending on the location of cell origin, membrane vesicles can be divided into three broad categories (Fig. 9.1).

9.3.1 Microvesicles

Microvesicles (MVs) are a group of membrane vesicles secreted by budding or shedding from the plasma membrane of most cell types. They usually have a diameter between 100 and 1,000 nm [33]. In the literature, they are also referred to as microparticles, ectosomes, shedding vesicles, and exovesicles depending on the cell type they originate from. Their release can be stimulated by calcium, upon ligand binding and activation, or stress either physical, e.g., shear stress, or chemical, e.g., hypoxia [34, 35]. Calcium can modify the asymmetric phospholipid distribution of the plasma membrane into cholesterol-rich microdomains by inhibiting translocase and activating scramblase causing phosphatidylserine to translocate from the inner facing membrane to be exposed on the outer at sites of MV shedding [36]. Small cytoplasmic protrusions bud from the plasma membrane due to curvature-mediated lateral redistribution of membrane components at these microdomains; a calcium influx activates calpain which alters the cytoskeletal structure causing the protrusions

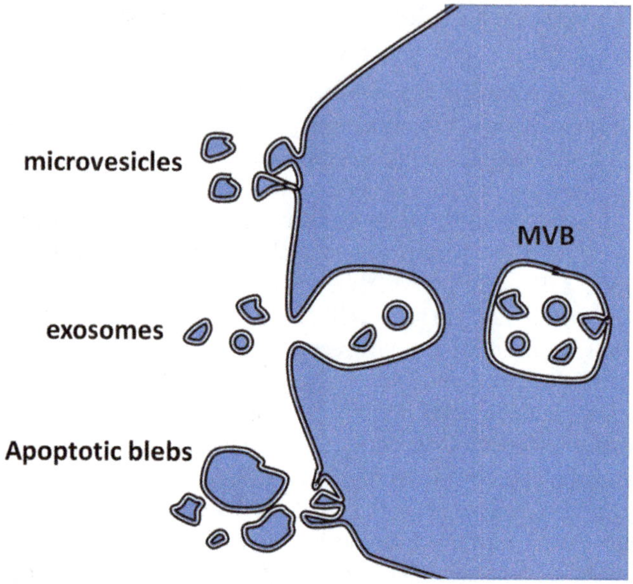

Fig. 9.1 Cellular origins of membrane vesicles. Diagram summarizing the cellular origin of membrane vesicles. Microvesicles arise from budding of the cell membrane. Exosomes are released from MVBs, while apoptotic blebs are released from the disintegrating membranes of apoptotic cells

to detach from cellular actin ending with fission of the stalk [33]. This shedding is not passive and requires energy input, RNA synthesis, and protein translation [37]. The resultant microvesicles contain cytosol and an extracellular membrane surface in the same orientation and topology as that of their donor cells so that their signaling properties reflect those of their cell of origin [38]. The membrane composition undergoes remodeling during the formation of these vesicles.

9.3.2 Apoptotic Blebs

Apoptotic blebs/bodies are released directly from the plasma membrane when it buds from disintegrating apoptotic cells. During the process of programmed cell death, there is an alteration in cell morphology, shrinkage of the cytoplasm, and nuclear condensation, and the final stage is cell fragmentation into apoptotic blebs, which enclose remnants of the shrinking cell. The number of apoptotic blebs correlates with the degree of apoptosis in cultured cells [39]. These are irregular in shape, and their size can vary between as low as 50 nm to a controversial maximum size of 500 nm [40] to 5 μm [41]. Phosphatidylserine is externalized as in MVs which encourages phagocytic recognition by other cells [42]. Although sharing some features with MVs, apoptotic blebs, in contrast to MVs, can contain cytosolic organelles and possible nuclear fragments [43].

Table 9.1 Characteristics of different membrane vesicles

	Microvesicles	Apoptotic blebs	Exosomes
Size	100–1,000 nm	50–5,000 nm	40–100 nm
Biogenesis	Released from budding of plasma membrane	Released as blebs from apoptotic cells	Released from MVBs
Methods for isolation	Centrifugation at 18,000–20,000 × g	Captured from cultures of apoptotic cells. Centrifugation at 1,200, 10,000 or 100,000 × g	Centrifugation at 100,000 × g, sucrose gradient
Methods for detection	Flow cytometry, capture-based assays	Flow cytometry	EM, western blotting, flow cytometry, latex bead flow cytometry
Markers	Integrins, selectins, CD40, tissue factor, and cell-specific markers	Contain DNA, histones	CD63, CD9, CD81, LAMP1, and TSG101
Characteristics	Most expose phosphatidylserine	Phosphatidylserine exposed	Membranes rich in ceramide, cholesterol and sphingomyelin, lipid rafts, phosphatidylserine exposed

Overall, there are unclear boundaries between the different types of membrane vesicles due to a lack of biochemical and biophysical validation among membrane vesicles and particularly among microvesicles where some research groups recognize different subtypes. Exosomes do not contain nuclear-, mitochondrial-, Golgi-, or endoplasmic reticulum-derived protein, while MV protein composition reflects that of the parent cell. Both contain mRNA and miRNA and can alter signaling in a recipient cell

9.3.3 Exosomes

Exosomes are small vesicles (40–100 nm) released from cells upon fusion of a multivesicular body (MVB) containing intraluminal vesicles (ILV) with the plasma membrane. MVBs form by inward budding of the endosomal membrane, generating multiple ILVs. Exosomes are released from most cell types in culture and are found in vivo in various biological fluids [44]. Exosomes contain an evolutionarily conserved set of protein families which can be used as markers, including CD63, CD9, Alix, and LAMP1. They also bear a set of proteins which reflect their source. Once considered to be merely transitional compartments destined for degradation of redundant cell constituents by the lysosome, it is now recognized that they are an important specific subcellular compartment described in mammals and invertebrates with an important role in both physiological and pathological processes. Table 9.1 provides an overview of the characteristics of the different membrane vesicles.

9.3.3.1 Exosome Biogenesis and Biological Role

Exosome release is part of the endosomal trafficking system, a mechanism by which the cell sorts incoming molecules for recycling, storing, or degradation by transportation to the lysosome. Endocytic vesicles form at the plasma membrane via different endocytic mechanisms and fuse to form the early endosome. This structure receives the endocytosed cargo from the plasma membrane and sorts it. Early endosomes survey the peripheral cytoplasm of the cell via salutatory movement along microtubules [45, 46].

Late endosomes are situated close to the nucleus and develop from early endosomes by acidification, altered protein content, and fusion of vesicles [47]. The limiting membrane, which is rich in LAMP1 to protect the membrane from acid hydrolases, continuously buds into the lumen trapping cytosol to enrich the population of ILVs [48]. Upon fusion with a lysosome, the endocytosed cargo is hydrolyzed, or on fusion with the plasma membrane, the ILVs are released as exosomes. A mechanism for selective sorting of membrane-associated cargo for degradation in lysosomes versus exportation from the cell in an exosome is regulated by the endosomal sorting complex required for transport (ESCRT) complex which recognizes and sequesters mono-ubiquitinated proteins in the endosomal membranes [49]. There is also an ESCRT-independent sorting mechanism involving lipid raft microdomains laterally segregating cargo with the involvement of ceramide that promotes vesicle budding [50].

The fate of exosomes after binding to their target cells remains to be fully elucidated, but they can activate cell surface receptors on a target cell or fuse with the target cell to transfer bioactive exosomal proteins and RNA or undergo internalization by endocytosis [51]. MVs and exosomes can signal at a distance and transfer proteins and mRNA horizontally, allowing them to coordinate processes such as inflammation and coagulation whereby their phosphatidylserine-enriched membranes act as a surface for the assembly of clotting factors [52] and contribute to antigen presentation as exosomes from antigen-presenting cells have functional MHC complexes on their surface [53]. They can regulate immune responses by stimulating or inhibiting T cells [54]. They are also implicated in the pathogenesis of different diseases. For example, exosomes derived from cancer cells can themselves be tumorigenic and are rich in metalloproteinases which degrade the extracellular matrix to facilitate the growth and spread of a tumor [55, 56]. The release of membrane vesicles from tumor cells correlates with poor prognosis [57]. MVs and exosomes are also involved in the spread of infectious agents including viruses such as HIV and prions [58, 59].

9.3.3.2 Exosomes as Biomarkers and Therapeutic Tools

Exosomes are being increasingly recognized as potential biomarkers for different disorders. They are easily obtainable from biological fluids and contain a unique mRNA and miRNA fingerprint reflective of the cell of origin [60]. Exosome numbers are elevated in different disease states, and the RNA expression profiles in

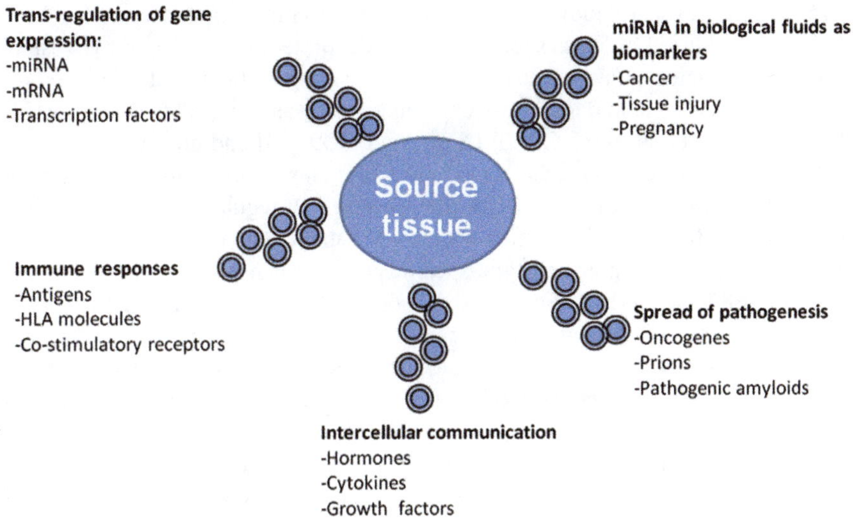

Fig. 9.2 Exosomes in health and disease. Summary of known biological functions of exosomes in trans-regulation of gene expression, immune responses and intercellular communication, their role in the spread of pathogenesis, and their exploitation as biological disease biomarkers

these circulating exosomes have been proposed to indicate survival and therapeutic outcome in human cancers [61].

As a natural biological product, exosomes hold great potential as a therapeutic tool. Exosomes from peritoneal cavity fluid in cancer patients can induce lysis of tumor cells as they prime T lymphocytes via an MHC I-dependent pathway involving dendritic cells and stimulate antigen-specific cytotoxic T cell responses [62]. Dendritic cell-derived exosomes loaded with antigenic peptides from human melanoma have been used as a cancer vaccine in a phase I clinical trial supporting their safety [63]. In addition, exosomes hold great promise as an alternative to stem cell treatment for cardiovascular disorders as injections of exosomes derived from mesenchymal stem cells are capable of ameliorating symptoms following myocardial infarction–reperfusion injury [64]. Direct exosome delivery could eliminate any safety and technical issues associated with stem cell transplantation. Figure 9.2 is a schematic diagram summarizing the known biological functions of exosomes in health and disease, as well as their therapeutic and diagnostic applications.

9.3.4 Membrane Vesicles as Natural Carriers of Protein and RNA

A variety of classes of protein and RNA have been identified in MVs/exosomes from various cell types and body fluids. An overview of identified molecules present in exosomes is provided in the ExoCarta database, an exosomal protein and RNA database (http://exocarta.ludwig.edu.au). The protein content of exosomes has been extensively analyzed by mass spectrometry-based proteomic studies or by western blotting. The content of exosomes may vary depending on their source. Some proteins are relatively abundant in all exosomes, such as Alix, TSG101, LAMP1, HSP70 and the tetraspanins CD63, CD81, and CD9 [65] and are thus frequently used as markers for exosome characterization. One class of cytosolic proteins commonly detected in exosomes includes the Rabs, which regulate exosome docking and membrane fusion, and annexins which aid in membrane trafficking and fusion [66]. Exosomes also contain cell-type-specific proteins such as A33, MHC II, CD86, and MFG-E8/lactadherin [66]. Other exosomal proteins include the metabolic enzymes, ribosomal proteins, transmembrane, signal transduction, adhesion, ATPases, cytoskeletal, and ubiquitin molecules [67]. In addition, the lipid composition of exosomes is characteristic of the cell origin and plays a critical role in exosome biogenesis [68].

MVs also contain genetic material in the form of mRNA and microRNA that may allow easy screening for disease genetic markers and offer new diagnostic and prognostic information [69]. The messages transmitted by intercellular communication may include those for survival, growth, division, differentiation, stress responses, apoptosis, etc. In addition, exosomes are present in body fluids such as saliva, urine, breast milk, and blood plasma, which are conveniently available from patients. For example, the discovery of microRNA in human salivary samples suggests a promising use of salivary exosomes as biomarkers for disease diagnosis. Hence, the horizontal transfer of RNA offers a new perspective on intercellular communication and has potential therapeutic applications, for example, in gene delivery [66].

9.4 Membrane Vesicles: Opportunities for RNAi Therapeutic Delivery and Insights into RNAi Biology

The multifunctionality of membrane vesicles in cellular transport and signaling underlies their potential as a valuable therapeutic resource. However, this potential is only just beginning to be unraveled. In this section, we will discuss the unique properties that render exosomes particularly useful in RNAi therapeutics and describe the seminal studies to first exploit exosomes in this context and the advantages of exosomes over existing RNAi delivery strategies. We will also highlight current and foreseeable obstacles in the route to full clinical translation of membrane vesicles in RNAi therapeutics.

9.4.1 Exosomes Possess Unique Properties That Render Them Useful for RNAi Therapeutics

Since they were first described in sheep reticulocytes [70], our knowledge of the biological functions of exosomes has been expanding, fuelling interest both in the potential of exosomes per se as drug delivery vehicles and in the design of novel "exosome-inspired" drug delivery vehicles. In this context, the most relevant feature of exosomes as aforementioned is that they are natural carriers of coding and non-coding RNA including miRNA between cells. Indeed, in 2006, Baj-Krzyworzeka et al. described the presence of tumor cell markers and mRNAs in tumor-derived microvesicles and demonstrated their transfer in vitro into monocytes [71]. Subsequently, in 2007, Valadi et al. discovered the presence of mRNA and miRNA inside exosomes derived from mast cells and suggested that the ability of exosomes to deliver nucleic acids to cells at a distance made them ideal candidates as RNA delivery vehicles [32]. At present, over 1,600 mRNAs and more than 700 miRNAs identified predominantly by microarrays have been reported in 134 studies. The profile of mRNAs observed in exosomes does not necessarily match the composition of the donor cells, suggesting that there is a selective loading of specific mRNA and miRNA molecules into exosomes [72]. Most importantly, it has been shown that exosome-delivered RNAs mediate de novo transcriptional and translational changes in the recipient cells, suggesting that exosomes are able to deliver their RNA cargo to the subcellular compartment appropriate for function [32, 73]. A study of the transfer of miRNA from activated T cells to APCs suggested the existence of immune synapse-mediated exosomal transport, providing evidence for the directed and unidirectional transfer of miRNA-carrying exosomes through the immune synapse between the donor T cells and the recipient APCs [74]. DC-derived exosomes were also shown to transport miRNA to recipient DCs and to consequently influence the regulation of the miRNA targets in recipient DCs [75]. Fluorescently labeled glioblastoma microvesicles have been shown to transfer Gaussia luciferase (Gluc) mRNA to human brain microvascular endothelial cells (HBMVECs), leading to expression of the Gluc protein in the recipient cells, demonstrating de novo translation of the transferred Gluc mRNA in the recipients [76]. In another study, incubation of human mast cells with exosomes derived from mouse cells resulted in the expression of the donor mouse proteins in the recipient human cells [32]. It was also shown that exosomes from cultured monocytes contained components of RNA-induced silencing complex (RISC), such as AGO2 and its interacting partner GW182 [77]. Importantly, mature miRNAs and their target mRNAs also co-localize to MVBs, but not to exosomes, raising the possibility that miRNA-loaded RISC assembly takes place in MVBs, and that exosomes are intrinsically implicated in miRNA-induced silencing, and more intriguingly, that silencing by miRNA may occur in trans in cells that do not express the miRNA (Fig. 9.3). The ability of exosomes to transfer RNA and miRNA between cells and to subsequently mediate changes in gene regulation in the recipient cells has now been demonstrated with a variety of donor and recipient cell types. This, together with their implication in the RISC-induced silencing, highlights their potential as RNAi delivery vehicles and raises the further opportunity for the design of "exosome-inspired" delivery vehicles.

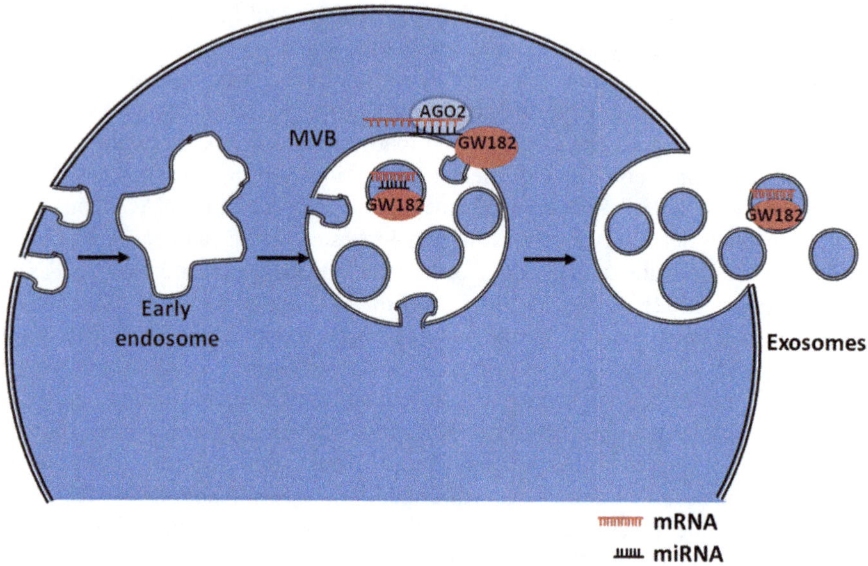

Fig. 9.3 Exosomes and miRNA-mediated silencing. miRNA RISC components associated with the membrane of late endosomes and are incorporated into intraluminal vesicles and then released into exosomes

9.4.2 Exosomes as Drug Delivery Vehicles: Proof-of-Concept Studies

The Wood Laboratory carried out the first study to harness the RNA-transporting capacity of exosomes and use them for delivery of exogenous RNAi in vivo, providing first proof-of-concept for biotechnological exploitation of membrane vesicles [78] (Fig. 9.4).

Immature dendritic cells (DCs) were derived from mouse bone marrow and used as a source of exosomes, as they are devoid of lymphocyte stimulatory molecules such as MHCII, CD80, and CD86. These were subsequently loaded by electroporation with exogenous siRNA for delivery, first in vitro then in vivo. The brain was selected as a target tissue, given the lack of efficient CNS-delivery vectors. To ensure that systemically injected exosomes targeted the brain in vivo and to reduce exosome homing to tissues of drug clearance, a novel targeting strategy was devised, utilizing the exosomal surface protein Lamp2b to display the previously mentioned brain-specific RVG peptide. By this method, we demonstrated specific delivery of siRNA to neurons in the brain, with up to 60% RNA and protein knockdown predominantly in the midbrain, cortex, and striatum. As well as efficient and specific delivery of siRNA, these exosomes produced little or no toxicity or immunogenicity.

A subsequent study by Zhuang et al. described the use of exosomes to deliver anti-inflammatory drugs to the brain through a noninvasive intranasal route [79].

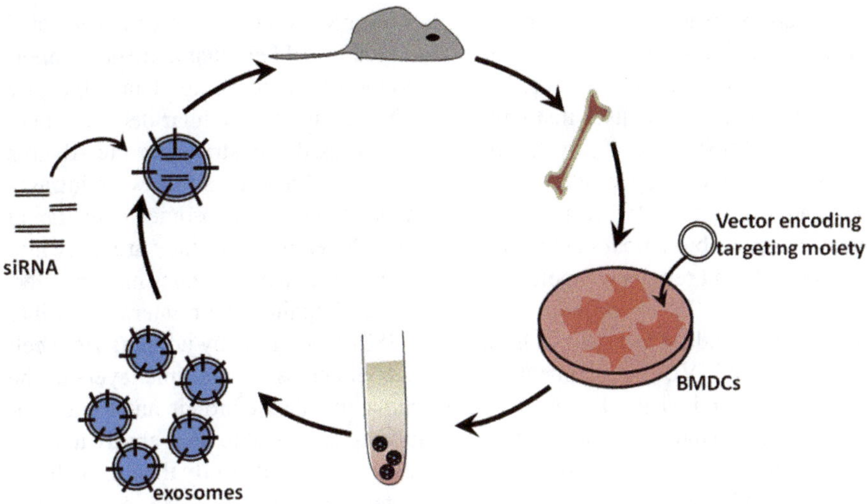

Fig. 9.4 Bone marrow dendritic cell-derived exosomes as siRNA delivery vehicles. BMDCs are generated from bone marrow cultures by differentiation with GM-CSF. BMDCs are transfected with a vector for expression of the targeting moiety (e.g., RVG) in fusion with an exosomal membrane protein (e.g., Lamp2b). Exosomes are harvested from supernatants by ultracentrifugation and loaded with siRNA by electroporation. Targeted and loaded exosomes are then injected into mice intravenously

The therapeutic value of this approach was demonstrated with exosome-complexed curcumin in LPS-induced inflammation and experimental allergic encephalomyelitis (EAE) and with an exosome-complexed stat3-inhibitor in a glioblastoma tumor model. Authors demonstrated that exosomes administered intranasally are potential delivery vehicles for small anti-inflammatory molecules, by increasing their biological stability and enabling them to bypass the BBB. This study represents further proof of concept for membrane vesicles as drug delivery vehicles with fast and selective homing to the brain and highlights the wider potential of membrane vesicles and exosomes in particular beyond oligonucleotide delivery, to include a potentially wider range of therapeutic cargoes.

9.4.3 *Advantages of Exosomes over Existing RNAi Delivery Vehicles*

These studies have also highlighted the advantages of using exosomes over viral vectors and liposomal formulations for RNAi delivery, particularly in vivo. In addition to being naturally derived vesicles, comprised of endogenously expressed proteins, RNAs, and lipids, exosomes appear capable of targeted delivery of exogenous RNAi cargoes to the site of silencing in the recipient cell, a feature that is an extension of their natural function in mRNA and miRNA transport and their association with RISC components. As such, it is likely that a greater proportion of exosome cargo is delivered directly to the site of silencing, than is the case with liposomes and viruses that have no such intrinsic intracellular delivery function. In addition, their ability to cross the BBB, as demonstrated in the Alvarez study, makes them highly attractive for targeting CNS disorders following intravenous administration. While it is not fully understood how exosomes enter target cells and cross biological barriers from the bloodstream, where they are known to be abundant and relatively stable, it has been proposed that exosomes may be internalized into MVBs of recipient cells then released again to be reinternalized into MVBs of secondary recipient cells, and so on [80]. Therefore, by jumping from cell to cell via the MVB compartment, exosomes could cross the multiple layers of the BBB. However, it is also known that their unique membrane protein and lipid composition, not found on liposomes or viral capsules, is critical to their ability to enter target cells. The membranes of exosomes are organized differently than the cellular plasma membrane and are enriched in cholesterol, ganglioside GM3, sphingomyelins, phosphatidylcholines, phosphatidylethanolamines, phosphatidylserines, prophatidylinositols, and lysophosphatidylcholines [81]. As phosphatidylserine is an important docking site for proteins involved in signaling and fusogenesis, phosphatidylserine may be involved in exosome homing and entry into target cells. It has also been shown that vesicles composed solely of lipids in the ratios encountered in exosomes were unable to achieve fusion with cells, indicating that proteins on exosomal membranes are as critical for their activity [81].

In addition to delivery of cargo to the site of silencing and the innate ability to cross biological barriers, exosomes appear to be relatively safe for in vivo use when compared to liposomes or viral vectors. In the Alvarez study, there was no evidence for immunogenicity or toxicity in animals even after repeated injections. There was also little evidence both in the Alvarez study and in the Zhuang study for exosome homing to the liver. It is likely that the immunological "inertness" of exosomes depends greatly on their cellular origin. Indeed exosomes used in the Alvarez study were derived from immature DCs that express minimal stimulatory molecules. However, the natural lipid and protein composition of their membranes and their rapid and efficient uptake into MVBs in host cells likely contribute to this immunological inertness and low toxicity.

9.5 Future Perspectives

The realization of the full potential of exosomes in drug delivery will be enabled by the establishment of a scalable and long-term source of well-characterized exosomes. In this context, the use of embryonic stem cells (ESC) and induced pluripotent stem cells (iPS) hold great promise. ESC cells can be differentiated into DCs, while iPS cells can be derived from patient skin fibroblasts by well-established methods, maintained in culture or frozen for much greater lengths of time than primary DCs, and differentiated into a chosen lineage, e.g., neurons and DCs [82–84]. As well as eliminating immunogenicity, this approach offers the option of using exosomes from stem cell-derived neurons or other brain cells, which are likely to display intrinsic neurotropic behavior and enhanced brain specificity. In addition, the applicability of exosome-delivered therapy to a large number of disease conditions would be greatly enhanced by broadening their cargo repertoire and improving the tissue-targeting strategies. Indeed, it would be important to explore other cargoes such as miRNA, miRNA antagonists, and shRNA expression plasmids. The delivery of miRNA mimics and antagonists by exosomes would be particularly beneficial where multiple pathways are targeted, as is the case in Alzheimer's (miR29) [85] and in cancer (e.g., tumor suppressor miR-7 and miR-128 replacement therapy in glioblastoma [86, 87]). These miRNAs would either be loaded directly into exosomes or enriched in them by expression in the exosome-producing parent cell providing a continuous source of loaded exosomes. The identification of novel targeting moieties, other than RVG, specific for the brain or other tissues of interest will further broaden the therapeutic applications of exosomes. Attractive candidates include monoclonal antibodies against receptors that are naturally expressed on the BBB or adhesion molecules expressed on endothelial cells in the lining of blood vessels. It would, therefore, be possible to achieve a degree of tissue specificity not previously achievable with viruses and liposomes, further enhancing the bioavailability of exosome cargo, reducing nonspecific homing to other tissues and to sites of clearance. These targeting moiety or moieties could either be expressed in the parent cells or inserted directly onto the exosomal membrane, allowing to control for levels of surface expression and the relative ratios of targeting moieties.

In addition to harnessing exosomes for RNAi delivery, lipidomic, proteomic, and transcriptomic studies promise to define more precisely the components that make exosomes competent in drug delivery. Defining these components will aid the design and development of (semi-)artificial drug delivery vehicles. Indeed, an alternative approach to obtaining membrane vesicles is to generate them synthetically by recapitulating the components that are essential for their function as a carrier system. Such synthetic exosome mimics would be a homogenous and reproducible type of drug delivery vehicles, devoid of other cargoes that are naturally found in exosomes and which otherwise mediate off-target effects. To mimic the exosomes' natural targeting properties and their ability to deliver RNAi to the site of tissue silencing, it is likely that artificial particles, e.g., liposomes, would have to comprise the specific lipids and proteins involved in exosomal cell trafficking. An attempt to produce such "exosome-like" vesicles has already been described by Martinez-Lostao et al. [88]. The effectiveness of APO2L/TRAIL (proteins that are decreased in the synovial fluid from patients with rheumatoid arthritis) conjugated with artificial lipid vesicles was evaluated in a rabbit model of antigen-induced arthritis (AIA) using artificial exosome-like liposomes which included phosphatidylcholine (PC) and sphingomyelin and were conjugated to APO2L/TRAIL. The authors observed that association of APO2L/TRAIL to the exosome-like liposomes increased its bioactivity and resulted in more effective treatment of AIA [88]. Other studies attempted to mimic the protein content of exosomes by including, for example, connexins. These components of cellular gap junctions had previously been identified on exosomes derived from various sources. Kaneda et al. showed efficacy of liposomes displaying connexin 43 in delivering the anti-inflammatory peptide, NEMO-binding domain peptide (NBDp) into Cx43-expressing U2OS cells and also demonstrated that such delivery was dependent on Cx43 expression in the recipient cells [89]. De La Pena et al. successfully coated liposomes consisting of exosomal phospholipids with an optimized number of MHC class I peptide complexes and other immunostimulatory ligands. These "exosome-like" liposomes recapitulated the function of antigen-presenting cells [90]. These studies demonstrate that artificial liposome-based delivery systems could be designed based on the properties derived from exosomes and that further understanding of exosome biology and trafficking will help the design and development of effective exosome-like delivery vehicles.

Despite the exciting progress with the discovery of the drug delivery potential of membrane vesicles, there exists a number of limiting factors in their clinical translation. First, membrane vesicles are known to play a role in pathology, for example, in spread of HIV or prion particles [91, 92]. There is also evidence that membrane vesicles may be involved in promoting tumor growth by inducing angiogenesis and spreading oncogenes [93]. Citrullinated proteins common in rheumatoid arthritis have been found in exosomes from synovial fluid and could be involved in stimulating the pathogenic autoimmune response [94]. It would be important to further characterize membrane vesicles intended for drug delivery by proteomic studies in order to identify any endogenous cargoes that may mediate potential unwanted side effects. Furthermore, gene expression studies in exosome-treated cells would identify pathways other than those targeted by the siRNA cargo, which are altered by exosome treatment. As well as eliminating potential pathological cargoes from membrane vesicles, it would be important to produce clinical-grade vesicles with a defined set of characteristics and a quantifiable level of drug cargo. However, this is currently limited by lack of suitable and scalable nanotechnologies for the purification, characterization, and loading of exosomes. Current ultracentrifugation protocols produce a heterogeneous mix of exosomes, other cellular vesicles, and macromolecular complexes. Therefore, novel purification methods based on the use of specific desired markers, such as the expression of the targeting moiety on the surface of exosomes, are required. In addition, siRNA loading into exosomes is relatively inefficient and cost-ineffective, highlighting the need for the development of novel transfection reagents tailored for nanoparticle applications. In summary, the clinical translation of membrane vesicles for drug delivery hinges upon better understanding of exosome biology and function in health and disease and the development of nanotechnologies for the specific purification and characterization of clinical-grade exosomes and their loading with a variety of therapeutic cargoes.

References

1. Zamore PD, Tuschl T, Sharp PA, Bartel DP (2000) RNAi: double-stranded RNA directs the ATP-dependent cleavage of mRNA at 21 to 23 nucleotide intervals. Cell 101:25–33
2. Bernstein E, Caudy AA, Hammond SM, Hannon GJ (2001) Role for a bidentate ribonuclease in the initiation step of RNA interference. Nature 409:363–366
3. Fire A, Xu S, Montgomery MK, Kostas SA, Driver SE, Mello CC (1998) Potent and specific genetic interference by double-stranded RNA in Caenorhabditis elegans. Nature 391: 806–811
4. Whitehead KA, Langer R, Anderson DG (2009) Knocking down barriers: advances in siRNA delivery. Nat Rev Drug Discov 8:129–138
5. Leuschner F, Dutta P, Gorbatov R, Novobrantseva TI, Donahoe JS, Courties G, Lee KM, Kim JI, Markmann JF, Marinelli B, Panizzi P, Lee WW, Iwamoto Y, Milstein S, Epstein-Barash H, Cantley W, Wong J, Cortez-Retamozo V, Newton A, Love K, Libby P, Pittet MJ, Swirski FK, Koteliansky V, Langer R, Weissleder R, Anderson DG, Nahrendorf M (2011) Therapeutic siRNA silencing in inflammatory monocytes in mice. Nat Biotechnol 29:1005–1010
6. Akinc A, Goldberg M, Qin J, Dorkin JR, Gamba-Vitalo C, Maier M, Jayaprakash KN, Jayaraman M, Rajeev KG, Manoharan M, Koteliansky V, Rohl I, Leshchiner ES, Langer R, Anderson DG (2009) Development of lipidoid-siRNA formulations for systemic delivery to the liver. Mol Ther 17:872–879
7. Ezzat K, El Andaloussi S, Abdo R, Langel U (2010) Peptide-based matrices as drug delivery vehicles. Curr Pharm Des 16:1167–1178
8. Andaloussi SE, Lehto T, Mager I, Rosenthal-Aizman K, Oprea II, Simonson OE, Sork H, Ezzat K, Copolovici DM, Kurrikoff K, Viola JR, Zaghloul EM, Sillard R, Johansson HJ, Said Hassane F, Guterstam P, Suhorutsenko J, Moreno PM, Oskolkov N, Halldin J, Tedebark U, Metspalu A, Lebleu B, Lehtio J, Smith CI, Langel U (2011) Design of a peptide-based vector, PepFect6, for efficient delivery of siRNA in cell culture and systemically in vivo. Nucleic Acids Res 39:3972–3987
9. Crombez L, Catherine Morris M, Dufort S, Aldrian-Herrada G, Nguyen Q, Mc Master G, Coll JL, Heitz F, Divita G (2009) Targeting cyclin B1 through peptide-based delivery of siRNA prevents tumour growth. Nucleic Acids Res 37(14):4559–4569
10. Zimmermann TS, Lee AC, Akinc A, Bramlage B, Bumcrot D, Fedoruk MN, Harborth J, Heyes JA, Jeffs LB, John M, Judge AD, Lam K, McClintock K, Nechev LV, Palmer LR, Racie T, Rohl I, Seiffert S, Shanmugam S, Sood V, Soutschek J, Toudjarska I, Wheat AJ, Yaworski E, Zedalis W, Koteliansky V, Manoharan M, Vornlocher HP, MacLachlan I (2006) RNAi-mediated gene silencing in non-human primates. Nature 441:111–114
11. Mae M, Armulik A, Betsholtz C (2011) Getting to know the cast – cellular interactions and signaling at the neurovascular unit. Curr Pharm Des 17:2750–2754
12. Nag S (2011) Morphology and properties of brain endothelial cells. Methods Mol Biol 686:3–47
13. Miller DS (2010) Regulation of P-glycoprotein and other ABC drug transporters at the blood-brain barrier. Trends Pharmacol Sci 31:246–254
14. Armulik A, Genove G, Mae M, Nisancioglu MH, Wallgard E, Niaudet C, He L, Norlin J, Lindblom P, Strittmatter K, Johansson BR, Betsholtz C (2010) Pericytes regulate the blood-brain barrier. Nature 468:557–561
15. Daneman R, Zhou L, Kebede AA, Barres BA (2010) Pericytes are required for blood-brain barrier integrity during embryogenesis. Nature 468:562–566
16. Araya R, Kudo M, Kawano M, Ishii K, Hashikawa T, Iwasato T, Itohara S, Terasaki T, Oohira A, Mishina Y, Yamada M (2008) BMP signaling through BMPRIA in astrocytes is essential for proper cerebral angiogenesis and formation of the blood-brain-barrier. Mol Cell Neurosci 38:417–430
17. Nimmerjahn A, Kirchhoff F, Helmchen F (2005) Resting microglial cells are highly dynamic surveillants of brain parenchyma in vivo. Science 308:1314–1318

18. Pakzaban P, Geller AI, Isacson O (1994) Effect of exogenous nerve growth factor on neurotoxicity of and neuronal gene delivery by a herpes simplex amplicon vector in the rat brain. Hum Gene Ther 5:987–995
19. Thorne RG, Pronk GJ, Padmanabhan V, Frey WH 2nd (2004) Delivery of insulin-like growth factor-I to the rat brain and spinal cord along olfactory and trigeminal pathways following intranasal administration. Neuroscience 127:481–496
20. Neuwelt EA, Barnett PA, McCormick CI, Frenkel EP, Minna JD (1985) Osmotic blood-brain barrier modification: monoclonal antibody, albumin, and methotrexate delivery to cerebrospinal fluid and brain. Neurosurgery 17:419–423
21. Hu J, Yuan X, Ko MK, Yin D, Sacapano MR, Wang X, Konda BM, Espinoza A, Prosolovich K, Ong JM, Irvin D, Black KL (2007) Calcium-activated potassium channels mediated blood-brain tumor barrier opening in a rat metastatic brain tumor model. Mol Cancer 6:22
22. Pardridge WM (2006) Molecular Trojan horses for blood-brain barrier drug delivery. Discov Med 6:139–143
23. Pardridge WM (2007) Drug targeting to the brain. Pharm Res 24:1733–1744
24. Boado RJ, Hui EK, Lu JZ, Pardridge WM (2010) Drug targeting of erythropoietin across the primate blood-brain barrier with an IgG molecular Trojan horse. J Pharmacol Exp Ther 333:961–969
25. Boado RJ, Lu JZ, Hui EK, Pardridge WM (2010) IgG-single chain Fv fusion protein therapeutic for Alzheimer's disease: expression in CHO cells and pharmacokinetics and brain delivery in the rhesus monkey. Biotechnol Bioeng 105:627–635
26. Pardridge WM (2010) Biologic TNF alpha-inhibitors that cross the human blood-brain barrier. Bioeng Bugs 1:231–234
27. Pardridge WM, Boado RJ (2009) Pharmacokinetics and safety in rhesus monkeys of a monoclonal antibody-GDNF fusion protein for targeted blood-brain barrier delivery. Pharm Res 26:2227–2236
28. Kumar P, Wu H, McBride JL, Jung KE, Kim MH, Davidson BL, Lee SK, Shankar P, Manjunath N (2007) Transvascular delivery of small interfering RNA to the central nervous system. Nature 448:39–43
29. Lentz TL (1990) Rabies virus binding to an acetylcholine receptor alpha-subunit peptide. J Mol Recognit 3:82–88
30. Moschos SA, Williams AE, Lindsay MA (2007) Cell-penetrating-peptide-mediated siRNA lung delivery. Biochem Soc Trans 35:807–810
31. Harris TJ, Green JJ, Fung PW, Langer R, Anderson DG, Bhatia SN (2010) Tissue-specific gene delivery via nanoparticle coating. Biomaterials 31:998–1006
32. Valadi H, Ekstrom K, Bossios A, Sjostrand M, Lee JJ, Lotvall JO (2007) Exosome-mediated transfer of mRNAs and microRNAs is a novel mechanism of genetic exchange between cells. Nat Cell Biol 9:654–659
33. Cocucci E, Racchetti G, Meldolesi J (2009) Shedding microvesicles: artefacts no more. Trends Cell Biol 19:43–51
34. Sanderson MP, Keller S, Alonso A, Riedle S, Dempsey PJ, Altevogt P (2008) Generation of novel, secreted epidermal growth factor receptor (EGFR/ErbB1) isoforms via metalloprotease-dependent ectodomain shedding and exosome secretion. J Cell Biochem 103: 1783–1797
35. Svensson KJ, Kucharzewska P, Christianson HC, Skold S, Lofstedt T, Johansson MC, Morgelin M, Bengzon J, Ruf W, Belting M (2011) Hypoxia triggers a proangiogenic pathway involving cancer cell microvesicles and PAR-2-mediated heparin-binding EGF signaling in endothelial cells. Proc Natl Acad Sci U S A 108:13147–13152
36. Zakharova L, Svetlova M, Fomina AF (2007) T cell exosomes induce cholesterol accumulation in human monocytes via phosphatidylserine receptor. J Cell Physiol 212:174–181
37. Dainiak N, Sorba S (1991) Intracellular regulation of the production and release of human erythroid-directed lymphokines. J Clin Invest 87:213–220
38. Enjeti AK, Lincz LF, Seldon M (2007) Detection and measurement of microparticles: an evolving research tool for vascular biology. Semin Thromb Hemost 33:771–779

39. Aupeix K, Hugel B, Martin T, Bischoff P, Lill H, Pasquali JL, Freyssinet JM (1997) The significance of shed membrane particles during programmed cell death in vitro, and in vivo, in HIV-1 infection. J Clin Invest 99:1546–1554
40. Thery C, Boussac M, Veron P, Ricciardi-Castagnoli P, Raposo G, Garin J, Amigorena S (2001) Proteomic analysis of dendritic cell-derived exosomes: a secreted subcellular compartment distinct from apoptotic vesicles. J Immunol 166:7309–7318
41. Hristov M, Erl W, Linder S, Weber PC (2004) Apoptotic bodies from endothelial cells enhance the number and initiate the differentiation of human endothelial progenitor cells in vitro. Blood 104:2761–2766
42. Ashman RF, Peckham D, Alhasan S, Stunz LL (1995) Membrane unpacking and the rapid disposal of apoptotic cells. Immunol Lett 48:159–166
43. Taylor MP, Kirkegaard K (2008) Potential subversion of autophagosomal pathway by picornaviruses. Autophagy 4:286–289
44. Simpson RJ, Jensen SS, Lim JW (2008) Proteomic profiling of exosomes: current perspectives. Proteomics 8:4083–4099
45. Hoepfner S, Severin F, Cabezas A, Habermann B, Runge A, Gillooly D, Stenmark H, Zerial M (2005) Modulation of receptor recycling and degradation by the endosomal kinesin KIF16B. Cell 121:437–450
46. Nielsen E, Severin F, Backer JM, Hyman AA, Zerial M (1999) Rab5 regulates motility of early endosomes on microtubules. Nat Cell Biol 1:376–382
47. Stoorvogel W, Strous GJ, Geuze HJ, Oorschot V, Schwartz AL (1991) Late endosomes derive from early endosomes by maturation. Cell 65:417–427
48. Futter CE, Collinson LM, Backer JM, Hopkins CR (2001) Human VPS34 is required for internal vesicle formation within multivesicular endosomes. J Cell Biol 155:1251–1264
49. Hurley JH (2008) ESCRT complexes and the biogenesis of multivesicular bodies. Curr Opin Cell Biol 20:4–11
50. Trajkovic K, Hsu C, Chiantia S, Rajendran L, Wenzel D, Wieland F, Schwille P, Brugger B, Simons M (2008) Ceramide triggers budding of exosome vesicles into multivesicular endosomes. Science 319:1244–1247
51. Morelli AE, Larregina AT, Shufesky WJ, Sullivan ML, Stolz DB, Papworth GD, Zahorchak AF, Logar AJ, Wang Z, Watkins SC, Falo LD Jr, Thomson AW (2004) Endocytosis, intracellular sorting, and processing of exosomes by dendritic cells. Blood 104:3257–3266
52. Zwaal RF, Comfurius P, Bevers EM (2004) Scott syndrome, a bleeding disorder caused by defective scrambling of membrane phospholipids. Biochim Biophys Acta 1636:119–128
53. Raposo G, Nijman HW, Stoorvogel W, Liejendekker R, Harding CV, Melief CJ, Geuze HJ (1996) B lymphocytes secrete antigen-presenting vesicles. J Exp Med 183:1161–1172
54. Taylor DD, Akyol S, Gercel-Taylor C (2006) Pregnancy-associated exosomes and their modulation of T cell signaling. J Immunol 176:1534–1542
55. Al-Nedawi K, Meehan B, Micallef J, Lhotak V, May L, Guha A, Rak J (2008) Intercellular transfer of the oncogenic receptor EGFRvIII by microvesicles derived from tumour cells. Nat Cell Biol 10:619–624
56. Liu C, Yu S, Zinn K, Wang J, Zhang L, Jia Y, Kappes JC, Barnes S, Kimberly RP, Grizzle WE, Zhang HG (2006) Murine mammary carcinoma exosomes promote tumor growth by suppression of NK cell function. J Immunol 176:1375–1385
57. Kim HK, Song KS, Park YS, Kang YH, Lee YJ, Lee KR, Ryu KW, Bae JM, Kim S (2003) Elevated levels of circulating platelet microparticles, VEGF, IL-6 and RANTES in patients with gastric cancer: possible role of a metastasis predictor. Eur J Cancer 39:184–191
58. Nguyen DG, Booth A, Gould SJ, Hildreth JE (2003) Evidence that HIV budding in primary macrophages occurs through the exosome release pathway. J Biol Chem 278:52347–52354
59. Robertson C, Booth SA, Beniac DR, Coulthart MB, Booth TF, McNicol A (2006) Cellular prion protein is released on exosomes from activated platelets. Blood 107:3907–3911
60. Koga K, Matsumoto K, Akiyoshi T, Kubo M, Yamanaka N, Tasaki A, Nakashima H, Nakamura M, Kuroki S, Tanaka M, Katano M (2005) Purification, characterization and biological significance of tumor-derived exosomes. Anticancer Res 25:3703–3707

61. Nilsson J, Skog J, Nordstrand A, Baranov V, Mincheva-Nilsson L, Breakefield XO, Widmark A (2009) Prostate cancer-derived urine exosomes: a novel approach to biomarkers for prostate cancer. Br J Cancer 100:1603–1607
62. Andre F, Schartz NE, Movassagh M, Flament C, Pautier P, Morice P, Pomel C, Lhomme C, Escudier B, Le Chevalier T, Tursz T, Amigorena S, Raposo G, Angevin E, Zitvogel L (2002) Malignant effusions and immunogenic tumour-derived exosomes. Lancet 360:295–305
63. Escudier B, Dorval T, Chaput N, Andre F, Caby MP, Novault S, Flament C, Leboulaire C, Borg C, Amigorena S, Boccaccio C, Bonnerot C, Dhellin O, Movassagh M, Piperno S, Robert C, Serra V, Valente N, Le Pecq JB, Spatz A, Lantz O, Tursz T, Angevin E, Zitvogel L (2005) Vaccination of metastatic melanoma patients with autologous dendritic cell (DC) derived-exosomes: results of the first phase I clinical trial. J Transl Med 3:10
64. Lai R, Arslan R, Lee M, Sze N, Choo A, Chrn T, Salto-Tellez M, Timmers L, Lee C, El Oalkey R, Pasterkamp G, de Kleijn D, Lim S (2010) Exosome secreted by MSC reduces myocardial ischemia/reperfusion injury. Stem Cell Res 4(3):214–222
65. Gyorgy B, Szabo TG, Pasztoi M, Pal Z, Misjak P, Aradi B, Laszlo V, Pallinger E, Pap E, Kittel A, Nagy G, Falus A, Buzas EI (2011) Membrane vesicles, current state-of-the-art: emerging role of extracellular vesicles. Cell Mol Life Sci 68:2667–2688
66. Ramachandran S, Palanisamy V (2012) Horizontal transfer of RNAs: exosomes as mediators of intercellular communication. Wiley Interdiscip Rev RNA 3:286–293
67. Pisitkun T, Shen RF, Knepper MA (2004) Identification and proteomic profiling of exosomes in human urine. Proc Natl Acad Sci U S A 101:13368–13373
68. de Gassart A, Geminard C, Fevrier B, Raposo G, Vidal M (2003) Lipid raft-associated protein sorting in exosomes. Blood 102:4336–4344
69. Camussi G, Deregibus MC, Bruno S, Grange C, Fonsato V, Tetta C (2011) Exosome/microvesicle-mediated epigenetic reprogramming of cells. Am J Cancer Res 1:98–110
70. Pan BT, Johnstone RM (1983) Fate of the transferrin receptor during maturation of sheep reticulocytes in vitro: selective externalization of the receptor. Cell 33:967–978
71. Baj-Krzyworzeka M, Szatanek R, Weglarczyk K, Baran J, Urbanowicz B, Branski P, Ratajczak MZ, Zembala M (2006) Tumour-derived microvesicles carry several surface determinants and mRNA of tumour cells and transfer some of these determinants to monocytes. Cancer Immunol Immunother 55:808–818
72. Ohshima K, Inoue K, Fujiwara A, Hatakeyama K, Kanto K, Watanabe Y, Muramatsu K, Fukuda Y, Ogura S, Yamaguchi K, Mochizuki T (2010) Let-7 microRNA family is selectively secreted into the extracellular environment via exosomes in a metastatic gastric cancer cell line. PLoS One 5:e13247
73. Nazarenko I, Rana S, Baumann A, McAlear J, Hellwig A, Trendelenburg M, Lochnit G, Preissner KT, Zoller M (2010) Cell surface tetraspanin Tspan8 contributes to molecular pathways of exosome-induced endothelial cell activation. Cancer Res 70:1668–1678
74. Mittelbrunn M, Gutierrez-Vazquez C, Villarroya-Beltri C, Gonzalez S, Sanchez-Cabo F, Gonzalez MA, Bernad A, Sanchez-Madrid F (2011) Unidirectional transfer of microRNA-loaded exosomes from T cells to antigen-presenting cells. Nat Commun 2:282
75. Montecalvo A, Larregina AT, Shufesky WJ, Beer Stolz D, Sullivan ML, Karlsson JM, Baty CJ, Gibson GA, Erdos G, Wang Z, Milosevic J, Tkacheva OA, Divito SJ, Jordan R, Lyons-Weiler J, Watkins SC, Morelli AE (2011) Mechanism of transfer of functional microRNAs between mouse dendritic cells via exosomes. Blood 119(3):756–766
76. Skog J, Wurdinger T, van Rijn S, Meijer DH, Gainche L, Sena-Esteves M, Curry WT Jr, Carter BS, Krichevsky AM, Breakefield XO (2008) Glioblastoma microvesicles transport RNA and proteins that promote tumour growth and provide diagnostic biomarkers. Nat Cell Biol 10:1470–1476
77. Gibbings DJ, Ciaudo C, Erhardt M, Voinnet O (2009) Multivesicular bodies associate with components of miRNA effector complexes and modulate miRNA activity. Nat Cell Biol 11:1143–1149

78. Alvarez-Erviti L, Seow Y, Yin H, Betts C, Lakhal S, Wood MJ (2011) Delivery of siRNA to the mouse brain by systemic injection of targeted exosomes. Nat Biotechnol 29:341–345
79. Zhuang X, Xiang X, Grizzle W, Sun D, Zhang S, Axtell RC, Ju S, Mu J, Zhang L, Steinman L, Miller D, Zhang HG (2011) Treatment of brain inflammatory diseases by delivering exosome encapsulated anti-inflammatory drugs from the nasal region to the brain. Mol Ther 19:1769–1779
80. Record M, Subra C, Silvente-Poirot S, Poirot M (2011) Exosomes as intercellular signalosomes and pharmacological effectors. Biochem Pharmacol 81:1171–1182
81. Subra C, Laulagnier K, Perret B, Record M (2007) Exosome lipidomics unravels lipid sorting at the level of multivesicular bodies. Biochimie 89:205–212
82. Fairchild PJ, Nolan KF, Waldmann H (2007) Genetic modification of dendritic cells through the directed differentiation of embryonic stem cells. Methods Mol Biol 380:59–72
83. Aasen T, Izpisua Belmonte JC (2010) Isolation and cultivation of human keratinocytes from skin or plucked hair for the generation of induced pluripotent stem cells. Nat Protoc 5:371–382
84. Karumbayaram S, Novitch BG, Patterson M, Umbach JA, Richter L, Lindgren A, Conway AE, Clark AT, Goldman SA, Plath K, Wiedau-Pazos M, Kornblum HI, Lowry WE (2009) Directed differentiation of human-induced pluripotent stem cells generates active motor neurons. Stem Cells 27:806–811
85. Bettens K, Brouwers N, Engelborghs S, Van Miegroet H, De Deyn PP, Theuns J, Sleegers K, Van Broeckhoven C (2009) APP and BACE1 miRNA genetic variability has no major role in risk for Alzheimer disease. Hum Mutat 30:1207–1213
86. Godlewski J, Nowicki MO, Bronisz A, Williams S, Otsuki A, Nuovo G, Raychaudhury A, Newton HB, Chiocca EA, Lawler S (2008) Targeting of the Bmi-1 oncogene/stem cell renewal factor by microRNA-128 inhibits glioma proliferation and self-renewal. Cancer Res 68:9125–9130
87. Kefas B, Godlewski J, Comeau L, Li Y, Abounader R, Hawkinson M, Lee J, Fine H, Chiocca EA, Lawler S, Purow B (2008) microRNA-7 inhibits the epidermal growth factor receptor and the Akt pathway and is down-regulated in glioblastoma. Cancer Res 68:3566–3572
88. Martinez-Lostao L, Garcia-Alvarez F, Basanez G, Alegre-Aguaron E, Desportes P, Larrad L, Naval J, Martinez-Lorenzo MJ, Anel A (2010) Liposome-bound APO2L/TRAIL is an effective treatment in a rabbit model of rheumatoid arthritis. Arthritis Rheum 62:2272–2282
89. Kaneda M, Nomura SM, Ichinose S, Kondo S, Nakahama K, Akiyoshi K, Morita I (2009) Direct formation of proteo-liposomes by in vitro synthesis and cellular cytosolic delivery with connexin-expressing liposomes. Biomaterials 30:3971–3977
90. De La Pena H, Madrigal JA, Rusakiewicz S, Bencsik M, Cave GW, Selman A, Rees RC, Travers PJ, Dodi IA (2009) Artificial exosomes as tools for basic and clinical immunology. J Immunol Methods 344:121–132
91. Lenassi M, Cagney G, Liao M, Vaupotic T, Bartholomeeusen K, Cheng Y, Krogan NJ, Plemenitas A, Peterlin BM (2010) HIV Nef is secreted in exosomes and triggers apoptosis in bystander CD4+ T cells. Traffic 11:110–122
92. Vella LJ, Sharples RA, Nisbet RM, Cappai R, Hill AF (2008) The role of exosomes in the processing of proteins associated with neurodegenerative diseases. Eur Biophys J 37:323–332
93. Schorey JS, Bhatnagar S (2008) Exosome function: from tumor immunology to pathogen biology. Traffic 9:871–881
94. Skriner K, Adolph K, Jungblut PR, Burmester GR (2006) Association of citrullinated proteins with synovial exosomes. Arthritis Rheum 54:3809–3814

Chapter 10
Aptamer-Mediated siRNA Targeting

Jiehua Zhou and John J. Rossi

Abstract RNA interference (RNAi) is a sequence-specific mechanism for post-transcriptional inhibition of gene expression. As such, it is an attractive approach for the therapeutic treatment of a wide variety of human maladies. Although conceptually elegant, there are key barriers to the widespread clinical application of this process. One of the most formidable impediments to clinical translation of RNAi is safe and effective delivery of the siRNAs to the desired target tissue at therapeutic doses. In this regard, the advent of versatile aptamer technology has prompted the development of aptamer-mediated cell-type-specific delivery for targeted RNAi triggers. In this chapter, we explore the developments of cell-type-specific aptamer applications. We also highlight recent advances of aptamers as functionalized nanocarriers for targeted siRNA delivery.

10.1 Introduction

Small interfering RNAs (siRNAs) are 21–22-base-long RNAs that guide the sequence-specific degradation of target mRNAs [1, 2]. The essential feature of the RNAi mechanism is the exquisite sequence specificity, which derives from

J. Zhou
Division of Molecular and Cellular Biology, Beckman Research Institute of City of Hope, City of Hope, Duarte, CA, USA

J.J. Rossi (✉)
Irell and Manella Graduate School of Biological Sciences, Division of Molecular and Cellular Biology, Beckman Research Institute of City of Hope, City of Hope, Duarte, CA, USA
e-mail: jrossi@coh.org

K.A. Howard (ed.), *RNA Interference from Biology to Therapeutics*, Advances in Delivery Science and Technology, DOI 10.1007/978-1-4614-4744-3_10,

complementary Watson–Crick base pairing of a target messenger RNA (mRNA) with the guide strand of the siRNA. Since double-stranded RNA-mediated RNAi was discovered in 1998 [3], the development of siRNA-based therapeutics has been an area of research [4, 5]. After more than a decade of development, there are over 20 siRNA-based preclinical and phase I/II clinical trials for a variety of human diseases which have continued to fuel the interest in RNAi clinical translation [4, 6]. For example, in 2004, OPKO Health (previous Acuity Pharmaceuticals) announced the first siRNA-related clinical trial in which bevasiranib, an unmodified siRNA targeting vascular endothelial growth factor (VEGF), was given to patients with wet age-related macular degeneration (AMD). The phase III trial of bevasiranib is the first and most advanced clinical study [7]. Additionally, RNAi therapy for respiratory syncytial virus (RSV) also has rapidly progressed from laboratory investigations to clinical trials [8–11]. In 2007, Alnylam Pharmaceuticals started phase II clinical studies for ALN-RSV01, which is an siRNA targeting a highly conserved region of the mRNA encoding the nucleocapsid (N) protein of RSV. The siRNA was shown to prophylactically reduce the incidence of RSV lung infections in adult patients [11]. The first in-human phase I clinical trial using a targeted nanoparticle-siRNA delivery system showed the first direct evidence for siRNA-mediated gene silencing in humans [12].

However, in similarity with the development of other drug types, some challenges/hurdles have resulted in setbacks in clinical translation of RNAi [5, 13]. Due to the negatively charged nature of nucleic acids, siRNAs cannot directly pass through the cell membrane and are vulnerable to RNase degradation in the absence of backbone modifications or protective delivery vehicles [14]. Although siRNA can be directly administered to a target (such as intratumoral delivery), for many diseases, systemic administration is required, which generally requires greater therapeutic doses, thus leading to higher costs and harmful side effects. On the other hand, when siRNAs are administered by systemic administration without some protective covering and/or appropriate chemical modifications, they are rapidly cleared through renal filtration, ultimately resulting in poor pharmacokinetics and marginal or no gene silencing. Before reaching the targeted cells, it is difficult for a nonformulated siRNA to pass through the blood vessel endothelial wall and multiple tissue barriers including liver, kidney, and lymphoid organs [14, 15]. Therefore, a nontargeting systemic administration of siRNA probably severely reduces the therapeutic efficacy and can even lead to harmful toxicities [15, 16]. In the case of bevasiranib, it was as administered as an unmodified siRNA without a delivery formulation that was given by intravitreal injection. In 2009, OPKO terminated the phase III clinical trial of bevasiranib because its primary end point of reducing vision loss was not achieved. Recently, another clinical trial for AMD using a chemically modified siRNA—Sirna-027 (also known as AGN 211745)—against a conserved region of the VEGF receptor-1 mRNA [17], eventually failed to meet the efficacy end points in its phase II trial. The specificity and mechanism of the aforementioned anti-VEGF siRNA drugs for treating AMD was called into question by a recent study [18], which showed that the siRNA-mediated inhibitory activity of neovascularization may be attributed to nonspecific immune response associated with activation of the cell-surface toll-like receptor 3 (TLR3), rather than to a target sequence-specific interaction.

Considering these setbacks, strategies to improve the therapeutic performance of siRNAs including appropriate chemical modifications and effective delivery formulations with targeting moieties, as well as combinatorial applications with other therapeutic agents, have been explored [19]. In particular, it is highly desirable to develop an intelligent delivery formulation for carrying siRNAs to the desired target tissues at therapeutically effective doses, especially by systemic administration [4]. With the intent of developing targeted delivery vehicles, nucleic acid-based aptamers which bind cell-surface proteins have been explored as targeting ligands or delivery vehicles for tissue and cell-type-specific delivery [20, 21]. Specific cellular internalization of the siRNAs and selective accumulation of the drug in the targeted tissues or cells can be achieved by conjugating cell-specific aptamers to siRNAs. This chapter focuses on cell-specific aptamer-mediated delivery of RNAi-based therapeutics.

10.2 Development of Cell-Specific Aptamers

Aptamers are evolved *in vitro* through a random combinatorial library, single-stranded nucleic acids that can specifically recognize and bind their cognate targets by means of well-defined stable, three-dimensional structures [22]. Over the past 20 years, numerous nucleic acid aptamers have been raised against a wide variety of targets by an *in vitro* procedure called systematic evolution of ligands by exponential enrichment (SELEX). The SELEX technique was first reported in 1990 [23–25] and has undergone several refinements and modifications to improve the selection efficiency and speed [26]. Many aptamer properties are comparable to those of protein monoclonal antibodies, but the nucleic acid aptamers with target binding affinities in the low to mid-nanomolar range offer some advantages over Mabs, including the potential for chemical synthesis and modification, stability, and lack of immunogenicity [26, 27]. Aptamers, therefore, have been utilized for a variety of applications ranging from diagnostics to therapeutics [28, 29].

Advances in the development of DNA or RNA aptamers that specifically target membrane receptors to deliver and enhance the efficacy of other therapeutics agents have created interest in exploiting cell-specific aptamers as targeted drug delivery carriers. Aptamers can be directly assembled with other therapeutic nucleic acids (like siRNAs) or be conjugated onto nanocarriers into versatile multifunctional molecular devices [21]. In order to generate a cell-specific aptamer, the selection procedure can vary from traditional purified membrane protein-based SELEX [25] (Fig. 10.1a) to live cell-based SELEX [30–32] (Fig. 10.1b). Alternatively, cell-specific aptamers also can be identified through a combinatorial SELEX strategy *via* switching between cells expressing the target protein with purified protein selection [33]. Generally, in a typical SELEX procedure, the initial single-stranded DNA/RNA pool contains a 20–60-nt random sequence, which largely guarantees the presence of active structures with high binding affinity to the target protein. By repetitive rounds of binding and selection, aptamers against virtually any given target can be routinely isolated from an initial combinatorial oligonucleotide library.

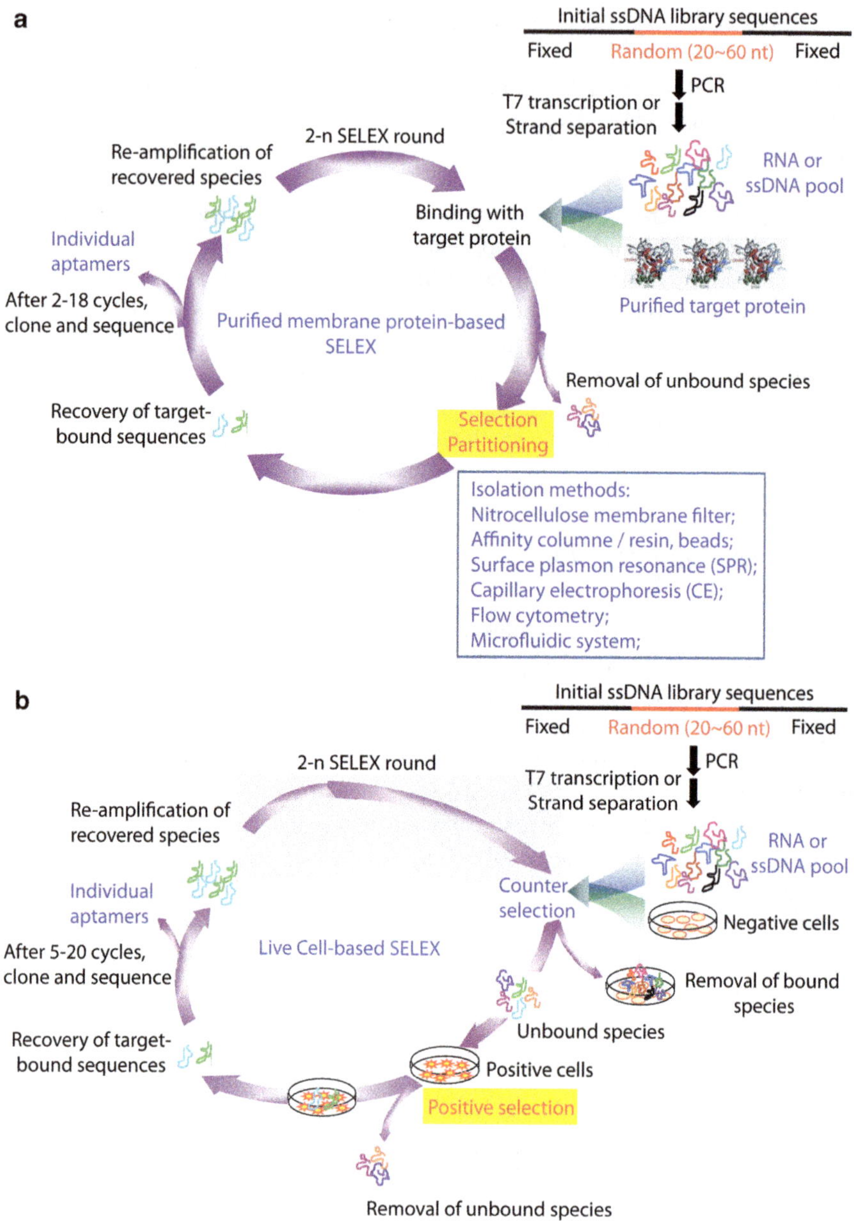

Fig. 10.1 In order to generate a cell-specific aptamer, the selection procedure can vary from traditional purified membrane protein-based SELEX (**a**) to live cell-based SELEX (**b**). Generally, in a typical SELEX procedure, the initial single-stranded DNA/RNA pool contains a 20–60-nt random sequence to provide a sequence space that facilitates presence of structures with high binding affinity to the target protein. By repeating selection rounds, aptamers against any given targets can be routinely isolated from an initial combinatorial oligonucleotide library. (**a**) A typical purified protein, membrane-based SELEX procedure consists of four main steps (1) binding to the target protein, (2) selective partitioning, (3) recovery of target-bound sequences, and (4) re-amplification of recovered species. (**b**) A live cell-based SELEX procedure also consists of four main steps: (1) counterselection by incubating library with negative cells that do not express the target protein, (2) positive selection by incubating recovered unbound sequences with positive cells expressing the target protein, (3) recovery of target-bound sequences, and finally (4) re-amplification of recovered species

10.2.1 Purified Membrane Protein-Based SELEX

As shown in Fig. 10.1a, a typical purified membrane protein-based SELEX procedure consists of four main steps (1) binding to the target protein, (2) selection partitioning, (3) recovery of target-bound sequences, and (4) re-amplification of recovered species [34]. The critical step in aptamer selection is the selection partitioning, which is to isolate target-bound sequences from unbound species in the library through various approaches, including traditional bead, resin, membrane, or chip-based segmentation approaches. Recently, novel isolation strategies have been developed to accelerate the selection procedure and improve the efficiency of aptamer selection. These include capillary gel electrophoresis [35–37], flow cytometry, and microfluidic devices [38, 39].

The purified membrane protein-based SELEX procedure has some advantages over other methods, including low nonspecific binding and facile control of the selection conditions [34]. Thus far, most of the cell-specific aptamers have been generated using soluble, purified membrane proteins/receptors. For example, a traditional nitrocellulose membrane filter-based isolation method was employed to generate anti-HIV-1 gp120 aptamers [40], anti-EGFR aptamers [41], and anti-TfR aptamers [42]. In addition, aptamers targeting CD4 were produced by immobilizing soluble, recombinant CD4 antigen onto Sepharose beads allowing elution of unbound sequences and retention of bound species [43]. Lupold et al. immobilized a purified fusion protein containing a modified extracellular form of PSMA on magnetic beads and successfully isolated two 2′-fluoro-modified RNase-resistant RNA aptamers (A-9 and A-10) with low nanomolar binding affinity [44]. Despite these successes, efficient generation of new aptamers as cell-specific homing agents still presents a major challenge due to the limited number of purified membrane proteins. A target protein that is insoluble or only functionalizes in a native conformation or in a multi-protein complex is unavailable for the purified protein-based SELEX procedure [45]. It was reported that some aptamers that were evolved to bind to a soluble, purified membrane protein failed to recognize the same targets in their native confirmation [46–48]. For such proteins, the traditional purified protein-based *in vitro* selection is not feasible.

10.2.2 Live Cell-Based SELEX

Live cell-based SELEX is the process in which live cells are used to select aptamers for target recognition (Fig. 10.1b) [31]. This method provides a promising alternative for the generation of aptamers that can recognize a particular target in its native cell conformation [49]. In principle, this strategy relies essentially on the differences between the target cell population (positive cells) relative to a control cell population (negative cells) which are used for counterselection. In similarity to the purified protein-based SELEX, a live cell-based SELEX procedure also consists of four main steps (1) counterselection by incubating the library with negative cells that do not express the target protein, (2) positive selection by incubating recovered unbound sequences with positive cells expressing the target protein, (3) recovery of target-bound sequences, and finally (4) re-amplification of the recovered species. In general, the number of cells, concentration of the library, ionic strength, incubation time, washing times, and temperature can all be varied to provide more stringent conditions favoring the selection of aptamers with the highest affinity.

The use of live cell-based SELEX to create aptamers that bind to target proteins on the surface of live cells has many advantages. In contrast to purified protein-based SELEX, it is not necessary to know the number or type of proteins on the cell membrane surface [49]. Since the selection target is whole cells, a panel of aptamers can be isolated against both known and unknown cell-surface proteins. Most importantly, the cell-surface proteins keep their native conformation throughout the selection process, thereby retaining their active biological functions. By using live cell-based SELEX, many new aptamers have been successfully generated that are capable of recognizing different cell populations, including lymphocytic leukemia cells [50, 51] and rat glioblastoma cells [52]. Employing the cell-based SELEX method, Tan and colleagues successfully isolated a panel of cell-specific ssDNA aptamers with high binding affinities to CCRF-CEM cells (a T cell acute lymphoblastic leukemia cell line) without the explicit knowledge of the cell's molecular signature [53]. The selected aptamers are able to distinguish molecular differences on cancer cells in patient samples. However, it should be noted that there are several pitfalls with the cell-based selection procedures [54], for example, nucleic acids are readily bound nonspecifically to dead cells, a problem which can impede effective enrichment of target-specific sequences and even result in failure to obtain the desired aptamers.

10.3 The Strategies of Aptamer–siRNA Targeting

Over the past few years, an increasing number of cell-specific RNA and DNA aptamers have been successfully adopted for targeted delivery of the various molecules of interest both in cell culture and *in vivo*, such as anticancer drugs [55–64], toxins [65], enzymes [42], radionuclides [66], virus [67], and RNAi-based therapeutics (siRNAs, shRNAs) [68]. Despite these validated examples, only a few RNA aptamers (PSMA, CD4, HIV-1 gp120, and CD30) have been demonstrated to have competence for targeted siRNA/shRNA delivery. In this section, we mainly focus on the application of cell-specific aptamers for cell-targeted RNA interference. Several representative examples are discussed below.

10.3.1 Noncovalent Aptamer/siRNA Conjugates

In order to achieve targeted RNAi effects, the desired siRNA molecules can be conjugated with a cell-specific aptamer through either covalent conjugation or noncovalent assembly. The most established and best characterized aptamers for siRNA targeting are the PSMA aptamers [44]. Several independent groups have demonstrated anti-PSMA aptamer-mediated siRNA delivery to PSMA-positive cells or human prostate cancer cells transplanted into nude mice. Chu et al. took advantage of the noncovalent interaction of streptavidin and biotin to create a modular streptavidin bridge for conjugating the anti-PSMA aptamer (A-9) with lamin A/C or GAPDH siRNA [69]. Two biotinylated aptamers and two biotinylated siRNAs were co-assembled, and effective PSMA receptor-mediated uptake of the aptamer–siRNA conjugates was observed along with siRNA-mediated specific silencing of the targeted transcripts in the cultured PSMA-positive cells.

We used a "sticky bridge" strategy to noncovalently assemble an HIV gp120 aptamer with various siRNAs [40]. The 3′-end of the aptamer and one of the two siRNA strands were chemically conjugated with complementary GC-rich sticky sequences, thereby, allowing the aptamer and siRNA portions to be noncovalently conjugated *via* Watson–Crick base pairing by simple mixing. Through mixing various siRNAs with different aptamers, the "sticky bridge" offers a facile conjugation approach for multiplex targeted delivery and target downregulation. For example, in the RNAi-based anti-HIV therapy, rapid emergence of viral escape mutants often abrogates RNAi efficacy of a single siRNA. Therefore, we combined three different siRNAs with a single gp120 aptamer: one against the HIV-1 *tat/rev* transcripts and two siRNAs targeting the HIV-1 host dependency factors [CD4 and transportin-3 (TNPO3)]. The resulting aptamer—"sticky bridge"—cocktail siRNA conjugates were specifically internalized into cells expressing the HIV-1 envelope protein and effectively suppressed viral loads in cell culture and in an HIV-1-infected humanized mouse model (Zhou et al. manuscript submitted).

10.3.2 Covalent Aptamer–siRNA Conjugates

Using a different, covalent conjugation approach, Giangrande and coworkers have developed a simple covalent anti-PSMA aptamer (A-10)–siRNA chimeric RNA [70]. In this approach, the aptamer was covalently attached to the sense strand of a 21-mer siRNA portion that was annealed with an unmodified antisense strand. The resulting chimeric RNA molecule was selectively internalized into cells expressing PSMA and effectively knocked down expression of the targeted tumor survival genes PLK1 and BCL2 both in cell culture and *in vivo*. Recently, this same group constructed an optimal second-generation chimera with an increased *in vivo* circulation time and bioavailability [71]. Based on their previous first-generation chimeric RNA design, the aptamer portion was truncated from 71 to 39 nt without loss of high binding affinity. The size of the resulting version (containing the truncated aptamer and sense strand of siRNA portion) was reduced to 64 nt, making this RNA amenable to milligram-scale chemical synthesis. Moreover, some rational structural changes of the siRNA portion were added, allowing more efficient incorporation of the siRNA by the cellular RNAi machinery and increasing the gene silencing efficacy. Additionally, the sense strand of the siRNA portion was appended with a 20-kDa polyethylene glycol (PEG) moiety, which substantially increased the circulating half-life and markedly improved the bioavailability of the aptamer–siRNA chimera. Compared with the swap-2′-F chimera, the *in vivo* circulating half-life of the PEG-conjugated chimera (the swap-2′-F-PEG chimera) was substantially increased (from <35 min to >30 h). Additionally, the therapeutic dose of the PEG-conjugated chimera was dramatically reduced from 10×1 nmol to 5×0.25 nmol per injection, minimizing both the cost of treatment and the risk for harmful side effects. Consequently, the optimized version resulted in pronounced regression of PSMA-expressing tumor growth after systemic administration in mice harboring human prostate cancer cells, and this was accomplished with a reduced therapeutic dose. More recently, Gliboa and coworkers also utilized the PSMA aptamer to specifically deliver NMD (nonsense-mediated messenger RNA decay) factor targeting siRNAs to tumor cells [72]. The systemic administration of PSMA aptamer–NMD siRNA chimeras resulted in significant suppression of tumor growth in subcutaneous and metastatic tumor animal models.

Using the same aptamer–siRNA fusion approach described by Giangrande, we successfully developed a dual-functional anti-gp120 aptamer–siRNA chimeric RNA [40, 73], in which both the aptamer and the siRNA portions have potent anti-HIV activities. By blocking the interaction of HIV-1 envelop gp120 and CD4 receptor, the aptamer portion can neutralize HIV-1 infectivity. The gp120 aptamer was covalently linked to a 27-mer siRNA against the HIV-1 *tat/rev* common exon. Treatment of HIV-1-infected T cells and PBMCs with these chimeras resulted in the selective gp120-mediated cellular internalization and the specific gene silencing. Furthermore, in an HIV-1-infected humanized Rag-hu mouse model, systemic administration of the gp120 aptamer–siRNA chimera provided several logs of inhibition of HIV-1 replication and complete inhibition of CD4+ T cell depletion normally mediated by viral infection [74]. Our results also demonstrated that the specific gene silencing effect and the siRNAs were only detectable in gp120 aptamer–siRNA chimera-treated mice, but not in animals treated with naked siRNA or a mutant aptamer–siRNA chimera, validating the aptamer-mediated cell-specific siRNA delivery and RNAi activity.

Similarly, an alternative aptamer that binds to the human CD4 receptor was also covalently fused with an siRNA to specifically induce gene silencing in CD4+ T cells and macrophages and in cervicovaginal tissue explants [75]. When the CD4 aptamer–siRNA chimeras (CD4-AsiCs) bearing siRNAs targeting HIV *gag* and *vif* or host CCR5 were administrated by the vaginal route to humanized mice, HIV vaginal transmission to cervicovaginal explants and to the mice was significantly prevented.

It has been reported that multivalent versions of aptamers can increase the potency and antitumor response and promote receptor activation [76–79]. In efforts to promote the binding affinity and internalization of chimeras and enhance the therapeutic potential, multivalent aptamer–siRNA conjugates have been generated. For example, the siRNA itself has been used to serve as a linker to join the two PSMA aptamers together or the siRNA was appended onto the 3′-ends of each aptamer to build bivalent aptamer–siRNA chimeras [80]. This bivalent design promoted internalization of the chimeras and resulted in an almost complete triggering of PSMA-positive cell death.

10.3.3 Aptamer-Nanoparticle Conjugates

In the past decade, the advances in nanotechnology have greatly accelerated developments in the material, life, and medical sciences [81, 82]. A variety of natural and synthetic nanocarriers for drug delivery, including liposomes, micelles, synthetic polymers, carbon nanotubes, and quantum dots, have been developed [83]. The integration of cell-specific aptamers with nanocarriers is expected to develop more versatile and multifunctional cell-type-specific delivery systems [55]. Through engineering of cell-specific aptamers with a nanoscale delivery vehicle, specific molecular recognition and cellular internalization of the therapeutic agents by the target tissue can be achieved *via* aptamer-mediated targeting [84]. In contrast to the average size of a single siRNA molecule which is well below 10 nm in diameter, aptamer-functionalized nanoparticles have an appropriate nanoscale size in the mid-nanometer range thereby allowing preferential accumulation in target tissues and organs *via* a passive targeting mode (EPR, enhanced permeability and retention effect) [85, 86] and facilitating cellular entry by endocytosis. These nanocarriers can also reduce renal clearance, improve circulation half-life, and biodistribution *in vivo*.

Self-assembling nanoparticles represent an attractive approach for siRNA targeting. The packaging RNA (pRNA) component of the bacteriophage phi29 DNA-packaging motor has been developed and manipulated to produce chimeric RNAs that form dimer, trimer, hexamer, and larger arrays ranging in size from nanometers to micrometers *via* interlocking right- and left-hand loops [87, 88]. pRNA monomers can fold into a stable and unique secondary structure that serves as the building blocks to form nanostructures *via* bottom-up assembly [89]. Fusion of the pRNA with a variety of therapeutic and chemical compounds does not impede the formation of dimers or interfere with function. By taking advantage of the self-assembling property of the pRNA, Guo and coworkers fabricated pRNA-based nanoparticles with a variety of structures and shapes [90]. For example, a CD4 aptamer and an siRNA against survivin, GFP, BAD, or lucoferase were, respectively, covalently fused with pRNAs. Subsequently, the pRNA–aptamer or pRNA–siRNA chimeras were incubated to form a dimer of 25 nm length as measured by atomic force microscopy (AFM) [91, 92]. The resulting nanoscale RNA dimer specifically bound to and was internalized in CD4-expressing T cells, delivering the siRNA to trigger specific knockdown of the target transcripts.

By taking advantage of straightforward chemical synthesis and modification, cell-specific aptamers can be conjugated to functional groups with relative ease, enabling their conjugation to the nanocarriers. The large surface areas of nanoparticles provide excellent platforms for conjugating multiple aptamers. For example, a PSMA aptamer was chemically conjugated with a synthetic branched polyethylenimine-grafted–polyethylene glycol (PEI-PEG) polymer [93]. The resulting PSMA aptamer-functionalized polymer specifically delivered to prostate cancer cells an shRNA targeting the anti-apoptotic gene (Bcl-xl).

The interior volumes of nanoparticles have been used to encapsulate large quantities of various drug molecules. Using a somewhat different strategy, a nanocomplex with a peak hydrodynamic diameter of ~140 nm was formulated by incorporating both ALK (anaplastic lymphoma kinase) siRNA and a CD30 RNA aptamer into nano-sized polyethylenimine-citrate carriers *via* noncovalent interaction [94]. The selective uptake of the nanocomplex and specific ALK gene silencing effects were observed in ALCL cells (anaplastic large cell lymphoma) expressing the CD30 receptor [95, 96].

10.4 Concluding Remarks and Future Perspectives

According to a PubMed search with "aptamer" as keyword, more than 2,700 publications indicate that the field of aptamer research has generated lots of interest in the scientific community. Aptamers have a number of advantages in their clinical usage over their antibody counterparts, and due to the relatively simple chemistry, aptamer-based applications will become more tunable and accessible compared to antibodies. However, several important issues limit the applications of aptamers. These include a reduced bioavailability compared to antibodies as well as high cost for the scale-up chemical syntheses of these RNA compounds. In this regard, increasing efforts are being made toward the development of efficacious selection methodologies to identify new aptamers with high affinity and novel chemical synthesis methods. For example, live cell-SELEX has offered very promising results in recent years and automated SELEX workstations are used to rapidly select aptamers for multiple targets. As these technologies are improved, we expect that the development and commercialization of aptamers will accelerate their clinical applications.

Interest in the therapeutic potential of nucleic acid-based agents, such as siRNAs, microRNAs, aptamers, antisense DNAs, mRNAs, and ribozymes has grown over the past two decades. With the technological maturation and increasing knowledge of RNAi as well as aptamers and their mechanism of action, it seems natural to partner multiple therapeutic nucleic acids (e.g., aptamers and siRNAs) to expand the potential for therapeutic efficacy. Because the building blocks for both are RNA molecules, combining cell-specific aptamers with RNAi-based therapeutic agents is facile and provides flexibility for achieving targeted delivery of siRNAs in the desired cells or tissue. Currently, siRNAs have been attached to cell-specific aptamers either through direct conjugation to the aptamer, or *via* functionalized groups appended on the aptamer. Additionally, nucleic acids also can be fused to polyethylene glycol (PEG) or assembled into a nanocarrier, thus reducing kidney clearance and increasing their propensity to circulate and distribute into tumor tissues. For example, a precisely engineered nanocarrier system functionalized with a cell-specific aptamer not only has an appropriate nanoscale size, thereby allowing preferential accumulation in the tumor in the passive target mode, but also can selectively recognize and bind to surface proteins on the targeted cells *via* the interaction of the ligand and cell-surface receptor, thus facilitating selective internalization. Therefore, the combined advantages of aptamers as targeting agents and the utility of nanoparticles as drug carriers make combining these into drug delivery vehicles very attractive.

Acknowledgments and Author Disclosure Statements This work is supported by grants from the National Institutes of Health AI29329, AI42552, and HL07470 awarded to J.J.R.

J.Z. drafted the article. J.J.R. revised it and gave final approval of the version to be published. All authors read and approved the final article.

The authors declare no competing financial interests.

References

1. Elbashir SM, Harborth J, Lendeckel W, Yalcin A, Weber K, Tuschl T (2001) Duplexes of 21-nucleotide RNAs mediate RNA interference in cultured mammalian cells. Nature 411:494–498
2. Zamore PD, Tuschl T, Sharp PA, Bartel DP (2000) RNAi: double-stranded RNA directs the ATP-dependent cleavage of mRNA at 21 to 23 nucleotide intervals. Cell 101:25–33
3. Fire A, Xu S, Montgomery MK, Kostas SA, Driver SE, Mello CC (1998) Potent and specific genetic interference by double-stranded RNA in Caenorhabditis elegans. Nature 391:806–811
4. Davidson BL, McCray PB Jr (2011) Current prospects for RNA interference-based therapies. Nat Rev Genet 12:329–340
5. Castanotto D, Rossi JJ (2009) The promises and pitfalls of RNA-interference-based therapeutics. Nature 457:426–433
6. Lares MR, Rossi JJ, Ouellet DL (2010) RNAi and small interfering RNAs in human disease therapeutic applications. Trends Biotechnol 28:570–579
7. Singerman L (2009) Combination therapy using the small interfering RNA bevasiranib. Retina 29:S49–S50
8. DeVincenzo J et al (2008) Evaluation of the safety, tolerability and pharmacokinetics of ALN-RSV01, a novel RNAi antiviral therapeutic directed against respiratory syncytial virus (RSV). Antiviral Res 77:225–231
9. DeVincenzo J et al (2010) A randomized, double-blind, placebo-controlled study of an RNAi-based therapy directed against respiratory syncytial virus. Proc Natl Acad Sci USA 107:8800–8805
10. Alvarez-Erviti L, Seow Y, Yin H, Betts C, Lakhal S, Wood MJ (2011) Delivery of siRNA to the mouse brain by systemic injection of targeted exosomes. Nat Biotechnol 29:341–345
11. Zamora MR et al (2011) RNA interference therapy in lung transplant patients infected with respiratory syncytial virus. Am J Respir Crit Care Med 183:531–538
12. Davis ME et al (2010) Evidence of RNAi in humans from systemically administered siRNA via targeted nanoparticles. Nature 464:1067–1070
13. Whitehead KA, Langer R, Anderson DG (2009) Knocking down barriers: advances in siRNA delivery. Nat Rev Drug Discov 8:129–138
14. Juliano R, Alam MR, Dixit V, Kang H (2008) Mechanisms and strategies for effective delivery of antisense and siRNA oligonucleotides. Nucleic Acids Res 36:4158–4171
15. Perez-Martinez FC, Guerra J, Posadas I, Cena V (2011) Barriers to non-viral vector-mediated gene delivery in the nervous system. Pharm Res 28:1843–1858
16. Wang J, Lu Z, Wientjes MG, Au JL (2010) Delivery of siRNA therapeutics: barriers and carriers. AAPS J 12:492–503
17. Kaiser PK et al (2010) RNAi-based treatment for neovascular age-related macular degeneration by Sirna-027. Am J Ophthalmol 150(33–39):e32
18. Kleinman ME et al (2008) Sequence- and target-independent angiogenesis suppression by siRNA via TLR3. Nature 452:591–597
19. Shim MS, Kwon YJ (2010) Efficient and targeted delivery of siRNA in vivo. FEBS J 277:4814–4827
20. Zhou J, Rossi JJ (2009) The therapeutic potential of cell-internalizing aptamers. Curr Top Med Chem 9:1144–1157

21. Zhou J, Rossi JJ (2011) Cell-specific aptamer-mediated targeted drug delivery. Oligonucleotides 21:1–10
22. Syed MA, Pervaiz S (2010) Advances in aptamers. Oligonucleotides 20:215–224
23. Ellington AD, Szostak JW (1990) In vitro selection of RNA molecules that bind specific ligands. Nature 346:818–822
24. Robertson DL, Joyce GF (1990) Selection in vitro of an RNA enzyme that specifically cleaves single-stranded DNA. Nature 344:467–468
25. Tuerk C, Gold L (1990) Systematic evolution of ligands by exponential enrichment: RNA ligands to bacteriophage T4 DNA polymerase. Science 249:505–510
26. Mayer G (2009) The chemical biology of aptamers. Angew Chem Int Ed Engl 48:2672–2689
27. Famulok M, Hartig JS, Mayer G (2007) Functional aptamers and aptazymes in biotechnology, diagnostics, and therapy. Chem Rev 107:3715–3743
28. Keefe AD, Pai S, Ellington A (2010) Aptamers as therapeutics. Nat Rev Drug Discov 9:537–550
29. Bunka DH, Platonova O, Stockley PG (2010) Development of aptamer therapeutics. Curr Opin Pharmacol 10:557–562
30. Phillips JA, Lopez-Colon D, Zhu Z, Xu Y, Tan W (2008) Applications of aptamers in cancer cell biology. Anal Chim Acta 621:101–108
31. Guo KT, Paul A, Schichor C, Ziemer G, Wendel HP (2008) CELL-SELEX: novel perspectives of aptamer-based therapeutics. Int J Mol Sci 9:668–678
32. Cerchia L, Giangrande PH, McNamara JO, de Franciscis V (2009) Cell-specific aptamers for targeted therapies. Methods Mol Biol 535:59–78
33. Hicke BJ et al (2001) Tenascin-C aptamers are generated using tumor cells and purified protein. J Biol Chem 276:48644–48654
34. Kulbachinskiy AV (2007) Methods for selection of aptamers to protein targets. Biochemistry (Mosc) 72:1505–1518
35. Berezovski M et al (2005) Nonequilibrium capillary electrophoresis of equilibrium mixtures: a universal tool for development of aptamers. J Am Chem Soc 127:3165–3171
36. Berezovski M, Musheev M, Drabovich A, Krylov SN (2006) Non-SELEX selection of aptamers. J Am Chem Soc 128:1410–1411
37. Mallikaratchy P, Stahelin RV, Cao Z, Cho W, Tan W (2006) Selection of DNA ligands for protein kinase C-delta. Chem Commun (Camb)30: 3229–3231
38. Farokhzad OC et al (2005) Microfluidic system for studying the interaction of nanoparticles and microparticles with cells. Anal Chem 77:5453–5459
39. Xu Y, Phillips JA, Yan J, Li Q, Fan ZH, Tan W (2009) Aptamer-based microfluidic device for enrichment, sorting, and detection of multiple cancer cells. Anal Chem 81:7436–7442
40. Zhou J et al (2009) Selection, characterization and application of new RNA HIV gp 120 aptamers for facile delivery of Dicer substrate siRNAs into HIV infected cells. Nucleic Acids Res 37(9):3094–3109
41. Li N, Larson T, Nguyen HH, Sokolov KV, Ellington AD (2010) Directed evolution of gold nanoparticle delivery to cells. Chem Commun (Camb) 46:392–394
42. Chen CH et al (2008) Aptamer-based endocytosis of a lysosomal enzyme. Proc Natl Acad Sci U S A 105:15908–15913
43. Kraus E, James W, Barclay AN (1998) Cutting edge: novel RNA ligands able to bind CD4 antigen and inhibit CD4+ T lymphocyte function. J Immunol 160:5209–5212
44. Lupold SE, Hicke BJ, Lin Y, Coffey DS (2002) Identification and characterization of nuclease-stabilized RNA molecules that bind human prostate cancer cells via the prostate-specific membrane antigen. Cancer Res 62:4029–4033
45. Cerchia L, de Franciscis V (2010) Targeting cancer cells with nucleic acid aptamers. Trends Biotechnol 28:517–525
46. Cerchia L et al (2005) Neutralizing aptamers from whole-cell SELEX inhibit the RET receptor tyrosine kinase. PLoS Biol 3:e123
47. Pestourie C et al (2006) Comparison of different strategies to select aptamers against a transmembrane protein target. Oligonucleotides 16:323–335

48. Liu Y et al (2009) Aptamers selected against the unglycosylated EGFRvIII ectodomain and delivered intracellularly reduce membrane-bound EGFRvIII and induce apoptosis. Biol Chem 390:137–144
49. Fang X, Tan W (2009) Aptamers generated from cell-SELEX for molecular medicine: a chemical biology approach. Acc Chem Res 43(1):48–57
50. Shangguan D et al (2008) Cell-specific aptamer probes for membrane protein elucidation in cancer cells. J Proteome Res 7:2133–2139
51. Shangguan D, Cao ZC, Li Y, Tan W (2007) Aptamers evolved from cultured cancer cells reveal molecular differences of cancer cells in patient samples. Clin Chem 53:1153–1155
52. Blank M, Weinschenk T, Priemer M, Schluesener H (2001) Systematic evolution of a DNA aptamer binding to rat brain tumor microvessels. Selective targeting of endothelial regulatory protein pigpen. J Biol Chem 276:16464–16468
53. Shangguan D et al (2006) Aptamers evolved from live cells as effective molecular probes for cancer study. Proc Natl Acad Sci U S A 103:11838–11843
54. Avci-Adali M, Metzger M, Perle N, Ziemer G, Wendel HP (2010) Pitfalls of cell-systematic evolution of ligands by exponential enrichment (SELEX): existing dead cells during in vitro selection anticipate the enrichment of specific aptamers. Oligonucleotides 20:317–323
55. Levy-Nissenbaum E, Radovic-Moreno AF, Wang AZ, Langer R, Farokhzad OC (2008) Nanotechnology and aptamers: applications in drug delivery. Trends Biotechnol 26:442–449
56. Bagalkot V, Farokhzad OC, Langer R, Jon S (2006) An aptamer-doxorubicin physical conjugate as a novel targeted drug-delivery platform. Angew Chem Int Ed Engl 45:8149–8152
57. Cheng J et al (2007) Formulation of functionalized PLGA-PEG nanoparticles for in vivo targeted drug delivery. Biomaterials 28:869–876
58. Dhar S, Gu FX, Langer R, Farokhzad OC, Lippard SJ (2008) Targeted delivery of cisplatin to prostate cancer cells by aptamer functionalized Pt(IV) prodrug-PLGA-PEG nanoparticles. Proc Natl Acad Sci USA 105:17356–17361
59. Farokhzad OC et al (2006) Targeted nanoparticle-aptamer bioconjugates for cancer chemotherapy in vivo. Proc Natl Acad Sci USA 103:6315–6320
60. Gu F et al (2008) Precise engineering of targeted nanoparticles by using self-assembled biointegrated block copolymers. Proc Natl Acad Sci USA 105:2586–2591
61. Zhang L et al (2007) Co-delivery of hydrophobic and hydrophilic drugs from nanoparticle-aptamer bioconjugates. ChemMedChem 2:1268–1271
62. Engels FK, Mathot RA, Verweij J (2007) Alternative drug formulations of docetaxel: a review. Anticancer Drugs 18:95–103
63. Cao Z et al (2009) Reversible cell-specific drug delivery with aptamer-functionalized liposomes. Angew Chem Int Ed Engl 48:6494–6498
64. Ferreira CS, Cheung MC, Missailidis S, Bisland S, Gariepy J (2009) Phototoxic aptamers selectively enter and kill epithelial cancer cells. Nucleic Acids Res 37:866–876
65. Chu TC et al (2006) Aptamer:toxin conjugates that specifically target prostate tumor cells. Cancer Res 66:5989–5992
66. Hicke BJ et al (2006) Tumor targeting by an aptamer. J Nucl Med 47:668–678
67. Tong GJ, Hsiao SC, Carrico ZM, Francis MB (2009) Viral capsid DNA aptamer conjugates as multivalent cell-targeting vehicles. J Am Chem Soc 131:11174–11178
68. Zhou J, Rossi JJ (2010) Aptamer-targeted cell-specific RNA interference. Silence 1:4
69. Chu TC, Twu KY, Ellington AD, Levy M (2006) Aptamer mediated siRNA delivery. Nucleic Acids Res 34:e73
70. McNamara JO 2nd et al (2006) Cell type-specific delivery of siRNAs with aptamer-siRNA chimeras. Nat Biotechnol 24:1005–1015
71. Dassie JP et al (2009) Systemic administration of optimized aptamer-siRNA chimeras promotes regression of PSMA-expressing tumors. Nat Biotechnol 27:839–849
72. Pastor F, Kolonias D, Giangrande PH, Gilboa E (2010) Induction of tumour immunity by targeted inhibition of nonsense-mediated mRNA decay. Nature 465:227–230
73. Zhou J, Li H, Li S, Zaia J, Rossi JJ (2008) Novel dual inhibitory function aptamer-siRNA delivery system for HIV-1 therapy. Mol Ther 16:1481–1489

74. Neff CP et al (2011) An aptamer-siRNA chimera suppresses HIV-1 viral loads and protects from helper CD4(+) T cell decline in humanized mice. Sci Transl Med 3:66ra66
75. Wheeler LA et al (2011) Inhibition of HIV transmission in human cervicovaginal explants and humanized mice using CD4 aptamer-siRNA chimeras. J Clin Invest 121:2401–2412
76. Shi H, Hoffman BE, Lis JT (1999) RNA aptamers as effective protein antagonists in a multicellular organism. Proc Natl Acad Sci USA 96:10033–10038
77. Santulli-Marotto S, Nair SK, Rusconi C, Sullenger B, Gilboa E (2003) Multivalent RNA aptamers that inhibit CTLA-4 and enhance tumor immunity. Cancer Res 63:7483–7489
78. McNamara JO et al (2008) Multivalent 4-1BB binding aptamers costimulate CD8+ T cells and inhibit tumor growth in mice. J Clin Invest 118:376–386
79. Dollins CM et al (2008) Assembling OX40 aptamers on a molecular scaffold to create a receptor-activating aptamer. Chem Biol 15:675–682
80. Wullner U, Neef I, Eller A, Kleines M, Tur MK, Barth S (2008) Cell-specific induction of apoptosis by rationally designed bivalent aptamer-siRNA transcripts silencing eukaryotic elongation factor 2. Curr Cancer Drug Targets 8:554–565
81. Guo P (2010) The emerging field of RNA nanotechnology. Nat Nanotechnol 5:833–842
82. De Rosa G, La Rotonda MI (2009) Nano and microtechnologies for the delivery of oligonucleotides with gene silencing properties. Molecules 14:2801–2823
83. Kim S, Kim JH, Jeon O, Kwon IC, Park K (2009) Engineered polymers for advanced drug delivery. Eur J Pharm Biopharm 71:420–430
84. Tan W et al (2011) Molecular aptamers for drug delivery. Trends Biotechnol 29:634–640
85. Matsumura Y, Maeda H (1986) A new concept for macromolecular therapeutics in cancer chemotherapy: mechanism of tumoritropic accumulation of proteins and the antitumor agent smancs. Cancer Res 46:6387–6392
86. Greish K (2007) Enhanced permeability and retention of macromolecular drugs in solid tumors: a royal gate for targeted anticancer nanomedicines. J Drug Target 15:457–464
87. Guo P (2005) RNA nanotechnology: engineering, assembly and applications in detection, gene delivery and therapy. J Nanosci Nanotechnol 5:1964–1982
88. Shu D, Huang LP, Hoeprich S, Guo P (2003) Construction of phi29 DNA-packaging RNA monomers, dimers, and trimers with variable sizes and shapes as potential parts for nanodevices. J Nanosci Nanotechnol 3:295–302
89. Guo P et al (2010) Engineering RNA for targeted siRNA delivery and medical application. Adv Drug Deliv Rev 62:650–666
90. Shu Y, Cinier M, Shu D, Guo P (2011) Assembly of multifunctional phi29 pRNA nanoparticles for specific delivery of siRNA and other therapeutics to targeted cells. Methods 54(2):204–214
91. Guo S, Tschammer N, Mohammed S, Guo P (2005) Specific delivery of therapeutic RNAs to cancer cells via the dimerization mechanism of phi29 motor pRNA. Hum Gene Ther 16:1097–1109
92. Hoeprich S, Guo P (2002) Computer modeling of three-dimensional structure of DNA-packaging RNA (pRNA) monomer, dimer, and hexamer of Phi29 DNA packaging motor. J Biol Chem 277:20794–20803
93. Kim E et al (2010) Prostate cancer cell death produced by the co-delivery of Bcl-xL shRNA and doxorubicin using an aptamer-conjugated polyplex. Biomaterials 31:4592–4599
94. Zhao N, Bagaria HG, Wong MS, Zu Y (2011) A nanocomplex that is both tumor cell-selective and cancer gene-specific for anaplastic large cell lymphoma. J Nanobiotechnology 9:2
95. Zhang P et al (2009) Using an RNA aptamer probe for flow cytometry detection of CD30-expressing lymphoma cells. Lab Invest 89:1423–1432
96. Duyster J, Bai RY, Morris SW (2001) Translocations involving anaplastic lymphoma kinase (ALK). Oncogene 20:5623–5637

Chapter 11
RNAi as Antiviral Therapy: The HIV-1 Case

Ben Berkhout and Julia J.M. Eekels

Abstract RNA interference (RNAi) is a cellular mechanism that mediates sequence-specific gene silencing. The RNAi mechanism also has therapeutic potential, and it can be used as an antiviral approach against infectious human pathogens. An attractive target for RNAi therapeutics is the RNA genome of the human immunodeficiency virus type 1 (HIV-1). In fact, the first clinical gene therapy trial with a lentiviral vector that encodes a single RNAi inhibitor in combination with other antiviral RNA molecules was initiated in early 2008. In this chapter, we will focus on basic mechanistic principles of an RNAi-based attack on HIV-1, which in some respects forms a formidable target. Among other items, we will discuss target site selection within the viral RNA genome, the phenomenon of viral escape, and therapeutic strategies to prevent such escape. Alternatively, one could target cofactors of the host that are essential for virus replication yet less important for cell physiology. The most promising anti-escape strategy is the implementation of a combinatorial RNAi attack on the virus.

11.1 RNAi: Transformation of the Natural Pathway into a Therapeutic Mode

The development of RNAi-based therapies against a wide variety of diseases, including cancer, neurological, autoimmune, and infectious diseases was triggered by the discovery of the RNAi mechanism and RNAi-mediated gene silencing in

B. Berkhout (✉) • J.J.M. Eekels
Laboratory of Experimental Virology, Department of Medical Microbiology, Center for Infection and Immunity Amsterdam, Academic Medical Center, University of Amsterdam, Amsterdam, The Netherlands
e-mail: b.berkhout@amc.uva.nl

K.A. Howard (ed.), *RNA Interference from Biology to Therapeutics*, Advances in Delivery Science and Technology, DOI 10.1007/978-1-4614-4744-3_11,

mammalian cells [1–7]. RNAi also holds promise for antiviral therapy against pathogenic viruses, either the ones that cause acute infections like respiratory syncytial virus (RSV) and influenza virus or the ones that cause chronic infections such as hepatitis B virus (HBV), hepatitis C virus (HCV), and HIV-1. We will focus on the chronic infections that require an RNAi-based gene therapy approach.

In the following, we will briefly describe the natural microRNA (miRNA) pathway. It is estimated that human cells express more than 500 miRNAs to control the process of cell differentiation and development by regulating gene expression at the posttranscriptional level [8–13]. The natural miRNA pathway uses RNA polymerase II or III to produce primary transcripts or pri-miRNAs that encode a single or multiple miRNAs as a polycistronic transcript [14]. The pri-miRNA is processed by the microprocessor complex, consisting of the Drosha enzyme and the dsRNA-binding protein DGCR8/Pasha [15–19], into a pre-miRNA with a 5′-monophosphate and 3′-hydroxyl 2-nucleotide (nt) overhang [20]. The pre-miRNA is processed in the nucleus and subsequently exported by Exportin-5 (Exp-5) to the cytoplasm [21–23]. The cytoplasmic Dicer endonuclease cleaves the base-paired stem approximately 22 base pairs (bp) away from its base, generating a 2-nt overhang at the 3′-end [24]. Dicer is associated with TAR RNA-binding protein (TRBP) that recruits Argonaute 2 (Ago2) [25]. The Ago2–RNA complex forms the minimal core of the RNA-induced silencing complex (RISC) [26, 27]. RISC unwinds the miRNA duplex and cleaves the passenger strand, such that only the guide strand remains in the complex to execute the subsequent mRNA silencing step [28–30].

Mammalian miRNA-mediated gene silencing occurs by translational repression of the target mRNA [9]. However, a recent study suggested mRNA cleavage as the favorite mode of miRNA action [31]. Degradation of the targeted mRNA is initiated by endonucleolytic mRNA cleavage opposite nucleotides 10 and 11 of the miRNA. The level of base-pairing complementarity in the miRNA–mRNA duplex is a major determinant, leading to mRNA cleavage with a perfect complementarity and translational repression with suboptimal complementarity [32–36]. The "seed" region of the miRNA (5′ terminal nucleotides 2–8) typically finds multiple partially complementary target sequences in the 3′ untranslated region (3′UTR) of the cellular mRNA target, and the silencing efficiency is determined by the overall arrangement of these 3′UTR targets, including their number and intermotif distance [37]. At least one human miRNA–mRNA set with perfect base-pairing complementarity has been reported [38].

In plants and insects, small interfering RNAs (siRNAs) arise from extended virus-derived double-stranded (ds) RNA molecules upon cleavage by Dicer. These siRNAs have antiviral activity and belong to an ancient antiviral mechanism of the host cell, which is less apparent in mammalian cells because the siRNAs are not readily detectable in infected human cells [39]. However, recent studies reported the accumulation of small virus-derived RNAs in cells infected with HIV-1, poliovirus, HCV, flock house virus (FHV), Drosophila C virus (DCV), Sindbis virus, and vesicular stomatitis virus (VSV) [40–42].

The most popular strategy for activation of RNAi-mediated gene silencing in mammalian cells is transfection of siRNAs either introduced directly into cells as

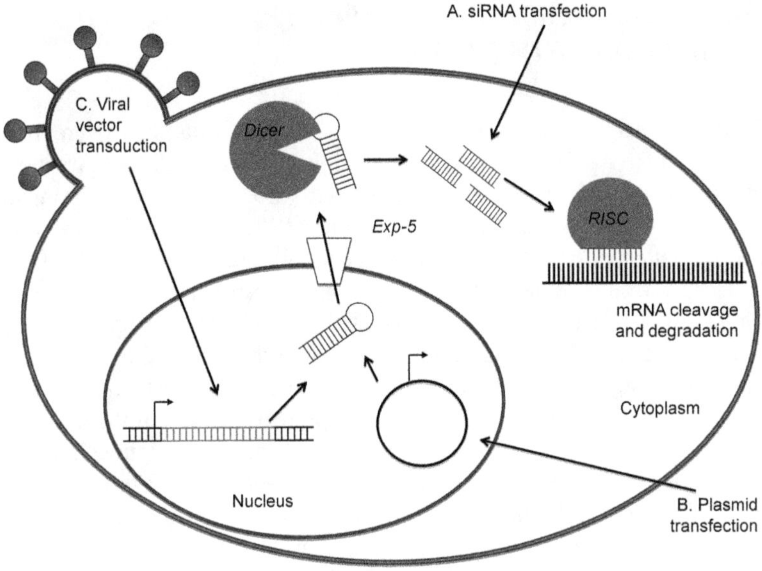

Fig. 11.1 The cellular RNAi pathway and delivery options for RNAi-based therapeutics. Several methods can be used to deliver RNAi inducer molecules into target cells. (A) Transient transfection of siRNAs will produce a silencing effect that ranges from days up to a week. The siRNAs are directly incorporated into the RISC complex residing in the cytoplasm, where one strand of the siRNA is degraded and one strand is used to guide the RISC complex to the target mRNA. The mRNA is subsequently cleaved, leading to sequence-specific gene silencing. This method is appropriate only for the attack on acute virus infections. (B) Plasmids encoding shRNAs are maintained longer in target cells (weeks to months), but the effect wanes due to dilution of the plasmid after each cell division or degradation of the plasmid. (C) Some viral vectors, in particular retroviral and lentiviral vectors, can integrate the shRNA-expression cassette into the host cell genome, ensuring constitutive expression of the therapeutic shRNA. This method is particularly useful in the fight against persistent virus infections like HIV-1. Plasmids and integrated viral vectors produce shRNAs in the nucleus. These RNA molecules are subsequently transported into the cytoplasm by Exportin-5 (Exp-5) and processed by Dicer into the active siRNAs

synthetic siRNA or vector-based intracellular expression (Fig. 11.1). Synthetic siRNAs are usually designed to have full base-pair complementarity and to direct mRNA cleavage. For this action, only a single target sequence is required, and this target can be positioned anywhere within the mRNA molecule. Synthetic siRNA introduced into cells reprogram the cytoplasmic RISC [43]. Short hairpin RNA (shRNA) transcripts [44–46] and man-made miRNA mimics can be expressed intracellularly from a plasmid or viral vector [47]. Such transcripts are expressed in the nucleus and will use the natural miRNA processing and transport pathways to instruct RISC for downregulation of a specific mRNA (Fig. 11.1). Whereas the siRNA approach is perfectly suited for a therapeutic attack on acute virus infections, the shRNA strategy using plasmid or viral vectors provides a more durable gene-silencing effect that is needed in the fight against chronic virus infections. We previously presented detailed protocols for the design of such RNAi constructs [48].

11.1.1 Towards a Durable Gene Therapy to Control HIV-1 Infection

Candidates for therapeutic intervention include diseases caused by overexpression of a specific mRNA. RNAi can also be designed to target the RNA genome of invading microbes such as HIV-1. HIV-1 is a retrovirus that causes a chronic infection that ultimately leads to disease progression, AIDS, and death. Disease progression can be halted effectively in most patients with antiviral drugs, but the virus readily escapes from a monotherapy. A combinatorial drug approach can avoid the evolution of drug-resistant HIV-1 variants. Despite this clinical success, some problems remain with multiple drug regimes, including drug-related toxicity that may show up during long-term follow-up. The goal of alternative gene therapy approaches against HIV-1 should, therefore, be to durably protect the cells of the immune system of an already infected individual against HIV-1 infection. "Intracellular immunization" of these cells, e.g., by means of RNAi technology, should prevent the depletion of immune cells like $CD4^+$ T cells, monocytes, macrophages, and dendritic cells. Maintenance of the immune function during chronic HIV-1 infection should prevent opportunistic infections and AIDS disease progression.

HIV-1 causes a persistent infection, and therefore, a continuously active treatment seems required. Repeated delivery of exogenous siRNA as anti-HIV therapy has been proposed based on experiments in a mouse with a humanized immune system [49]. Effective virus inhibition was observed with concomitant prevention of human $CD4^+$ T cell loss. The systemic delivery of siRNA was recently reported in a human phase I clinical trial via targeted nanoparticles to patients with solid cancers [50]. We, however, seriously doubt whether such an siRNA approach would be suitable in the setting of an HIV-infected patient, where the prevention of viral escape requires the continuous presence of an effective dose of multiple siRNAs in all human cell types that can be infected, which are in fact located in many different tissues and body compartments.

A more effective gene therapy approach would instead trigger the continuous expression of anti-HIV molecules after a single transduction of HIV-susceptible cells with an appropriate delivery vector. We believe that the lentiviral vector system, which is based on the HIV-1 genome, is particularly attractive for this anti-HIV action. The lentiviral vector is made replication-incompetent by removal of essential and pathogenic genes, which were replaced by novel control and therapeutic sequences. This vector can efficiently infect most target cell types, and transduction of dividing and nondividing cell types is possible. A major asset for an anti-HIV approach is the stable integration of the lentiviral vector in the host cell genome, thus yielding permanent transduction [51, 52]. This same property raises obvious safety concerns that will be discussed later.

Use of the HIV-based lentiviral vector system to target the HIV-1 RNA genome with RNAi reagents can cause some unwanted side effects [53]. This includes self-targeting of the vector RNA by the antiviral shRNAs, thus triggering a severe vector production problem. One can of course select only those anti-HIV shRNAs for which the target sequences were deleted from the vector genome [54]. Also miRNA cassettes may raise specific problems such as Drosha-mediated cleavage of the vector RNA genome, but these problems can be countered by appropriate vector design [53, 55]. Alternative viral vectors are available for delivery of a therapeutic RNAi transgene, and these systems have been reviewed previously [55–57].

Another major question relates to the cell types that should be targeted in a gene therapy for HIV-infected individuals. Hematopoietic CD34⁺ stem cells seed the different lineages of immune cells in the blood and organs and are, therefore, interesting targets for such a gene therapy. This should be an ex vivo procedure, followed by autologous transplantation of the cells into the same patient. The transduced hematopoietic stem cells should durably supply all derived immune cells with the antiviral arsenal, even after a single transduction event. In the presence of HIV-1, one expects the preferential survival of these shRNA-expressing immune cells over untreated cells. The latter cells are likely to become infected and will subsequently be recognized and removed by the immune system. This survival benefit should result in a gradual increase in the percentage of protected cells, which should eventually result in partial reconstitution of the immune system such that AIDS disease progression is delayed or stopped.

Hematopoietic stem cells (HSC) transduced with a retroviral vector that encodes an anti-HIV-1 ribozyme have already been evaluated in clinical trials [58, 59]. These studies demonstrate the feasibility and safety of the proposed stem cell approach, although little therapeutic effect was scored for the ribozyme. A recent study demonstrated the safety of the lentiviral vector in combination with ex vivo targeting of $CD34^+$ cells [60]. Another option is the treatment of the mature $CD4^+$ T cell population, which represents the major target cell population for HIV-1. However, the gene therapy should be applied repetitively in this scenario because the T cells have a restricted life span [61].

Treatment of hematopoietic stem cells has a disadvantage as well, as the number of cells that can be obtained for manipulation is limited. The use of induced pluripotent stem cells (iPSC) could provide an elegant solution. The iPSC technology makes it possible to produce patient-specific pluripotent stem cells by treatment of somatic cells (e.g., fibroblasts or T cells) with specific reprogramming transcription factors. These iPS cells are similar to human embryonic stem cells and are able to differentiate into a wide variety of cell types. By transducing the iPSCs with a lentiviral vector encoding the antiviral genes, a continuous supply of therapeutic cells could become available for transplantation. For instance, macrophages that express a shRNA against CCR5 in combination with a gene encoding the restriction factor TRIM5α were produced with this technology, and HIV-1 replication was substantially inhibited in these cells [62].

11.1.2 Where to Target the HIV-1 RNA Genome?

Several criteria can be used to identify the optimal targets for an RNAi attack on the 9-kb HIV-1 RNA genome. First, one could reason to select targets in the multiply spliced HIV-1 mRNAs that are synthesized early upon virus infection and that encode the early viral proteins Tat, Rev, and Nef. The idea behind this strategy is that an early block in viral gene expression will severely impact the subsequent expression of unspliced and singly spliced mRNAs that encode the structural HIV-1 proteins. Second, one could target HIV-1 genomic regions that are represented in all spliced viral mRNAs, which is the case for small sequence stretches in the 5′UTR and 3′UTR [63]. Third, as target RNA structure can effectively neutralize an RNAi attack [64, 65], focusing on the "accessible" RNA domains is important. For HIV-1, this strategy is clearly facilitated by the recent description of the secondary structure of the complete HIV-1 RNA genome [66, 67]. Fourth, an important selection criterion when targeting a variable virus like HIV-1 concerns the variability of the target sequence among different virus isolates. One should ideally select targets that are highly conserved because one wants to inhibit as many virus strains as possible. We discovered that targeting of highly conserved genome regions may also restrict the evolution of viral escape mutants because well-conserved sequences exhibit an important function in HIV-1 biology, as RNA signal and/or protein-coding sequence. RNAi-induced sequence variation may, thus, be expected to have an impact on the viral replication capacity and fitness [68].

We performed an extensive RNAi screen against highly conserved HIV-1 genome sequences, which yielded about 20 potent shRNA inhibitors [69]. Stable shRNA-expressing T cell lines were generated that were subsequently infected with HIV-1, which yielded four durable shRNA inhibitors that restricted virus replication in a transformed T cell line for at least 3 months [70]. Other groups have also screened large sets of anti-HIV shRNAs [71], and effective inhibitors were described that target regulatory sequences [68, 71] or protein-coding sequences in the gag [69, 73–75], pol [73, 76, 77], vif [72], tat [69, 77–79], rev [69, 78, 79], vpu [73], env [75], and nef genes [72]. Follow-up analyses should include prolonged culturing of stably transduced T cells to score the impact on cell viability. To address safety in more detail, the off-target effects of the antiviral shRNAs on human mRNAs can be evaluated [80].

An important lesson to be learned from various siRNA investigations is the necessity for inclusion of appropriate controls. Several preclinical antiviral studies used EGFP-specific siRNA as a control. Favorable therapeutic effects were observed; however, these were later attributed to nonspecific induction of TLR7/8 interferon pathway innate immunity triggered by the antiviral-specific sequence but not the EGFP [81]. Another lesson comes from a study on an siRNA therapeutic for the treatment of age-related macular degeneration in the eye. The siRNA exhibited a therapeutic effect, but this was not likely elicited by the RNAi mechanism since the charged siRNA molecule cannot easily penetrate cells. Instead, the clinical effect occurred through TLR-3 signaling [82]. These two examples illustrate the importance of selecting the correct controls to ensure RNAi-specificity. For HIV-1 therapies that target the viral genome, exclusive specificity can be demonstrated by the selection of escape variants with a single point mutation in the target sequence. The sequence-specificity of a particular RNAi effector must be demonstrated in vitro and subsequently in appropriate in vivo models before translation into the clinical test phase [4, 83, 84]. Guidelines for the proper testing and selection of potent and safe shRNA inhibitors against HIV-1 have been formulated [70].

11.2 Viral Escape and Combinatorial RNAi Approaches

Potent and sequence-specific HIV-1 inhibition has been reported with RNAi-inducing reagents in infected cell cultures, but it soon became apparent that HIV-1 is prone to viral escape when a single shRNA inhibitor is applied [64, 69, 85–89]. Prolonged culturing in the presence of HIV-1 should be performed to test the likelihood of viral escape, which is difficult to predict. Care should be taken not to misinterpret the results of such escape studies. For instance, the appearance of point mutations in the viral target sequence forms definitive proof of viral escape, and in fact, it demonstrates the exquisite sequence-specificity of the RNAi mechanism. However, sometimes viral breakthrough is observed without the acquisition of apparent escape mutations. First, mutations may be selected outside the actual target sequence. These changes can nevertheless cause resistance by triggering a conformational switch in the RNA such that the target becomes inaccessible to the RNAi machinery [64, 65]. Second, no changes may be apparent despite phenotypic viral escape, which may in fact represent false breakthrough replication due to suboptimal virus inhibition [70, 84]. We previously discussed the complexities of properly interpreting HIV-1 evolution studies [90–94]. Thus, both detailed phenotypic and genotypic analyses are required to satisfactorily address the issue of viral escape.

The ease of HIV-1 escape mimics what occurs in patients treated with a single antiretroviral drug, but we know that combinatorial drug regimes can prevent viral escape and therapy failure. Thus, also an RNAi therapeutic attempt should tackle the virus with multiple shRNA inhibitors at the same time. Such a combinatorial RNAi attack can target the virus at multiple genome positions [83], but one can add an attack against host-encoded cofactors, as will be discussed later. One could also combine RNAi molecules with other RNA effector molecules such as decoys and ribozymes [95, 96]. Different RNA-based inhibitors can also be combined in a single transcript such as the conjugate of an antiviral aptamer that binds the HIV-1 envelope protein and an antiviral siRNA [97]. The aptamer not only blocks the envelope protein on virion particles, but it also selectively ferries the siRNA to HIV-infected cells that express the envelope protein on their surface. This conjugate demonstrated good antiviral activity in the preclinical model of the humanized mouse, although uncertainty remains about the efficiency of the intracellular delivery of the siRNA [98]. Another elegant solution to avoid viral escape is the use of the second-generation shRNAs that specifically target viral escape variants [99]. However, the shear endless number of possible viral escape routes may limit the feasibility of this approach [84]. In fact, we recently demonstrated the power of the second-generation concept by effectively blocking common viral escape routes, but little therapeutic benefit was achieved because HIV-1 selected alternative escape routes [100]. RNAi can also be used to specifically block the evolution of drug-resistant virus variants that evolve under pressure by the regular antiretroviral drugs, e.g., inhibitors of the viral protease enzyme [101].

Combinatorial drug/RNAi approaches are essential to restrict HIV-1 evolution and to prevent viral escape that consequently will lead to therapeutic failure [102]. A variety of strategies have been described for multiplexing of shRNA cassettes in a single therapeutic vector. As repeat sequences should be avoided in the lentiviral vector to prevent recombination-mediated deletions, the multiple shRNA cassettes generally use separate polymerase III promoters or a combination of polymerase II and III promoters [103]. Multiplexed siRNAs can also be expressed from a single transcript. We and others have developed extended-shRNAs that are processed into two or maximally three functional siRNAs [104–106]. Another strategy uses truly long-hairpin RNAs (lhRNAs) that should encode numerous siRNAs [107–109]. A disadvantage of the lhRNA approach is that it is a priori unknown what siRNA molecules will be produced and whether these molecules will be active inhibitors [99], and it was recently demonstrated that such constructs express a very low siRNA level [105]. Polycistronic miRNA transcripts have also been developed [110]. Various groups have reported toxicity of shRNAs [111–114], which can perhaps be solved by inserting the siRNA sequence into a natural miRNA backbone [112].

Conditional expression of the siRNA molecules will increase the safety of a therapeutic vector. For instance, one would like to avoid shRNA expression in transduced hematopoietic stem cells that still have to undergo hematopoiesis, a process that is particularly susceptible to changes in the RNAi machinery. Tissue-specific miRNA expression has been described for several organs, including the liver [115]. Another option is the design of constructs that are induced by HIV-1 infection [89]. Selective expression in HIV-1 susceptible cells would be an elegant way to restrict putative saturation and off-target effects. Another option is the use of inducible gene expression systems such as the doxycycline-controlled Tet system [116, 117]. While shRNAs are generally expressed from polymerase III promoters, miRNAs are expressed from polymerase II promoters. These polymerase II systems are better equipped for tissue-specific or drug-regulated gene expression.

Comprehensive reviews on combinatorial RNA approaches are available [118, 119]. Other types of inhibitory RNA molecules can be added to the RNAi-inducing antiviral regimes, and we already mentioned the anti-HIV aptamer-siRNA conjugate transcript [97]. The currently ongoing phase I clinical trial at the City of Hope employs a lentiviral vector that encodes a TAR-decoy, CCR5-ribozyme, and shRNA that targets the HIV-1 genome in the tat–rev region [95, 96]. The TAR-decoy is a small nucleolar RNA molecule that absorbs the viral Tat protein, which will prevent the Tat–TAR interaction that is essential for enhanced viral promoter activity [120]. The ribozyme cleaves the CCR5-encoding mRNA to cause reduced expression of this important HIV-1 receptor on the cell surface [121]. Alternative antiviral RNA molecules include antisense transcripts [122, 123], decoys [124], ribozymes [121], and aptamers [125]. A new addition to this arsenal is an antisense molecule that can elicit transcriptional gene silencing of the viral LTR promoter [126]. The novel RNAu method is based on the expression of a modified U1 small nuclear RNA that blocks polyadenylation of the targeted mRNA, which is subsequently degraded [127, 128].

11.3 HIV-1 Inhibition by Knockdown of Critical Cellular Cofactors

A theoretical advantage of targeting host cell proteins that are important for HIV-1 replication is the reduced chance of escape by viral mutation. Silencing of several of these cofactors resulted in HIV-1 inhibition: nuclear factor kappa B [76], CD4 [73, 129], CXCR4 [129–132], DDX3 [133], LEDGF/p75 [134], CCR5 [129, 131, 135], and stable expression of shRNA against several cofactors could inhibit HIV-1 replication in vitro up to 2 months [136]. CCR5 is a critical receptor for HIV-1 entry and is one of the most promising cofactor targets. First, individuals with the delta-32 mutation in the CCR5 gene do not express the receptor and are not susceptible to HIV-1 infection. Second, this gene inactivation does not cause any health problems, except for an increased risk for infection with the West Nile virus [137]. Third, potent shRNA targeting the mRNA for this host cell factor have already been described [129, 135]. Fourth, the therapeutic potential of downregulation of CCR5 is supported by the cure of an HIV-1 infected patient who had leukemia in addition to AIDS. This patient received a bone marrow transplantation of a matching donor who was homozygous for the 32-bp deletion in the CCR5 gene. Surprisingly, HIV-1 has remained undetectable in the patient's plasma for at least 600 days post transfusion [138], and the patient has recently been declared "cured" of HIV-1 [139].

Viral escape is theoretically possible in the CCR5 case. CCR5-tropic viruses are generally responsible for HIV-1 transmission, although the virus can also use the alternative CXCR4 receptor. Downregulation of the CCR5 receptor will potentially set the stage for selection of CXCR4-tropic HIV-1 variants, but this evolutionary route was not observed in the Berlin patient. The same virus escape route was discussed when CCR5-blocking drugs were developed, and such receptor-switch escape routes have indeed been witnessed in patients treated with the CCR5-antagonist maraviroc [140, 141]. In general, the concept that targeting of a cellular cofactor prevents viral escape should be verified experimentally [136].

Many cellular targets cannot be considered for silencing because they are essential for cell metabolism and the host. For instance, the alternative HIV-1 coreceptor CXCR4 is required for homing of hematopoietic stem cells to the bone marrow and subsequent T cell differentiation [142]. Although HIV-1 replication has been studied extensively, many details of the viral replication cycle remain elusive, including many of the involved cellular cofactors. Recently, three high-throughput RNAi gene knockdown screens were published that identified numerous new candidate cofactors [143–145]. Although each of the screens reported hundreds of new candidates, overlap between them was surprisingly small with only three proteins: MED7, MED8, and RELA. A number of reasons for this enormous variation in experimental results have been discussed (reviewed in [146, 147]). For instance, the three studies used rather different experimental setups, including different cell lines (293T versus HeLa cells) and lab-adapted viral strains versus viral vectors. A meta-analysis of these RNAi screens confirmed that experimental variation contributed to the many discrepancies between the screens [148]. Although the overlap between individual cofactors identified in these screens was remarkably small, the overlap of the cellular pathways involved seems more significant. Interfering with these pathways could, thus, be of therapeutic use.

Important candidate cofactors may have been missed because several well-known cofactors were not found in any of the screens. First, cofactors of which the function is redundant with that of other cellular proteins will not easily be detected. Second, cofactors that require a nearly complete expression knockdown before an impact on HIV-1 replication is scored will also be missed. Third, some important cofactors may simply have been missed because the siRNAs were excluded because of toxic off-target effects.

The attack on cellular cofactors may have a more general advantage for the attack on other human pathogenic viruses that may use the same cellular pathways and/or cellular cofactors. Similar genome-wide screens have been performed for other human viruses, including influenza virus A, HCV, dengue virus, and West Nile virus [149–152]. When comparing all screens performed for the different viruses, proteins involved in cytoskeleton complexes invariably present themselves as candidate cofactors with a broad antiviral impact [150]. Ten genes that were identified in at least one of the HIV-1 screens were also picked up in the HCV screen by Li et al. [151] and could, thus, be of double therapeutic value, as an estimated 25–30% of HIV-infected individuals are coinfected with HCV [153].

The use of RNAi against cellular cofactors is not restricted to the gene therapy setting. A recent publication described the use of an miRNA targeting the human gene PERK to enhance the immunogenicity of an HIV-1 Env DNA vaccine [154]. Normally, PERK has an antiviral function by preventing, in an indirect manner, high expression levels of viral proteins such as the HIV-1 envelope. Coexpression of HIV-1 envelope and an engineered miRNA targeting PERK resulted in increased envelope expression, and a higher Env-specific $CD8^{+}$ T cell response was measured in vaccinated mice.

11.4 Appropriate Preclinical Test Systems

Programming of the cellular miRNA pathway with new siRNA specificity is associated with certain risks. A general problem is that the artificial siRNA molecules can compete with the endogenous siRNAs, and siRNA overexpression may lead to saturation of the miRNA pathway. As the miRNA pathway is important in the control of cellular gene expression, one could expect a disturbance in the cellular differentiation program, possibly cell death or even cancer [155]. Saturation of the miRNA pathway was reported to cause death when high doses of shRNAs were delivered by an adeno-associated virus (AAV) vector in mice [4, 111–114, 156]. Thus, exogenous RNAi inducers should be expressed at a moderate level. Off-target effects on unintended mRNAs can occur because an miRNA requires only a seed sequence complementarity of 7–8 base pairs with a given mRNA [157]. Furthermore, off-targeting can not only be elicited by the guide strand but also by the passenger strand [80, 158, 159]. Such off-target effects are notoriously difficult to predict, and one should carefully screen for adverse effects in appropriate experimental systems. Another problem relates to the induction of an immune response by transfected siRNA and intracellularly expressed shRNA duplexes [160, 161], but this effect can be minimized by optimal design of the si/shRNA molecule [162]. We recently developed an ultrasensitive assay to score minor cell growth retardation effects in stably transduced shRNA-expressing cell lines [163].

When potent antiviral shRNAs have been identified in cell culture experiments, one can move to relevant preclinical models to critically assess the safety and efficacy of the proposed therapy. A simple and efficient in vitro test system to measure the impact of shRNA expression on cell viability is to perform a coculture of the transduced GFP^+ cells and nontransduced GFP^- cells (Eekels and Berkhout, submitted for publication). A reduction in the percentage of GFP^+ cells over time forms an indication of delayed cell growth and RNAi toxicity. Outgrowth of the transduced and thereby protected cells should occur in the presence of HIV-1, which can also be screened for by using simple FACS analysis of a mixed cell culture.

The SIV/macaque model [164] is used extensively for vaccination studies but could also be considered for testing of an RNAi gene therapy. However, this model has several limitations. First, anti-HIV shRNA cannot easily be tested against SIV because of sequence dissimilarity. Thus, one should either convert the anti-HIV into anti-SIV shRNA, which may affect their inhibitory power, or HIV-1 target sequences should be incorporated into the SIV test genome. Second, transduction of the HIV-based lentiviral vector is restricted by TRIM5α in macaque cells [165]. Third, macaque experiments are rather expensive, and the number of animals may be restricted because of budgetary or ethical reasons.

Most of these limitations do not apply to humanized immune system (HIS) mouse models [166, 167]. All major human myeloid and lymphoid cellular compartments develop and mature from input human stem cells in the most recent HIS mouse [168–170]. This model provides access to in vivo and ex vivo experimentation on human T cells [171]. HIS mice can be infected by intravenous injection of the virus but also via rectal and vaginal transmission routes. Infection results in viremia and depletion of human $CD4^+$ cells [172–177]. We used this model to test the safety and efficacy of a lentiviral-based gene therapy in hematopoietic stem cells [4]. These animal models and their advantages and disadvantages have been reviewed [178].

11.5 Safety Concerns of Gene Therapy

By now, more than 1,700 clinical trials involving a gene therapy have been performed (http://www.wiley.co.uk/genmed/clinical), and trials up to 2007 are reviewed in [179]. The first gene therapy patient was treated in 1989 for adenosine deaminase deficiency, a severe form of combined immunodeficiency (SCID) [180]. Another gene therapy patient was treated in 1999 for a genetic liver disease and ornithine transcarbamylase deficiency and received an adenovirus vector encoding the wild-type gene. This patient died 4 days later of a massive immune response, most likely triggered by the high viral vector load [181]. SCID patients received a gamma-retroviral gene transfer with the wild-type interleukin 2 gene in a trial that started in 2000. Although this procedure improved the condition of all patients, a true success, two patients developed leukemia due to clonal lymphocyte proliferation [182]. In both cases, the retroviral vector was integrated near the LMO2 proto-oncogene, leading to enhanced expression of the LMO2 protein that has a crucial role in hematopoietic development [183]. Such insertional oncogenesis was described for a few more patients. After nearly 10 years of follow-up, gene therapy seems highly successful in correcting the immunodeficiency associated with SCID [184]. Hematopoietic stem-cell transplantation will remain the standard therapy, but gene therapy may be an option in the absence of donors with a compatible HLA-type [185].

The retroviral vectors have gradually been replaced by lentiviral vectors that are considered more safe because they tend to integrate in genes distant from the promoter region. In addition, all transcriptional enhancer motifs have been removed from the so-called third-generation lentiviral vectors "self-inactivating" design [186]. Experiments with a lentiviral vector and hematopoietic stem cells in tumor-prone mice did not, in contrast to the retroviral vector, show signs of insertional oncogenesis [187]. Other safety and regulatory issues concerning lentiviral vectors are addressed in a comprehensive review based on the expertise gained in the first lentiviral trial [188]. The power of the lentiviral vector system to safely transduce $CD34^+$ hematopoietic precursor cells ex vivo was recently demonstrated in a gene therapy trial for children with adrenoleukodystrophy [60].

11.6 Future Perspectives

An overview of ongoing gene therapy trials for treatment of HIV-1 infection has been published [189]. Thus far, these clinical trials have failed to demonstrate a therapeutic benefit. One of the bottlenecks is effective gene delivery to a clinically relevant number of cells, in particular in the early studies that treated T cells or hematopoietic stem cells with retroviral vectors [59]. The use of lentiviral vectors allows a much higher transduction efficiency of a variety of cell types.

The first clinical trial with a lentiviral vector was, in fact, directed against HIV-1 by expression of an extended antisense transcript against the viral RNA genome. Persistent in vivo expression of the therapeutic antisense molecule was documented by the VirXsys company [123]. In addition, vector integration sites in blood cells revealed a preference for gene-rich regions, which is typical for a lentivirus, and no signs of insertional oncogenesis were observed. Another anti-HIV gene therapy trial that uses a triple RNA payload (ribozyme, decoy, shRNA) was performed at the City of Hope by the team of John Rossi. AIDS patients undergoing autologous transplantation for lymphoma were treated with gene-modified

CD34+ hematopoietic progenitor cells. The transplant consisted, for safety reasons, of a mixture of genetically modified and untreated cells. No infusion-related toxicities were reported. Persistent gene marking and sustained shRNA and ribozyme expression were documented in blood cells for up to 24 months [96]. Future optimization of the transplant procedure, in particular the preferential infusion of transduced cells, should provide the setting for delivery of therapeutic levels of HIV-resistant cells. Combined with the power of antiviral RNAi approaches, this strategy remains a promising candidate for development of a durable anti-HIV-1 gene therapy, especially for a minority patient group that exhausted all treatment options due to drug toxicity or viral resistance.

We reviewed the current status of the development of an RNAi-based gene therapy to control HIV-1 infection and AIDS disease progression. The potent and sequence-specific inhibition observed for such RNAi action forms the cornerstone for such a therapy. The superior transduction of hematopoietic stem cells with lentiviral vectors provides the means to deliver the transgene. The availability of several lentiviral production facilities is another promising development in the field. We are currently developing a candidate clinical vector that yielded very potent antiviral effects in prolonged in vitro cell cultures. It is currently being evaluated for safety in a humanized mouse model, and we expect to initiate a clinical trial within the next 3 years.

Acknowledgments RNAi research in the Berkhout lab is sponsored by ZonMw (Translational gene therapy program) and the Dutch AIDS Fund (grant number 2006006). We also thank the Belgian Federal Government for financial support through the Inter-University Attraction Pole grant no. P6/41.

References

1. McCaffrey AP, Nakai H, Pandey K, Huang Z, Salazar FH, Xu H, Wieland SF, Marion PL, Kay MA (2003) Inhibition of hepatitis B virus in mice by RNA interference. Nat Biotechnol 21:639–644
2. Kapadia SB, Brideau-Andersen A, Chisari FV (2003) Interference of hepatitis C virus RNA replication by short interfering RNAs. Proc Natl Acad Sci USA 100:2014–2018
3. Banerjea A, Li MJ, Bauer G, Remling L, Lee NS, Rossi J, Akkina R (2003) Inhibition of HIV-1 by lentiviral vector-transduced siRNAs in T lymphocytes differentiated in SCID-hu mice and CD34+ progenitor cell-derived macrophages. Mol Ther 8:62–71
4. Ter Brake O, Legrand N, von Eije KJ, Centlivre M, Spits H, Weijer K, Blom B, Berkhout B (2009) Evaluation of safety and efficacy of RNAi against HIV-1 in the human immune system (Rag-2(−/−)(c)(−/−)) mouse model. Gene Ther 16:148–153
5. Ding H, Schwarz DS, Keene A, Affar EB, Fenton L, Xia X, Shi Y, Zamore PD, Xu Z (2003) Selective silencing by RNAi of a dominant allele that causes amyotrophic lateral sclerosis. Aging Cell 2:209–217
6. Takeshita F, Ochiya T (2006) Therapeutic potential of RNA interference against cancer. Cancer Sci 97:689–696
7. Davidson BL, Paulson HL (2004) Molecular medicine for the brain: silencing of disease genes with RNA interference. Lancet Neurol 3:145–149

8. Carrington JC, Ambros V (2003) Role of microRNAs in plant and animal development. Science 301:336–338
9. Ambros V (2004) The functions of animal microRNAs. Nature 431:350–355
10. Ambros V (2001) MicroRNAs: tiny regulators with great potential. Cell 107:823–826
11. Bartel DP (2004) MicroRNAs: genomics, biogenesis, mechanism, and function. Cell 116:281–297
12. Baehrecke EH (2003) miRNAs: micro managers of programmed cell death. Curr Biol 13:R473–R475
13. McManus MT (2004) Small RNAs and immunity. Immunity 21:747–756
14. Lee Y, Jeon K, Lee JT, Kim S, Kim VN (2002) MicroRNA maturation: stepwise processing and subcellular localization. EMBO J 21:4663–4670
15. Denli AM, Tops BB, Plasterk RH, Ketting RF, Hannon GJ (2004) Processing of primary microRNAs by the microprocessor complex. Nature 432:231–235
16. Gregory RI, Yan KP, Amuthan G, Chendrimada T, Doratotaj B, Cooch N, Shiekhattar R (2004) The microprocessor complex mediates the genesis of microRNAs. Nature 432:235–240
17. Han J, Lee Y, Yeom KH, Kim YK, Jin H, Kim VN (2004) The Drosha-DGCR8 complex in primary microRNA processing. Genes Dev 18:3016–3027
18. Landthaler M, Yalcin A, Tuschl T (2004) The human DiGeorge syndrome critical region gene 8 and its D. melanogaster homolog are required for miRNA biogenesis. Curr Biol 14:2162–2167
19. Lee Y, Ahn C, Han J, Choi H, Kim J, Yim J, Lee J, Provost P, Radmark O, Kim S et al (2003) The nuclear RNase III Drosha initiates microRNA processing. Nature 425:415–419
20. Bernstein E, Caudy AA, Hammond SM, Hannon GJ (2001) Role for a bidentate ribonuclease in the initiation step of RNA interference. Nature 409:363–366
21. Bohnsack MT, Czaplinski K, Gorlich D (2004) Exportin 5 is a RanGTP-dependent dsRNA-binding protein that mediates nuclear export of pre-miRNAs. RNA 10:185–191
22. Lund E, Guttinger S, Calado A, Dahlberg JE, Kutay U (2004) Nuclear export of microRNA precursors. Science 303:95–98
23. Yi R, Qin Y, Macara IG, Cullen BR (2003) Exportin-5 mediates the nuclear export of pre-microRNAs and short hairpin RNAs. Genes Dev 17:3011–3016
24. Zhang H, Kolb FA, Jaskiewicz L, Westhof E, Filipowicz W (2004) Single processing center models for human Dicer and bacterial RNase III. Cell 118:57–68
25. Chendrimada TP, Gregory RI, Kumaraswamy E, Norman J, Cooch N, Nishikura K, Shiekhattar R (2005) TRBP recruits the Dicer complex to Ago2 for microRNA processing and gene silencing. Nature 436:740–744
26. Gregory RI, Chendrimada TP, Cooch N, Shiekhattar R (2005) Human RISC couples microRNA biogenesis and posttranscriptional gene silencing. Cell 123:631–640
27. Maniataki E, Mourelatos Z (2005) A human, ATP-independent RISC assembly machine fueled by pre-miRNA. Genes Dev 19:2979–2990
28. Hammond SM, Boettcher S, Caudy AA, Kobayashi R, Hannon GJ (2001) Argonaute2, a link between genetic and biochemical analyses of RNAi. Science 293:1146–1150
29. Tomari Y, Matranga C, Haley B, Martinez N, Zamore PD (2004) A protein sensor for siRNA asymmetry. Science 306:1377–1380
30. Tomari Y, Zamore PD (2005) Perspective: machines for RNAi. Genes Dev 19:517–529
31. Cohen MS, Hellmann N, Levy JA, DeCock K, Lange J (2008) The spread, treatment, and prevention of HIV-1: evolution of a global pandemic. J Clin Invest 118:1244–1254
32. Brennecke J, Hipfner DR, Stark A, Russell RB, Cohen SM (2003) Bantam encodes a developmentally regulated microRNA that controls cell proliferation and regulates the proapoptotic gene hid in Drosophila. Cell 113:25–36
33. Doench JG, Sharp PA (2004) Specificity of microRNA target selection in translational repression. Genes Dev 18:504–511
34. Lewis BP, Shih IH, Jones-Rhoades MW, Bartel DP, Burge CB (2003) Prediction of mammalian microRNA targets. Cell 115:787–798

35. Kiriakidou M, Nelson PT, Kouranov A, Fitziev P, Bouyioukos C, Mourelatos Z, Hatzigeorgiou A (2004) A combined computational-experimental approach predicts human microRNA targets. Genes Dev 18:1165–1178
36. Lai EC (2002) Micro RNAs are complementary to 3' UTR sequence motifs that mediate negative post-transcriptional regulation. Nat Genet 30:363–364
37. Saetrom P, Heale BS, Snove O Jr, Aagaard L, Alluin J, Rossi JJ (2007) Distance constraints between microRNA target sites dictate efficacy and cooperativity. Nucleic Acids Res 35:2333–2342
38. Yekta S, Shih IH, Bartel DP (2004) MicroRNA-directed cleavage of HOXB8 mRNA. Science 304:594–596
39. Pfeffer S, Zavolan M, Grasser FA, Chien M, Russo JJ, Ju J, John B, Enright AJ, Marks D, Sander C et al (2004) Identification of virus-encoded microRNAs. Science 304:734–736
40. Schopman NC, Willemsen M, Liu YP, Bradley T, van Kampen A, Baas F, Berkhout B, Haasnoot J (2012) Deep sequencing of virus-infected cells reveals HIV-encoded small RNAs. Nucleic Acids Res 40:414–427
41. Parameswaran P, Sklan E, Wilkins C, Burgon T, Samuel MA, Lu R, Ansel KM, Heissmeyer V, Einav S, Jackson W et al (2010) Six RNA viruses and forty-one hosts: viral small RNAs and modulation of small RNA repertoires in vertebrate and invertebrate systems. PLoS Pathog 6:e1000764
42. Yeung ML, Bennasser Y, Watashi K, Le SY, Houzet L, Jeang KT (2009) Pyrosequencing of small non-coding RNAs in HIV-1 infected cells: evidence for the processing of a viral-cellular double-stranded RNA hybrid. Nucleic Acids Res 37:6575–6586
43. Elbashir SM, Harborth J, Lendeckel W, Yalcin A, Weber K, Tuschl T (2001) Duplexes of 21-nucleotide RNAs mediate RNA interference in cultured mammalian cells. Nature 411:494–498
44. Brummelkamp TR, Bernards R, Agami R (2002) A system for stable expression of short interfering RNAs in mammalian cells. Science 296:550–553
45. Paddison PJ, Caudy AA, Bernstein E, Hannon GJ, Conklin DS (2002) Short hairpin RNAs (shRNAs) induce sequence-specific silencing in mammalian cells. Genes Dev 16:948–958
46. Schopman NC, Liu YP, Konstantinova P, Ter Brake O, Berkhout B (2010) Optimization of shRNA inhibitors by variation of the terminal loop sequence. Antiviral Res 86:204–211
47. Zeng Y, Wagner EJ, Cullen BR (2002) Both natural and designed micro RNAs can inhibit the expression of cognate mRNAs when expressed in human cells. Mol Cell 9:1327–1333
48. Liu YP, Westerink JT, Ter Brake O, Berkhout B (2011) RNAi-inducing lentiviral vectors for anti-HIV-1 gene therapy. Methods Mol Biol 721:293–311
49. Kumar P, Ban HS, Kim SS, Wu H, Pearson T, Greiner DL, Laouar A, Yao J, Haridas V, Habiro K et al (2008) T cell-specific siRNA delivery suppresses HIV-1 infection in humanized mice. Cell 134:577–586
50. Davis ME, Zuckerman JE, Choi CH, Seligson D, Tolcher A, Alabi CA, Yen Y, Heidel JD, Ribas A (2010) Evidence of RNAi in humans from systemically administered siRNA via targeted nanoparticles. Nature 464:1067–1070
51. Naldini L, Blomer U, Gallay P, Ory D, Mulligan R, Gage FH, Verma IM, Trono D (1996) In vivo gene delivery and stable transduction of nondividing cells by a lentiviral vector. Science 272:263–267
52. Nguyen TH, Oberholzer J, Birraux J, Majno P, Morel P, Trono D (2002) Highly efficient lentiviral vector-mediated transduction of nondividing, fully reimplantable primary hepatocytes. Mol Ther 6:199–209
53. Liu YP, Vink MA, Westerink JT, Ramirez de Arellano E, Konstantinova P, Ter Brake O, Berkhout B (2010) Titers of lentiviral vectors encoding shRNAs and miRNAs are reduced by different mechanisms that require distinct repair strategies. RNA 16:1328–1339
54. Ter Brake O, Berkhout B (2007) Lentiviral vectors that carry anti-HIV shRNAs: problems and solutions. J Gene Med 9:743–750
55. Liu YP, Berkhout B (2011) miRNA cassettes in viral vectors: problems and solutions. Biochim Biophys Acta 1809:732–745

56. Nguyen T, Menocal EM, Harborth J, Fruehauf JH (2008) RNAi therapeutics: an update on delivery. Curr Opin Mol Ther 10:158–167
57. de Fougerolles AR (2008) Delivery vehicles for small interfering RNA in vivo. Hum Gene Ther 19:125–132
58. Amado RG, Mitsuyasu RT, Rosenblatt JD, Ngok FK, Bakker A, Cole S, Chorn N, Lin LS, Bristol G, Boyd MP et al (2004) Anti-human immunodeficiency virus hematopoietic progenitor cell-delivered ribozyme in a phase I study: myeloid and lymphoid reconstitution in human immunodeficiency virus type-1-infected patients. Hum Gene Ther 15:251–262
59. Mitsuyasu RT, Merigan TC, Carr A, Zack JA, Winters MA, Workman C, Bloch M, Lalezari J, Becker S, Thornton L et al (2009) Phase 2 gene therapy trial of an anti-HIV ribozyme in autologous CD34(+) cells. Nat Med 15:285–292
60. Cartier N, Hacein-Bey-Abina S, Bartholomae CC, Veres G, Schmidt M, Kutschera I, Vidaud M, Abel U, Dal-Cortivo L, Caccavelli L et al (2009) Hematopoietic stem cell gene therapy with a lentiviral vector in X-linked adrenoleukodystrophy. Science 326:818–823
61. Dropulic B (2001) Lentivirus in the clinic. Mol Ther 4:511–512
62. Kambal A, Mitchell G, Cary W, Gruenloh W, Jung Y, Kalomoiris S, Nacey C, McGee J, Lindsey M, Fury B et al (2011) Generation of HIV-1 resistant and functional macrophages from hematopoietic stem cell-derived induced pluripotent stem cells. Mol Ther 19:584–593
63. Muesing MA, Smith DH, Cabradilla CD, Benton CV, Lasky LA, Capon DJ (1985) Nucleic acid structure and expression of the human AIDS/lymphadenopathy retrovirus. Nature 313:450–458
64. Westerhout EM, Ooms M, Vink M, Das AT, Berkhout B (2005) HIV-1 can escape from RNA interference by evolving an alternative structure in its RNA genome. Nucleic Acids Res 33:796–804
65. Westerhout EM, Berkhout B (2007) A systematic analysis of the effect of target RNA structure on RNA interference. Nucleic Acids Res 35:4322–4330
66. Watts JM, Dang KK, Gorelick RJ, Leonard CW, Bess JW Jr, Swanstrom R, Burch CL, Weeks KM (2009) Architecture and secondary structure of an entire HIV-1 RNA genome. Nature 460:711–716
67. Low JT, Knoepfel SA, Watts JM, ter Brake O, Berkhout B, Weeks KM (2012) SHAPE-directed discovery of potent shRNA inhibitors of HIV-1 Mol Ther: 20(4):820–828
68. Kafri T, Blomer U, Peterson DA, Gage FH, Verma IM (1997) Sustained expression of genes delivered directly into liver and muscle by lentiviral vectors. Nat Genet 17:314–317
69. Ter Brake O, Konstantinova P, Ceylan M, Berkhout B (2006) Silencing of HIV-1 with RNA interference: a multiple shRNA approach. Mol Ther 14:883–892
70. von Eije KJ, Ter Brake O, Berkhout B (2009) Stringent testing identifies highly potent and escape-proof anti-HIV short hairpin RNAs. J Gene Med 11:459–467
71. McIntyre GJ, Groneman JL, Yu YH, Jaramillo A, Shen S, Applegate TL (2009) 96 shRNAs designed for maximal coverage of HIV-1 variants. Retrovirology 6:55
72. Jacque JM, Triques K, Stevenson M (2002) Modulation of HIV-1 replication by RNA interference. Nature 418:435–438
73. Chang LJ, Liu X, He J (2005) Lentiviral siRNAs targeting multiple highly conserved RNA sequences of human immunodeficiency virus type 1. Gene Ther 12:1133–1144
74. Novina CD, Murray MF, Dykxhoorn DM, Beresford PJ, Riess J, Lee SK, Collman RG, Lieberman J, Shankar P, Sharp PA (2002) siRNA-directed inhibition of HIV-1 infection. Nat Med 8:681–686
75. Park WS, Miyano-Kurosaki N, Hayafune M, Nakajima E, Matsuzaki T, Shimada F, Takaku H (2002) Prevention of HIV-1 infection in human peripheral blood mononuclear cells by specific RNA interference. Nucleic Acids Res 30:4830–4835
76. Berkhout B, Ter Brake O (2008) Towards an RNAi-based gene therapy. BIOforum Europe 4:35–37
77. Surabhi RM, Gaynor RB (2002) RNA interference directed against viral and cellular targets inhibits human immunodeficiency virus type 1 replication. J Virol 76:12963–12973

78. Coburn GA, Cullen BR (2002) Potent and specific inhibition of human immunodeficiency virus type 1 replication by RNA interference. J Virol 76:9225–9231
79. Lee NS, Dohjima T, Bauer G, Li H, Li MJ, Ehsani A, Salvaterra P, Rossi J (2002) Expression of small interfering RNAs targeted against HIV-1 rev transcripts in human cells. Nat Biotechnol 20:500–505
80. Jackson AL, Bartz SR, Schelter J, Kobayashi SV, Burchard J, Mao M, Li B, Cavet G, Linsley PS (2003) Expression profiling reveals off-target gene regulation by RNAi. Nat Biotechnol 21:635–637
81. Robbins M, Judge A, Ambegia E, Choi C, Yaworski E, Palmer L, McClintock K, Maclachlan I (2008) Misinterpreting the therapeutic effects of siRNA caused by immune stimulation. Hum Gene Ther 19:991–999
82. Kleinman ME, Yamada K, Takeda A, Chandrasekaran V, Nozaki M, Baffi JZ, Albuquerque RJ, Yamasaki S, Itaya M, Pan Y et al (2008) Sequence- and target-independent angiogenesis suppression by siRNA via TLR3. Nature 452:591–597
83. Ter Brake O, 't Hooft K, Liu YP, Centlivre M, von Eije KJ, Berkhout B (2008) Lentiviral vector design for multiple shRNA expression and durable HIV-1 inhibition. Mol Ther 16:557–564
84. von Eije KJ, Ter Brake O, Berkhout B (2008) Human immunodeficiency virus type 1 escape is restricted when conserved genome sequences are targeted by RNA interference. J Virol 82:2895–2903
85. Boden D, Pusch O, Lee F, Tucker L, Ramratnam B (2003) Human immunodeficiency virus type 1 escape from RNA interference. J Virol 77:11531–11535
86. Das AT, Brummelkamp TR, Westerhout EM, Vink M, Madiredjo M, Bernards R, Berkhout B (2004) Human immunodeficiency virus type 1 escapes from RNA interference-mediated inhibition. J Virol 78:2601–2605
87. Nishitsuji H, Kohara M, Kannagi M, Masuda T (2006) Effective suppression of human immunodeficiency virus type 1 through a combination of short- or long-hairpin RNAs targeting essential sequences for retroviral integration. J Virol 80:7658–7666
88. Sabariegos R, Gimenez-Barcons M, Tapia N, Clotet B, Martinez MA (2006) Sequence homology required by human immunodeficiency virus type 1 to escape from short interfering RNAs. J Virol 80:571–577
89. Unwalla HJ, Li HT, Bahner I, Li MJ, Kohn D, Rossi JJ (2006) Novel Pol II fusion promoter directs human immunodeficiency virus type 1-inducible coexpression of a short hairpin RNA and protein. J Virol 80:1863–1873
90. Leonard JN, Shah PS, Burnett JC, Schaffer DV (2008) HIV evades RNA interference directed at TAR by an indirect compensatory mechanism. Cell Host Microbe 4:484–494
91. Berkhout B (2009) A new Houdini act: multiple routes for HIV-1 escape from RNAi-mediated inhibition. Future Microbiol 4:151–154
92. Hache G, Abbink TE, Berkhout B, Harris RS (2009) Optimal translation initiation enables Vif-deficient human immunodeficiency virus type 1 to escape restriction by APOBEC3G. J Virol 83:5956–5960
93. Hache G, Shindo K, Albin JS, Harris RS (2008) Evolution of HIV-1 isolates that use a novel Vif-independent mechanism to resist restriction by human APOBEC3G. Curr Biol 18:819–824
94. Das AT, Berkhout B (2010) HIV-1 evolution: frustrating therapies, but disclosing molecular mechanisms. Philos Trans R Soc Lond B Biol Sci 365:1965–1973
95. Li MJ, Kim J, Li S, Zaia J, Yee JK, Anderson J, Akkina R, Rossi JJ (2005) Long-term inhibition of HIV-1 infection in primary hematopoietic cells by lentiviral vector delivery of a triple combination of anti-HIV shRNA, anti-CCR5 ribozyme, and a nucleolar-localizing TAR decoy. Mol Ther 12:900–909
96. DiGiusto DL, Krishnan A, Li L, Li H, Li S, Rao A, Mi S, Yam P, Stinson S, Kalos M et al (2010) RNA-based gene therapy for HIV with lentiviral vector-modified CD34(+) cells in patients undergoing transplantation for AIDS-related lymphoma. Sci Transl Med 2:36ra43
97. Neff CP, Zhou J, Remling L, Kuruvilla J, Zhang J, Li H, Smith DD, Swiderski P, Rossi JJ, Akkina R (2011) An aptamer-siRNA chimera suppresses HIV-1 viral loads and protects from helper CD4(+) T cell decline in humanized mice. Sci Transl Med 3:66ra6

98. Wang GP, Levine BL, Binder GK, Berry CC, Malani N, McGarrity G, Tebas P, June CH, Bushman FD (2009) Analysis of lentiviral vector integration in HIV+study subjects receiving autologous infusions of gene modified CD4+ T cells. Mol Ther 17:844–850
99. Ter Brake O, Berkhout B (2005) A novel approach for inhibition of HIV-1 by RNA interference: counteracting viral escape with a second generation of siRNAs. J RNAi Gene Silencing 1:56–65
100. Schopman NC, Ter Brake O, Berkhout B (2010) Anticipating and blocking HIV-1 escape by second generation antiviral shRNAs. Retrovirology 7:52
101. Schopman NC, Braun A, Berkhout B (2012) Directed HIV-1 evolution of protease inhibitor resistance by second-generation short hairpin RNAs. Antimicrob Agents Chemother 56:479–486
102. Berkhout B, Sanders RW (2011) Molecular strategies to design an escape-proof antiviral therapy. Antiviral Res 92:7–14
103. Ter Brake O, Berkhout B (2008) Development of an RNAi-based gene therapy against HIV-1. In: Kurreck J (ed) Therapeutic oligonucleotides. RSC Publishing, Cambridge, pp 296–311
104. Liu YP, Haasnoot J, Berkhout B (2007) Design of extended short hairpin RNAs for HIV-1 inhibition. Nucleic Acids Res 35:5683–5693
105. Liu YP, von Eije KJ, Schopman NC, Westerink JT, Ter Brake O, Haasnoot J, Berkhout B (2009) Combinatorial RNAi against HIV-1 using extended short hairpin RNAs. Mol Ther 17:1712–1723
106. Saayman S, Barichievy S, Capovilla A, Morris KV, Arbuthnot P, Weinberg MS (2008) The efficacy of generating three independent anti-HIV-1 siRNAs from a single U6 RNA Pol III-expressed long hairpin RNA. PLoS One 3:e2602
107. Barichievy S, Saayman S, von Eije KJ, Morris KV, Arbuthnot P, Weinberg MS (2007) The inhibitory efficacy of RNA POL III-expressed long hairpin RNAs targeted to untranslated regions of the HIV-1 5' long terminal repeat. Oligonucleotides 17:419–431
108. Sano M, Li H, Nakanishi M, Rossi JJ (2008) Expression of long anti-HIV-1 hairpin RNAs for the generation of multiple siRNAs: advantages and limitations. Mol Ther 16:170–177
109. Konstantinova P, de Vries W, Haasnoot J, Ter Brake O, de Haan P, Berkhout B (2006) Inhibition of human immunodeficiency virus type 1 by RNA interference using long-hairpin RNA. Gene Ther 13:1403–1413
110. Liu YP, Haasnoot J, Ter Brake O, Berkhout B, Konstantinova P (2008) Inhibition of HIV-1 by multiple siRNAs expressed from a single microRNA polycistron. Nucleic Acids Res 36:2811–2824
111. Grimm D, Streetz KL, Jopling CL, Storm TA, Pandey K, Davis CR, Marion P, Salazar F, Kay MA (2006) Fatality in mice due to oversaturation of cellular microRNA/short hairpin RNA pathways. Nature 441:537–541
112. McBride JL, Boudreau RL, Harper SQ, Staber PD, Monteys AM, Martins I, Gilmore BL, Burstein H, Peluso RW, Polisky B et al (2008) Artificial miRNAs mitigate shRNA-mediated toxicity in the brain: implications for the therapeutic development of RNAi. Proc Natl Acad Sci USA 105:5868–5873
113. Castanotto D, Sakurai K, Lingeman R, Li H, Shively L, Aagaard L, Soifer H, Gatignol A, Riggs A, Rossi JJ (2007) Combinatorial delivery of small interfering RNAs reduces RNAi efficacy by selective incorporation into RISC. Nucleic Acids Res 35:5154–5164
114. Vickers TA, Lima WF, Nichols JG, Crooke ST (2007) Reduced levels of Ago2 expression result in increased siRNA competition in mammalian cells. Nucleic Acids Res 35:6598–6610
115. Snyder LL, Esser JM, Pachuk CJ, Steel LF (2008) Vector design for liver-specific expression of multiple interfering RNAs that target hepatitis B virus transcripts. Antiviral Res 80:36–44
116. Wiznerowicz M, Trono D (2003) Conditional suppression of cellular genes: lentivirus vector-mediated drug-inducible RNA interference. J Virol 77:8957–8961
117. Zhou X, Symons J, Hoppes R, Krueger C, Berens C, Hillen W, Berkhout B, Das AT (2007) Improved single-chain transactivators of the Tet-On gene expression system. BMC Biotechnol 7:6

118. Grimm D, Kay MA (2007) Combinatorial RNAi: a winning strategy for the race against evolving targets? Mol Ther 15:878–888
119. Liu YP, Berkhout B (2008) Combinatorial RNAi strategies against HIV-1 and other escape-prone viruses. Int J Biosci Technol 1:1–10
120. Lisziewicz J, Sun D, Smythe J, Lusso P, Lori F, Louie A, Markham P, Rossi J, Reitz M, Gallo RC (1993) Inhibition of human immunodeficiency virus type 1 replication by regulated expression of a polymeric Tat activation response RNA decoy as a strategy for gene therapy in AIDS. Proc Natl Acad Sci USA 90:8000–8004
121. Sarver N, Cantin EM, Chang PS, Zaia JA, Ladne PA, Stephens DA, Rossi JJ (1990) Ribozymes as potential anti-HIV-1 therapeutic agents. Science 247:1222–1225
122. Chatterjee S, Johnson PR, Wong KK Jr (1992) Dual-target inhibition of HIV-1 in vitro by means of an adeno-associated virus antisense vector. Science 258:1485–1488
123. Levine BL, Humeau LM, Boyer J, MacGregor RR, Rebello T, Lu X, Binder GK, Slepushkin V, Lemiale F, Mascola JR et al (2006) Gene transfer in humans using a conditionally replicating lentiviral vector. Proc Natl Acad Sci USA 103:17372–17377
124. Kohn DB, Bauer G, Rice CR, Rothschild JC, Carbonaro DA, Valdez P, Hao Q, Zhou C, Bahner I, Kearns K et al (1999) A clinical trial of retroviral-mediated transfer of a rev-responsive element decoy gene into CD34(+) cells from the bone marrow of human immunodeficiency virus-1-infected children. Blood 94:368–371
125. Symensma TL, Giver L, Zapp M, Takle GB, Ellington AD (1996) RNA aptamers selected to bind human immunodeficiency virus type 1 Rev in vitro are Rev responsive in vivo. J Virol 70:179–187
126. Weinberg MS, Villeneuve LM, Ehsani A, Amarzguioui M, Aagaard L, Chen ZX, Riggs AD, Rossi JJ, Morris KV (2006) The antisense strand of small interfering RNAs directs histone methylation and transcriptional gene silencing in human cells. RNA 12:256–262
127. Abad X, Vera M, Jung SP, Oswald E, Romero I, Amin V, Fortes P, Gunderson SI (2008) Requirements for gene silencing mediated by U1 snRNA binding to a target sequence. Nucleic Acids Res 36:2338–2352
128. Knoepfel SA, Abad A, Abad X, Fortes P, Berkhout B (2012) Design of modified U1i molecules against HIV-1 RNA. Antiviral Res.94(3):208–216.
129. Anderson J, Banerjea A, Akkina R (2003) Bispecific short hairpin siRNA constructs targeted to CD4, CXCR4, and CCR5 confer HIV-1 resistance. Oligonucleotides 13:303–312
130. Martinez J, Patkaniowska A, Urlaub H, Luhrmann R, Tuschl T (2002) Single-stranded antisense siRNAs guide target RNA cleavage in RNAi. Cell 110:563–574
131. Anderson J, Akkina R (2005) CXCR4 and CCR5 shRNA transgenic CD34+ cell derived macrophages are functionally normal and resist HIV-1 infection. Retrovirology 2:53
132. Anderson J, Banerjea A, Planelles V, Akkina R (2003) Potent suppression of HIV type 1 infection by a short hairpin anti-CXCR4 siRNA. AIDS Res Hum Retroviruses 19:699–706
133. Ishaq M, Hu J, Wu X, Fu Q, Yang Y, Liu Q, Guo D (2008) Knockdown of cellular RNA helicase DDX3 by short hairpin RNAs suppresses HIV-1 viral replication without inducing apoptosis. Mol Biotechnol 39:231–238
134. Vandekerckhove L, Christ F, Van Maele B, De Rijck J, Gijsbers R, Van den HC, Witvrouw M, Debyser Z (2006) Transient and stable knockdown of the integrase cofactor LEDGF/p75 reveals its role in the replication cycle of human immunodeficiency virus. J Virol 80:1886–1896
135. An DS, Donahue RE, Kamata M, Poon B, Metzger M, Mao SH, Bonifacino A, Krouse AE, Darlix JL, Baltimore D et al (2007) Stable reduction of CCR5 by RNAi through hematopoietic stem cell transplant in non-human primates. Proc Natl Acad Sci USA 104: 13110–13115
136. Eekels JJM, Geerts D, Jeeninga RE, Berkhout B (2011) Long-term inhibition of HIV-1 replication with RNA interference against cellular cofactors. Antiviral Res 89:43–53
137. Lim JK, Glass WG, McDermott DH, Murphy PM (2006) CCR5: no longer a "good for nothing" gene–chemokine control of West Nile virus infection. Trends Immunol 27:308–312

138. Liu Z, Yang F, Robotham JM, Tang H (2009) Critical role of cyclophilin A and its prolyl-peptidyl isomerase activity in the structure and function of the hepatitis C virus replication complex. J Virol 83:6554–6565
139. Allers K, Hutter G, Hofmann J, Loddenkemper C, Rieger K, Thiel E, Schneider T (2010) Evidence for the cure of HIV infection by CCR5{Delta}32/{Delta}32 stem cell transplantation. Blood 117(10):2791–2799
140. Bangsberg DR, Charlebois ED, Grant RM, Holodniy M, Deeks SG, Perry S, Conroy KN, Clark R, Guzman D, Zolopa A et al (2003) High levels of adherence do not prevent accumulation of HIV drug resistance mutations. AIDS 17:1925–1932
141. Menendez-Arias L (2010) Molecular basis of human immunodeficiency virus drug resistance: an update. Antiviral Res 85:210–231
142. Lapidot T (2001) Mechanism of human stem cell migration and repopulation of NOD/SCID and B2mnull NOD/SCID mice. The role of SDF-1/CXCR4 interactions. Ann N Y Acad Sci 938:83–95
143. Yang F, Robotham JM, Nelson HB, Irsigler A, Kenworthy R, Tang H (2008) Cyclophilin A is an essential cofactor for hepatitis C virus infection and the principal mediator of cyclosporine resistance in vitro. J Virol 82:5269–5278
144. Gaspar HB, Thrasher AJ (2005) Gene therapy for severe combined immunodeficiencies. Expert Opin Biol Ther 5:1175–1182
145. Parkin NT, Chamorro M, Varmus HE (1992) Human immunodeficiency virus type 1 gag-pol frameshifting is dependent on downstream mRNA secondary structure: demonstration by expression in vivo. J Virol 66:5147–5151
146. Goff SP (2008) Knockdown screens to knockout HIV-1. Cell 135:417–420
147. Eekels JJM, Berkhout B (2011) Toward a durable treatment of HIV-1 infection using RNA interference. In: Grimm D (ed) Cellular RNA interference mechanisms, vol 102. Elsevier, London, pp 141–163
148. Bushman FD, Malani N, Fernandes J, D'Orso I, Cagney G, Diamond TL, Zhou H, Hazuda DJ, Espeseth AS, Konig R et al (2009) Host cell factors in HIV replication: meta-analysis of genome-wide studies. PLoS Pathog 5:e1000437
149. Karlas A, Machuy N, Shin Y, Pleissner KP, Artarini A, Heuer D, Becker D, Khalil H, Ogilvie LA, Hess S et al (2010) Genome-wide RNAi screen identifies human host factors crucial for influenza virus replication. Nature 463:818–822
150. Konig R, Stertz S, Zhou Y, Inoue A, Hoffmann HH, Bhattacharyya S, Alamares JG, Tscherne DM, Ortigoza MB, Liang Y et al (2010) Human host factors required for influenza virus replication. Nature 463:813–817
151. Li Q, Brass AL, Ng A, Hu Z, Xavier RJ, Liang TJ, Elledge SJ (2009) A genome-wide genetic screen for host factors required for hepatitis C virus propagation. Proc Natl Acad Sci USA 106:16410–16415
152. Sessions OM, Barrows NJ, Souza-Neto JA, Robinson TJ, Hershey CL, Rodgers MA, Ramirez JL, Dimopoulos G, Yang PL, Pearson JL et al (2009) Discovery of insect and human dengue virus host factors. Nature 458:1047–1050
153. van den Berg CH, Smit C, Bakker M, Geskus RB, Berkhout B, Jurriaans S, Coutinho RA, Wolthers KC, Prins M (2007) Major decline of hepatitis C virus incidence rate over two decades in a cohort of drug users. Eur J Epidemiol 22:183–193
154. Wheatley AK, Kramski M, Alexander MR, Toe JG, Center RJ, Purcell DF (2011) Co-expression of miRNA targeting the expression of PERK, but not PKR, enhances cellular immunity from an HIV-1 Env DNA vaccine. PLoS One 6:e18225
155. Liu Z, Robida JM, Chinnaswamy S, Yi G, Robotham JM, Nelson HB, Irsigler A, Kao CC, Tang H (2009) Mutations in the hepatitis C virus polymerase that increase RNA binding can confer resistance to cyclosporine A. Hepatology 50(1):25–33
156. Boudreau RL, Monteys AM, Davidson BL (2008) Minimizing variables among hairpin-based RNAi vectors reveals the potency of shRNAs. RNA 14:1834–1844
157. Brennecke J, Stark A, Russell RB, Cohen SM (2005) Principles of microRNA-target recognition. PLoS Biol 3:e85

158. Fedorov Y, Anderson EM, Birmingham A, Reynolds A, Karpilow J, Robinson K, Leake D, Marshall WS, Khvorova A (2006) Off-target effects by siRNA can induce toxic phenotype. RNA 12:1188–1196
159. Jackson AL, Burchard J, Schelter J, Chau BN, Cleary M, Lim L, Linsley PS (2006) Widespread siRNA "off-target" transcript silencing mediated by seed region sequence complementarity. RNA 12:1179–1187
160. Bridge AJ, Pebernard S, Ducraux A, Nicoulaz AL, Iggo R (2003) Induction of an interferon response by RNAi vectors in mammalian cells. Nat Genet 34:263–264
161. Sledz CA, Holko M, de Veer MJ, Silverman RH, Williams BR (2003) Activation of the interferon system by short-interfering RNAs. Nat Cell Biol 5:834–839
162. Marques JT, Williams BR (2005) Activation of the mammalian immune system by siRNAs. Nat Biotechnol 23:1399–1405
163. Eekels JJM, Pasternak AO, Schut AM, Geerts D, Jeeninga RE, Berkhout B (2012) A competitive cell growth assay for the detection of subtle effects of gene transduction on cell proliferation. Gene Ther. Advance online publication. doi: 10.1038/gt.2011.191.
164. Lackner AA, Veazey RS (2007) Current concepts in AIDS pathogenesis: insights from the SIV/macaque model. Annu Rev Med 58:461–476
165. Stremlau M, Owens CM, Perron MJ, Kiessling M, Autissier P, Sodroski J (2004) The cytoplasmic body component TRIM5alpha restricts HIV-1 infection in Old World monkeys. Nature 427:848–853
166. Traggiai E, Chicha L, Mazzucchelli L, Bronz L, Piffaretti JC, Lanzavecchia A, Manz MG (2004) Development of a human adaptive immune system in cord blood cell-transplanted mice. Science 304:104–107
167. Gimeno R, Weijer K, Voordouw A, Uittenbogaart CH, Legrand N, Alves NL, Wijnands E, Blom B, Spits H (2004) Monitoring the effect of gene silencing by RNA interference in human CD34+ cells injected into newborn RAG2-/- gammac-/- mice: functional inactivation of p53 in developing T cells. Blood 104:3886–3893
168. Legrand N, Weijer K, Spits H (2006) Experimental models to study development and function of the human immune system in vivo. J Immunol 176:2053–2058
169. Shultz LD, Ishikawa F, Greiner DL (2007) Humanized mice in translational biomedical research. Nat Rev Immunol 7:118–130
170. Manz MG (2007) Human-hemato-lymphoid-system mice: opportunities and challenges. Immunity 26:537–541
171. Legrand N, Cupedo T, van Lent AU, Ebeli MJ, Weijer K, Hanke T, Spits H (2006) Transient accumulation of human mature thymocytes and regulatory T cells with CD28 superagonist in "human immune system" Rag2(-/-)gammac(-/-) mice. Blood 108:238–245
172. Baenziger S, Tussiwand R, Schlaepfer E, Mazzucchelli L, Heikenwalder M, Kurrer MO, Behnke S, Frey J, Oxenius A, Joller H et al (2006) Disseminated and sustained HIV infection in CD34+ cord blood cell-transplanted Rag2-/-gamma c-/- mice. Proc Natl Acad Sci USA 103:15951–15956
173. Zhang L, Kovalev GI, Su L (2007) HIV-1 infection and pathogenesis in a novel humanized mouse model. Blood 109:2978–2981
174. Watanabe S, Terashima K, Ohta S, Horibata S, Yajima M, Shiozawa Y, Dewan MZ, Yu Z, Ito M, Morio T et al (2007) Hematopoietic stem cell-engrafted NOD/SCID/IL2Rgamma null mice develop human lymphoid systems and induce long-lasting HIV-1 infection with specific humoral immune responses. Blood 109:212–218
175. Berges BK, Wheat WH, Palmer BE, Connick E, Akkina R (2006) HIV-1 infection and CD4 T cell depletion in the humanized Rag2-/-gamma c-/- (RAG-hu) mouse model. Retrovirology 3:76
176. Berges BK, Akkina SR, Folkvord JM, Connick E, Akkina R (2008) Mucosal transmission of R5 and X4 tropic HIV-1 via vaginal and rectal routes in humanized Rag2(-/-)gammac(-/-) (RAG-hu) mice. Virology 373:342–351
177. An DS, Poon B, Ho Tsong FR, Weijer K, Blom B, Spits H, Chen IS, Uittenbogaart CH (2007) Use of a novel chimeric mouse model with a functionally active human immune system to study human immunodeficiency virus type 1 infection. Clin Vaccine Immunol 14:391–396

178. Goldstein H (2008) Summary of presentations at the NIH/NIAID New Humanized Rodent Models 2007 Workshop. AIDS Res Ther 5:3
179. Edelstein ML, Abedi MR, Wixon J (2007) Gene therapy clinical trials worldwide to 2007–an update. J Gene Med 9:833–842
180. Culliton BJ (1990) Gene therapy begins. Science 249:1372
181. Marshall E (1999) Gene therapy death prompts review of adenovirus vector. Science 286:2244–2245
182. Hacein-Bey-Abina S, Le Deist F, Carlier F, Bouneaud C, Hue C, de Villartay JP, Thrasher AJ, Wulffraat N, Sorensen R, Dupuis-Girod S et al (2002) Sustained correction of X-linked severe combined immunodeficiency by ex vivo gene therapy. N Engl J Med 346:1185–1193
183. Hacein-Bey-Abina S, von Kalle C, Schmidt M, McCormack MP, Wulffraat N, Leboulch P, Lim A, Osborne CS, Pawliuk R, Morillon E et al (2003) LMO2-associated clonal T cell proliferation in two patients after gene therapy for SCID-X1. Science 302:415–419
184. Cavazzana-Calvo M, Fischer A (2007) Gene therapy for severe combined immunodeficiency: are we there yet? J Clin Invest 117:1456–1465
185. Robotham JM, Nelson HB, Tang H (2009) Selection and characterization of drug-resistant HCV replicons in vitro with a flow cytometry-based assay. Methods Mol Biol 510:227–242
186. Bushman F, Lewinski M, Ciuffi A, Barr S, Leipzig J, Hannenhalli S, Hoffmann C (2005) Genome-wide analysis of retroviral DNA integration. Nat Rev Microbiol 3:848–858
187. Montini E, Cesana D, Schmidt M, Sanvito F, Ponzoni M, Bartholomae C, Sergi SL, Benedicenti F, Ambrosi A, Di Serio C et al (2006) Hematopoietic stem cell gene transfer in a tumor-prone mouse model uncovers low genotoxicity of lentiviral vector integration. Nat Biotechnol 24:687–696
188. Manilla P, Rebello T, Afable C, Lu X, Slepushkin V, Humeau LM, Schonely K, Ni Y, Binder GK, Levine BL et al (2005) Regulatory considerations for novel gene therapy products: a review of the process leading to the first clinical lentiviral vector. Hum Gene Ther 16:17–25
189. Rossi JJ, June CH, Kohn DB (2007) Genetic therapies against HIV. Nat Biotechnol 25:1444–1454

Chapter 12
Genome-Wide RNAi Screening to Identify Human Host Factors Crucial for Influenza Virus Replication

Katharina Ahrens and Alexander Karlas

Abstract Influenza A virus has developed strategies to exploit and in some cases subvert cellular proteins and pathways to promote its own replication and to suppress antiviral immune responses. Identification of these host factors would expand the number of potential drug targets far beyond the 11 proteins encoded in the viral genome. Recently, several laboratories have set out to provide an insight into the interface between influenza virus and its host by performing genome-wide siRNA silencing screens to characterize these host proteins and to monitor the effects on viral infectivity. Initial hits from each study were used to search databases of protein–protein interactions, allowing prediction of host-cell pathways likely to be involved either in the viral replicative cycle or in the immune response to viral infection. The results of these screens will promote our understanding of influenza virus biology as well as identify potential targets for the rational design of broad-spectrum antiviral drugs such as siRNA and small molecules.

12.1 Influenza

12.1.1 Life Cycle and Disease

Influenza is an infectious disease caused by RNA viruses of the family Orthomyxoviridae. Antigenic differences in virus nucleoprotein (NP) and matrix proteins (M1) have led to a classification of influenza viruses into types A, B, or C;

K. Ahrens • A. Karlas (✉)
Max Planck Institute for Infection Biology, Charitéplatz 1, 10117 Berlin, Germany
e-mail: karlas@mpiib-berlin.mpg.de

K.A. Howard (ed.), *RNA Interference from Biology to Therapeutics*, Advances in Delivery Science and Technology, DOI 10.1007/978-1-4614-4744-3_12,

however, only A and B cause significant disease in humans, with influenza A virus showing the highest degree of genetic variability. Influenza A viruses (IAV) are further subtyped according to their surface antigens, hemagglutinin (HA) and neuraminidase (NA), of which 16 HA subtypes and 9 NA subtypes have been identified to date [1, 2]. Influenza type A (IAV) is responsible for seasonal epidemics and reoccurring pandemics, such as the so-called Spanish flu (1918–1919) caused by an H1N1 virus and responsible for the deaths of at least 40 million people [3, 4]. The first influenza pandemic of the twenty-first century was caused by the H1N1 virus in 2009, resulting in over 80 million cases of infection worldwide. The highly pathogenic avian H5N1 has already spread throughout Asia, Europe, and Africa, overcoming host species barriers to infect humans with many fatal outcomes [5]. The current panel of preventive and therapeutic measures against influenza virus infections rests on (a) active vaccination and (b) the use of antiviral drugs directed against virus proteins as can be seen in the depiction of the viral life cycle (Fig. 12.1). Both strategies have their intrinsic limitations owing to the high variability of influenza viruses.

12.1.2 Vaccine and Variability

The most commonly employed method of controlling influenza A and B is by vaccination focused toward non-conserved surface proteins. However, several caveats exist when using this type of immunoprophylaxis: (a) overall rates of vaccination within the population still remain low; (b) efficacy of the vaccination is reduced or even contraindicated in high-risk groups, such as the elderly and the immunocompromised; and (c) antigenic drift and antigenic shift demand ongoing development of new vaccines every season [7].

Antigenic drift is the gradual evolution of viral strains due to frequent (point) mutations within the HA protein, the NA protein, or both [8]. Antigenic shift is only seen in influenza A viruses and results from the substitution of HA (and less frequently NA) subtypes, with novel ones resulting in new viruses unknown to the human immune system or last circulated decades before. Moreover, if such viral recombinants are derived from animal reservoirs, they may exhibit surface antigens completely unknown to the human immune system, hence increasing the pandemic potential of such a virus.

12.1.3 Drugs and Resistance

Existing antiviral drugs are typically directed against *bona fide* viral targets, so-called direct antiviral targets, bearing the usual risk of generating therapeutic resistance. Adamantane compounds, such as amantadine and rimantadine, have a type-specific inhibitory effect on type A but not on type B influenza viruses by

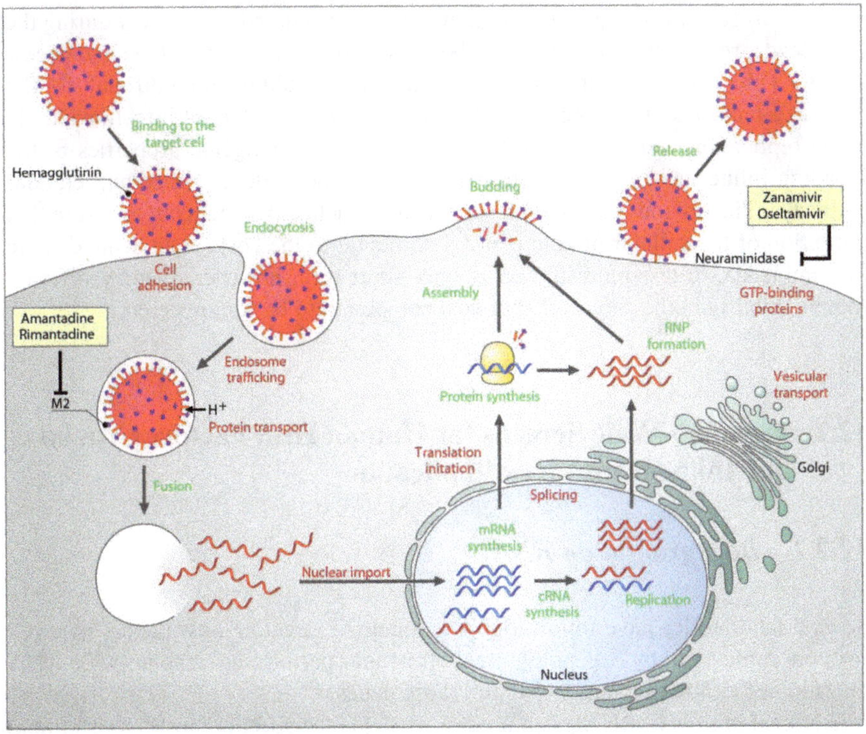

Fig. 12.1 Life cycle of influenza A virus and possible targets for therapeutic intervention. IAV enters the cell via receptor-mediated endocytosis and is internalized in endosomes. Endosomal acidification alters the conformation of the viral hemagglutinin, leading to the release of viral ribonucleoproteins (vRNPs) into the cytoplasm. Viral RNAs enter the nucleus via nuclear import, where they are transcribed to generate mRNA required for the synthesis of viral proteins, and are replicated by the viral RNA-dependent RNA polymerase complex. Newly generated viral vRNPs shuttle to the cytoplasm and assemble with viral proteins, forming mature viruses that bud from the cell surface. Two classes of antivirals are currently licensed for the treatment of influenza: M2 inhibitors, such as amantadine and rimantadine, which interfere with the viral M2 protein, thereby inhibiting the release of vRNPs out of the endosomes; and inhibitors such as zanamivir and oseltamivir, which target the viral neuraminidase and prevent the release of newly generated virions. Cellular pathways that are required for the IAV replication cycle and that have recently been identified in global siRNA screens are shown in *red*. Adapted from Min and Subbarao [6]

blocking the viral M2 protein [7]. A single mutation in human and avian influenza viruses, however, can confer resistance to amantadine [3, 7]. Therefore, the potential of amantadine and amantadine derivatives for controlling pandemics might be limited. In contrast, anti-influenza neuraminidase (NA) inhibitors, such as oseltamivir and zanamivir preventing viral release from the cell membrane, have been shown to efficiently inhibit influenza virus A and B strains in clinical studies [9]. Although resistance to these compounds has been shown to emerge less frequently than resistance to amantadine, resistant viruses have been identified *in vitro* [10] and *in vivo* (Fig. 12.1).

Whereas functions of the viral proteins have been studied extensively during the last decade, relatively little is known about the cellular factors involved and necessary for viral infection. A missing cellular function would be more difficult for the virus to adapt to, so targeting a host factor essential for influenza infection should affect replication, independent of the type, strain, and antigenic properties of the invading influenza virus, since viruses may share dependence on certain cellular pathways. The obvious caveat with such an approach is the necessity to insure that inhibition of the cellular protein is not detrimental to the host. In an acute disease like influenza, drug administration is only short term however, thereby offering more flexibility in the choice of chemical compound to be administered.

12.2 Genome-Wide Screens for Human Host Factors Crucial for Influenza A Virus Replication

12.2.1 Background on RNAi

Several laboratories have followed this rationale of targeting host factors required for viral replication by expending great effort into performing genome-wide RNA interference (RNAi) screens to identify these genes.

RNA interference (RNAi) is a process of sequence-specific, posttranscriptional gene silencing in animals and plants by degradation of mRNA. The RNA interference pathway was first described in 1998 by Fire and Mellow, who injected double-stranded (dsRNA) into the nematode *Caenorhabditis elegans*, initiating sequence-specific degradation of cytoplasmic mRNAs containing the same sequence as the dsRNA trigger [11]. Early applications using long dsRNA were not effective in most mammalian cells, however, because they induced an antiviral interferon (IFN) response hereby functioning as the host's first line of defense against viral infection but ultimately leading to cell death [12]. Further investigations regarding the RNA silencing mechanism in different organisms revealed that the mediators of sequence-specific messenger RNA degradation were 21- and 22-nucleotide small interfering RNAs (siRNAs) generated by ribonuclease III cleavage from longer dsRNAs by the enzyme termed Dicer [13]. Components of the RNAi machinery specifically recognize these small interfering RNA (siRNA) duplexes and incorporate a single siRNA strand into the so-called RISC complex (RNA-induced silencing complex) [14]. After integration into the RISC, the antisense "guide" strand base-pairs to the cognate site of the mRNA and induces cleavage of the mRNA, thereby preventing it from being used as a translation template. This mechanism, therefore, enables strong reduction of *de novo* protein synthesis for the corresponding mRNA in a sequence-specific manner [14]. The natural function of RNAi appears to be the protection of the genome against invasion of either mobile genetic elements such as transposons or of viruses, which exploit the RNA-producing machinery of the host cell upon infection to produce aberrant RNA or dsRNA.

Interestingly, these high-impact early studies describing RNAi in mammalians often disregarded the fact that chemically synthesized and virally transcribed siRNAs are still capable of activating components of the innate immune response in certain *in vitro* culture systems, as demonstrated by several later reports [15–19]. This immune responsiveness for siRNA was shown to be highly cell type- and sequence-specific, however. siRNAs capable of activating the mammalian immune response do so primarily through endosomal activation of the Toll-like receptor 7/8 (TLR7/8) pathway, resulting in induction of interferons (IFNs) and inflammatory cytokines from immune cells [19, 20]. Cytokine production via siRNA was shown to be eliminated by the selective incorporation of 2′-*O*-methyl (2′-OMe) modified ribonucleotides into the constituent RNA oligonucleotides [16, 21]. Furthermore, Judge et al. were able to demonstrate that the innate immune system is able to recognize immunostimulatory RNA motifs within both ssRNA and dsRNA via protein members of the TLR family, a feature analogous to the recognition of CpG DNA motifs by TLR9 [19, 22]. siRNA duplexes containing several 5′-UGU-3′ motifs were found to be highly immunostimulatory, and conversely, functional siRNA sequences lacking these GU-rich regions were shown to have inherently low immunostimulatory capacity [19].

Chemically synthesized siRNA has become the method of choice to manipulate gene expression in mammalian cell culture in low- and in high-throughput approaches. The potential influence of siRNA-mediated immune response on key readouts for *in vivo* RNAi studies is a critical consideration; however, responsibility rests on the individual researcher to appropriately choose cell type and siRNA library and to include prudent controls and hit identification criteria. Putatively immunostimulatory motifs within siRNA should be considered when designing synthetic siRNA to avoid or minimize immunotoxicity and to reduce potential off-target gene effects.

12.2.2 Genome-Wide RNAi Screens

Cell-based RNAi screening technology allows for genome-wide loss-of-function studies and is broadly used in the identification of genes associated with specific biological phenotypes. The basic methodology of RNAi screening includes the following steps:

- Choice of an RNAi library, usually consisting of several thousand siRNAs
- Selection of a robust and suitable cell line, as well as of an adequate assay readout system
- Transfection of cells with RNAi agents from the chosen RNAi library
- Treatment and/or incubation of cells
- Signal detection and statistical and bioinformatical analysis

Five screens for host factors required for influenza virus replication were performed in the manner described above and are presented and further discussed in the following.

The first genome-wide screen to identify host factors involved in IAV infection was by Hao et al. using *Drosophila* RNAi technology [23]. Hao and coworkers first modified IAV to allow for infection of *Drosophila* cells and for expression of a reporter gene product in infected cells. They then tested an RNAi library against 13,071 genes (90% of the *Drosophila* genome), identifying over 100 genes whose suppression in *Drosophila* cells significantly inhibited or stimulated reporter gene (*Renilla* luciferase) expression from an IAV-derived vector. Among the over 100 candidate genes found to be crucial to IAV replication in *Drosophila*, they selected the human homologues of several encoding components in host pathways/machineries known to be involved in the life cycle of IAV, such as *ATP6V0D1* (an ATPase involved in the endocytosis pathway), *COX6A1* (a cytochrome c oxidase subunit with mitochondrial function), and *NXF1* (a nuclear RNA export factor) for further analysis in mammalian cells. The validated factors were found to play key roles in the replication of H5N1 and H1N1 influenza A viruses, but not in vesicular stomatitis virus or vaccinia virus, in human HEK293 cells [23].

With the rapid development of more advanced RNAi-based screening technology including more sophisticated, commercially available siRNA libraries as well as improved robotics, image acquisition, and analysis instruments and software, comprehensive analysis of mammalian host cell functions was made possible.

Recently, three different groups have performed genome-wide RNAi screens for influenza virus host factors in human cells [4, 24, 25].

Experiments by Brass and coworkers used A/PR/8/34 virus-infected human osteosarcoma cells (U2OS) and replication efficiency was detected by microscopy-based quantification of the HA signal. This system allowed for the identification of host gene products involved in the early and mid-stages of IAV infection only since readout of the assay was the measurement of a viral protein. The primary screen identified 312 human genes regulating susceptibility or resistance to influenza PR8 virus infection and confirmed 129 required host factors for IAV infection. Notably, they discovered that the interferon-inducible transmembrane proteins (IFITM) 1, 2, and 3 restrict an early step in influenza A viral replication [24]. The IFITM proteins were shown to be induced by interferons type I and II, subsequently leading to basal resistance against influenza A virus. In addition, it was shown that IFITM proteins inhibit the early replication of flaviviruses, including Dengue virus and West Nile virus [24].

König et al. performed a genome-wide siRNA screen with human lung epithelial (A549) cells [25]. The coding region for the influenza A/WSN/33 virus hemagglutinin (HA) protein was replaced with that of *Renilla* luciferase to facilitate the readout of the screen. Since no HA was produced, the recombinant virus could not complete its replication cycle. Similar to the readout system employed by Brass et al., this RNAi screen also focused on the cellular factors required for viral entry, uncoating, vRNP nuclear import, genome transcription, and viral protein translation, excluding late events such as virus assembly, budding, and release. König et al. identified 295 cellular cofactors required for early-stage influenza virus replication, 219 of which were confirmed to be required for efficient wild-type influenza virus growth. Further analysis of a subset of genes showed 23 factors necessary for viral entry, including members of the vacuolar ATPase (vATPase) and COPI-protein families, fibroblast growth factor receptor (FGFR) proteins, and glycogen synthase kinase 3 (GSK3)-beta [25].

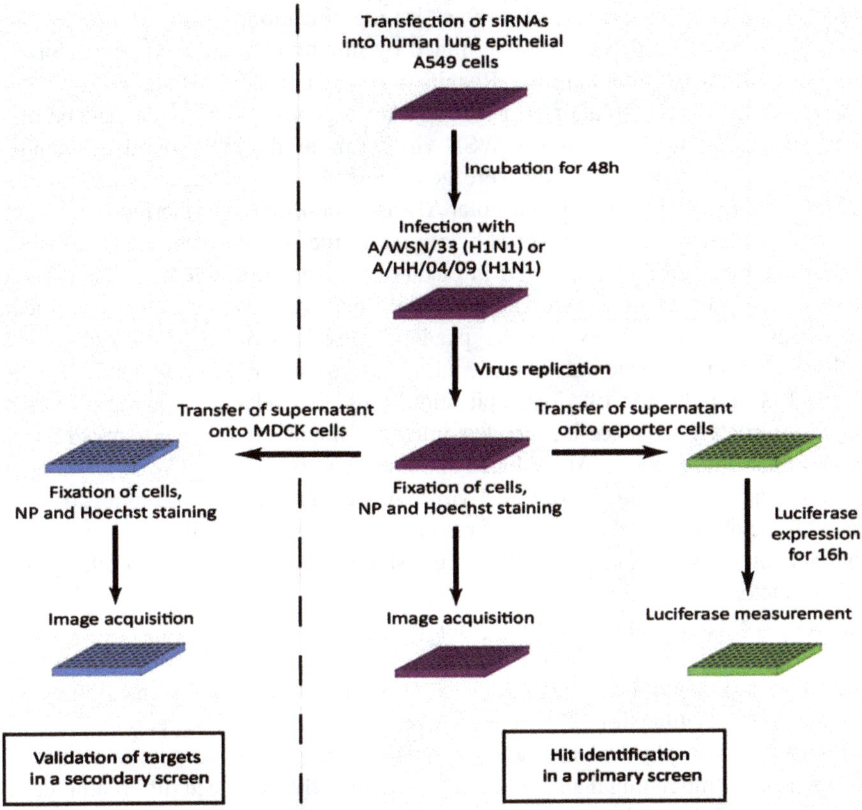

Fig. 12.2 Outline of screening and validation procedure. A549 cells were reverse transfected with a library of siRNAs and further incubated for 48 h to allow maximal gene knockdown. Subsequently, cells were infected with IAV (either A/WSN/33 or the new pandemic H1N1 virus A/HH/04/09), and 24 h (primary screen) or 36 h (secondary screen) later virus replication was quantified using different techniques. The infection rate within transfected and infected A549 cells was detected using immunofluorescence and automated microscopy. The numbers of infectious virus particles were monitored by addition of A549 cell supernatants onto 293 T reporter cells that had been transfected with an influenza-virus-specific luciferase construct (*FlaA*) [26], followed by luciferase measurement 16 h post-reinfection. For the validation of primary hits, the numbers of infectious particles in the supernatants of A549 cells were quantified by reinfection of MDCK cells for 6 h, followed by immunofluorescence and automated microscopy

In contrast, our group [4] set out to study the entire viral replication cycle from viral entry to viral budding in a genome-wide RNAi screen. Human A549 lung epithelial cells were first transfected with a genome-wide library of siRNA and subsequently infected with influenza A/WSN/33 virus. The readout was then divided into two parts: (1) Infected A549 cells were stained with a virus-specific antibody at 24 h postinfection to monitor cell infection rates, and (2) virus supernatants from these transfected A549 cells were transferred onto 293 T human embryonic kidney reporter cells, containing an inducible, influenza-virus-specific luciferase construct (*FlaA*) [26] for quantification of released virus (Fig. 12.2). Ultimately, three parameters were included in the process of primary hit identification: luciferase expression, percentage of infected cells (as determined by immunofluorescence microscopy), and total number of infected cells. Results from each of the three parameters were combined, and from a total of over 22,000 human genes, 287 were designated primary hits, affecting replication of WSN virus. Among the high-confidence candidates pivotal in IAV replication were genes encoding for nuclear export factors *NXF1* and *XPO1*, as well as the vacuolar ATPase *ATP6V0D1*. Furthermore, our data set was markedly enriched in genes belonging to the spliceosome, the small ribosomal subunit, and those involved in eukaryotic translation initiation and coated vesicle transport. One hundred and sixty-eight hits were confirmed in subsequent analysis, using A/WSN/33 and the pandemic H1N1 virus A/Hamburg/04/2009 influenza strains. A gene was considered a hit if two or more individual siRNAs showed clear reduction of virus replication, thus also selecting against off-target effects by siRNA-sequence-independent immunostimulation. In order to avoid false-positive hits caused by RNAi-mediated toxicity, we also performed WST-1 assays to monitor viability of cells. Also, H5N1 virus (A/Vietnam/1203/2004) replication was found to be reduced by two orders of magnitude by a subset of siRNAs, suggesting that influenza A viruses of different subtypes share a dependency on a specific group of host factors.

Shapira et al. used an integrative genomics approach to investigate the IAV-host cell relationship. They used a combination of yeast two-hybrid analysis, genome-wide expression profiling, and RNAi screen to determine genes transcriptionally regulated by IAV infection [27]. First, they performed a systematic yeast two-hybrid screen to build a physical interaction map for viral proteins and human cellular genes using computational analysis. They subsequently examined differential gene expression in primary human bronchial epithelial cells (HBECs) exposed to either IAV, or to viral RNA, IAV lacking the NS1 gene (ΔNS1), or to interferon beta (IFN-β). These different approaches yielded 1,745 candidate genes, 1,056 of which were shown to be transcriptionally regulated by influenza virus infection upon further analysis. Further validation using siRNA pools to mediate knockdown resulted in a list of 616 human genes affecting virus replication and/or IFN production (Table 12.1).

12.2.3 Results and Interpretation

The genome-wide siRNA screens have provided an overabundance of likely influenza host factors required for steps throughout all stages of the infection cycle; however, only a limited number of factors actually overlap among the individual screens when analyzed at gene level. Only ~8% of genes identified are common to at least two of the five siRNA screens discussed above, with no gene shared by all five. If the respective data is analyzed at the level of cellular function instead of that of gene name, however, more matches appear, suggesting that the screens agree at the level of functional pathway, just not by individual gene name. Also, while

Table 12.1 RNAi screens performed for the identification of human host factors associated in IAV infection and replication, as described in [4, 23–25, 27] and adapted from [5, 6, 28]

	Karlas et al. [4]	König et al. [25]	Brass et al. [24]	Hao et al. [23]	Shapira et al. [27]
Cell type	Human lung (A549)	Human lung (A549)	Human osteosarcoma (U2OS)	Drosophila (DL1)	Human (HBEC)
Virus strain	WSN	Recombinant WSN	PR8	Recombinant WSN	PR8
Readout method	NP expression and luciferase activity	Luciferase activity	HA expression on cell surface	Luciferase activity	Infectious virus yield by luciferase activity
Events in viral life cycle covered by screen	Attachment until budding/release	Attachment until protein expression	Attachment until HA surface trafficking	Uncoating until protein expression	Attachment until budding/release
siRNA library used	Qiagen	Qiagen	Dharmacon	Ambion	Dharmacon
Number of genes targeted	22,843	19,628	17,877	13,071	1,745[a]
Number of filtered hits	287	295	312	110	Not determined

[a]1,745 candidate genes identified via Y2H and genome-wide microarray analyses subjected to siRNA screen for validation

screens are initially unbiased, the necessity to establish criteria by which to select hits automatically introduces bias: Scores and hit lists are dependent on experimental setup of the screening systems employed (cell types, virus strains, reporter assays, and variability among them) as well as on data evaluation methods, i.e., by way of different analysis algorithms, Z-score cutoffs, and further in- or exclusion criteria (see Table 12.1). Moreover, knockdown efficiency can vary among different siRNAs used and can also be due to different protein expression levels and stabilities among cell types. In addition, the screens cannot differentiate among factors with redundant functions; these functions, thus, stay unaffected by single-gene knockdown.

Several cross-comparative reviews have attempted to find common patterns among the host target genes identified from the results of the genome-wide screens discussed [5, 6, 28]. A very comprehensive, comparative review was performed by Watanabe et al. which identified a total of 1,449 human genes potentially involved during the influenza A infection and replication process. Watanabe et al. then narrowed down this number to 128 candidates by pair-wise comparisons of the respective candidate genes and by selection of those found in at least two screens. They include a screen by Sui et al. [29] into their bioinformatic assessment which employs the so-called random homozygous gene perturbation (RHGP) technique based on insertion of a lentiviral genetic element at a single site in one allele of the genome in either the sense or antisense orientation. Unlike RNAi screens, the RHGP system can theoretically knock down or overexpress any gene without any prior knowledge or annotation of that gene, resulting in alteration of the phenotype.

In their metareview, Watanabe et al. found the number of genes common between pairs of screens to be relatively small (from 0 to 32 genes in the 15 pair-wise comparisons performed). This low incidence of overlap between hits of the different screens is similar to that observed in three RNAi screens aimed at identifying factors required for HIV replication [30–32]. In both cases, the most likely explanation for this low degree of overlaps originates from differences within the screening systems, such as cell type, virus strain, siRNA libraries employed, and/or individual readout system that could result in bias among the screens. This explanation is supported by the fact that the number of common hits among the IAV screens was highest among those performed by König et al. and by our group. Both of these screens were performed using the same siRNA library and both used the same human epithelial cell line.

Watanabe et al. use the PANTHER Classification System [33] as well as analysis by Reactome [34] to assess for enriched gene categories among the 128 candidate genes. These map the influenza virus-host interaction and respective enriched gene targets to individual steps of the influenza virus life cycle, based on the known functions of the identified host proteins. The greatest overlap among the hits discovered in the individual screens is found among the following four groups of host factors: (a) coating on vesicles-mediating retrograde transport from the Golgi to the ER and endosomal trafficking mediated by coatomer 1 (COPI) transport complex (*ARCN1*, *COPA*, *COPB2*, and *COPG*), (b) genes encoding the vacuolar ATPases (i.e., *ATP6V0C*, *ATP6V0D1*, and *ATP6V1A*), (c) nuclear transport of mRNAs and proteins (i.e., *KPNB1*, *Nup98*, and *NXF1*), and (d) splicing of cellular pre-mRNA (i.e., *PTBP1*, *NHP2L1*, *SNRP70*, *SF3B1*, *SF3A1*, *P14*, and *PRPF8*) [5].

12.3 Applications

The human genes ostensibly involved in the influenza A infection and replication process as identified in the screens discussed constitute potential future targets for antiviral strategies. Therapeutically targeting a host factor pivotal for IAV infection is anticipated to minimize the likelihood of developing drug-resistant variants and, thus, also to protect from new emerging influenza viruses. There are two main strategies of how to apply this information with respect to drug development which will be discussed in more detail below: The first strategy focuses on RNAi-based post-transcriptional prevention of expression of the respective human genes. The second approach is based upon posttranslational binding and inactivation of the corresponding human gene products (proteins) by small molecule inhibitors or new chemical entities. By targeting cellular proteins required by the virus, both approaches reduce the chances of drug-resistances arising and also enhance the possibility of identifying a broad-spectrum antiviral blocking certain cellular pathways shared by different viruses.

12.3.1 Therapeutic Applications of siRNA Directed Against Influenza

Since siRNAs are regulators of gene expression in cells, and are specific, efficient, and considered nontoxic, silencing of host genes essential for viral replication not only holds great therapeutic promise but also avoids the problem of resistance development often encountered when employing the traditional approach of inhibiting expression of viral genes.

Although great progress in the drug-like use of siRNAs has been made in recent years, the success of siRNAs as therapeutics to block and subsequently degrade mRNA of a specific host gene still depends to a great extent on the development of efficient *in vivo* delivery methods. siRNAs need to cross the cell membranes and to enter cells at the site of influenza infection to then be incorporated into the silencing complex. In addition to delivery, specificity and stability are two major obstacles in the development of therapeutic siRNA. The two most important caveats encountered in the siRNA selection process, with respect to specificity, are off-target effects caused by unsolicited gene homologies and immune stimulation due to recognition of specific, unmodified siRNA by the innate immune system [35]. Off-target effects induced by mismatch tolerance are influenced mainly by the thermodynamic stability of the first few base pairs of both ends of an siRNA duplex, crucial for the determination of which siRNA strand will be incorporated into the RISC. A common method utilized to improve stability and to avoid off-target effects is the inclusion of various chemical modifications of the ribose at the 2′ position of the sense strand [36–38].

It was demonstrated in two preclinical investigations targeting viral components inhibiting IAV replication that antiviral siRNAs provided both prophylactic and therapeutic benefits during influenza infection in mice [39]. Also, siRNAs targeting

IAV viral nucleoprotein or polymerase protected against an otherwise lethal infection [40]. Robbins et al., however, were able to demonstrate that the effects observed in both of these studies were most likely misinterpreted, owing to the use of a green fluorescent protein (GFP) control siRNA identified as having unusually low immunostimulatory activity compared with the active anti-influenza-formulated siRNAs [41]. Remarkably, this GFP siRNA has served as a negative control for a number of groups reporting therapeutic effects of siRNAs [42]. Thus, the existence of putatively immunostimulatory motifs within functional siRNAs must be considered when attempting to minimize immunotoxicity and to reduce the potential for off-target gene effects.

An important consideration especially when intending to use siRNA targeting a cellular host factor in a therapeutic application for influenza virus infection is that knockdown kinetics of siRNA will always lag behind IAV replication kinetics and spread in the body. This could interfere with treatment implementation since a potential antiviral siRNA therapeutic would have to be administered within hours after virus infection. Nonetheless, therapeutic siRNAs can be very well envisioned for prophylactic treatment when attempting, e.g., to protect family members of an infected person. Further studies will show if the benefits of such treatment outweigh the conceivable risks.

12.3.2 Small Molecule Inhibitors

An alternative to siRNA to block the human target genes identified and pivotal for IAV propagation is the use of small molecule inhibitors. Protein–protein interactions are central to many biological processes in the human body and in the course of viral infection. Therefore, inhibiting these interactions by small organic molecule antagonists offers an attractive and feasible opportunity for therapeutic intervention. However, development of small molecule inhibitors is a complex task due to factors such as a) the general lack of small molecule starting points for drug design; b) difficulty of small molecule compounds in competing against the large surface area typically involved in a protein–protein interaction, which furthermore tends to be fairly flat and devoid of small molecule binding pockets; c) difficulty of distinguishing real binding from that present only as an artifact; and d) the size and quality and makeup of small molecule libraries [43].

The development of a suitable assay system is a prerequisite for screening compound libraries for the identification of small molecule ligands able to bind and to interfere with IAV essential targets and to thereby reduce virus pathology.

Despite these hurdles, administering small molecule antagonist targeting a cellular protein required by the virus appears to hold great therapeutic promise, ensuring of course that inhibition of the cellular protein is not detrimental to the host. A striking example in support of the idea is the protection of humans from HIV infection conferred from the homozygous deletion in the HIV co-receptor gene

CCR5. This subgroup of people, however, do not suffer from any known pathology, and an anti-HIV drug was successfully developed on this basis [44, 45].

Since influenza viruses cause acute infections, the transient reduction of certain cellular proteins by small molecule inhibitors (as well as by siRNAs) might be even better tolerated as compared to the complete and permanent lack of host determinants. In addition, a drug inhibiting virus host factors would have reduced chances of drug resistances arising, and since different viruses might share dependence on similar cellular pathways, this type of drug could potentially possess broad-spectrum antiviral activity.

12.4 Future Perspectives

All viruses depend on the host cell machinery for their replication. While there have been tremendous advances in the study of influenza A virus and the protein functions it encodes, knowledge of how the IAV-usurped host cell promotes virus replication is limited. However, therapeutically targeting a host factor pivotal for IAV infection is anticipated to minimize the likelihood of developing drug-resistant variants and thus to also protect from novel emerging influenza viruses. The siRNA screens discussed here have followed this rationale in order to define host factors which could serve as targets for antiviral drug development. Interestingly, whereas most of the potential host factors appear to function by facilitating, directly or indirectly, virus replication, those restricting virus replication were also detected in the genome-wide screens. Notably, Brass et al. were able to demonstrate that interferon-inducible transmembrane proteins (IFITM) function as first line antagonists of viral infection. These findings support the notion described above that different viruses might be therapeutically targeted by one antiviral which blocks the same essential cellular gene product they all mutually depend on.

Acknowledgments We thank T.F. Meyer for providing constructive comments and K. Holden-Dye for critical reading of the manuscript. AK and KA are supported by European Commission funding under the 7th Framework Programme HEALTH.2010.2.3.3-4 Novel therapeutics against influenza; FP7-INFLUENZA-2010.

References

1. Zambon MC (1999) Epidemiology and pathogenesis of influenza. J Antimicrob Chemother 44(Suppl B):3–9
2. Corti D, Voss J, Gamblin SJ, Codoni G, Macagno A, Jarrossay D et al (2011) A neutralizing antibody selected from plasma cells that binds to group 1 and group 2 influenza A hemagglutinins. Science 333(6044):850–856
3. Horimoto T, Kawaoka Y (2005) Influenza: lessons from past pandemics, warnings from current incidents. Nat Rev Microbiol 3(8):591–600

4. Karlas A, Machuy N, Shin Y, Pleissner KP, Artarini A, Heuer D et al (2010) Genome-wide RNAi screen identifies human host factors crucial for influenza virus replication. Nature 463(7282):818–822
5. Watanabe T, Watanabe S, Kawaoka Y (2010) Cellular networks involved in the influenza virus life cycle. Cell Host Microbe 7(6):427–439
6. Min JY, Subbarao K (2010) Cellular targets for influenza drugs. Nat Biotechnol 28(3):239–240
7. Ludwig S, Planz O, Pleschka S, Wolff T (2003) Influenza-virus-induced signaling cascades: targets for antiviral therapy? Trends Mol Med 9(2):46–52
8. Both GW, Sleigh MJ, Cox NJ, Kendal AP (1983) Antigenic drift in influenza virus H3 hemagglutinin from 1968 to 1980: multiple evolutionary pathways and sequential amino acid changes at key antigenic sites. J Virol 48(1):52–60
9. Gubareva LV, Kaiser L, Hayden FG (2000) Influenza virus neuraminidase inhibitors. Lancet 355(9206):827–835
10. Gubareva LV (2004) Molecular mechanisms of influenza virus resistance to neuraminidase inhibitors. Virus Res 103(1–2):199–203
11. Fire A, Xu S, Montgomery MK, Kostas SA, Driver SE, Mello CC (1998) Potent and specific genetic interference by double-stranded RNA in Caenorhabditis elegans. Nature 391(6669):806–811
12. Stark GR, Kerr IM, Williams BR, Silverman RH, Schreiber RD (1998) How cells respond to interferons. Annu Rev Biochem 67:227–264
13. Elbashir SM, Lendeckel W, Tuschl T (2001) RNA interference is mediated by 21- and 22-nucleotide RNAs. Genes Dev 15(2):188–200
14. Hammond SM, Bernstein E, Beach D, Hannon GJ (2000) An RNA-directed nuclease mediates post-transcriptional gene silencing in Drosophila cells. Nature 404(6775):293–296
15. Bridge AJ, Pebernard S, Ducraux A, Nicoulaz AL, Iggo R (2003) Induction of an interferon response by RNAi vectors in mammalian cells. Nat Genet 34(3):263–264
16. Judge A, MacLachlan I (2008) Overcoming the innate immune response to small interfering RNA. Hum Gene Ther 19(2):111–124
17. Sledz CA, Holko M, de Veer MJ, Silverman RH, Williams BR (2003) Activation of the interferon system by short-interfering RNAs. Nat Cell Biol 5(9):834–839
18. Kleinman ME, Yamada K, Takeda A, Chandrasekaran V, Nozaki M, Baffi JZ et al (2008) Sequence- and target-independent angiogenesis suppression by siRNA via TLR3. Nature 452(7187):591–597
19. Judge AD, Sood V, Shaw JR, Fang D, McClintock K, MacLachlan I (2005) Sequence-dependent stimulation of the mammalian innate immune response by synthetic siRNA. Nat Biotechnol 23(4):457–462
20. Hornung V, Guenthner-Biller M, Bourquin C, Ablasser A, Schlee M, Uematsu S et al (2005) Sequence-specific potent induction of IFN-alpha by short interfering RNA in plasmacytoid dendritic cells through TLR7. Nat Med 11(3):263–270
21. Judge AD, Bola G, Lee AC, MacLachlan I (2006) Design of noninflammatory synthetic siRNA mediating potent gene silencing in vivo. Mol Ther 13(3):494–505
22. Krieg AM (2002) CpG motifs in bacterial DNA and their immune effects. Annu Rev Immunol 20:709–760
23. Hao L, Sakurai A, Watanabe T, Sorensen E, Nidom CA, Newton MA et al (2008) Drosophila RNAi screen identifies host genes important for influenza virus replication. Nature 454(7206):890–893
24. Brass AL, Huang IC, Benita Y, John SP, Krishnan MN, Feeley EM et al (2009) The IFITM proteins mediate cellular resistance to influenza A H1N1 virus, West Nile virus, and dengue virus. Cell 139(7):1243–1254
25. König R, Stertz S, Zhou Y, Inoue A, Hoffmann HH, Bhattacharyya S et al (2010) Human host factors required for influenza virus replication. Nature 463(7282):813–817
26. Lutz A, Dyall J, Olivo PD, Pekosz A (2005) Virus-inducible reporter genes as a tool for detecting and quantifying influenza A virus replication. J Virol Methods 126(1–2):13–20

27. Shapira SD, Gat-Viks I, Shum BO, Dricot A, de Grace MM, Wu L et al (2009) A physical and regulatory map of host-influenza interactions reveals pathways in H1N1 infection. Cell 139(7):1255–1267
28. Stertz S, Shaw ML (2011) Uncovering the global host cell requirements for influenza virus replication via RNAi screening. Microbes Infect 13(5):516–525
29. Sui B, Bamba D, Weng K, Ung H, Chang S, Van DJ et al (2009) The use of random homozygous gene perturbation to identify novel host-oriented targets for influenza. Virology 387(2):473–481
30. Brass AL, Dykxhoorn DM, Benita Y, Yan N, Engelman A, Xavier RJ et al (2008) Identification of host proteins required for HIV infection through a functional genomic screen. Science 319(5865):921–926
31. König R, Zhou Y, Elleder D, Diamond TL, Bonamy GM, Irelan JT et al (2008) Global analysis of host-pathogen interactions that regulate early-stage HIV-1 replication. Cell 135(1):49–60
32. Zhou H, Xu M, Huang Q, Gates AT, Zhang XD, Castle JC et al (2008) Genome-scale RNAi screen for host factors required for HIV replication. Cell Host Microbe 4(5):495–504
33. Mi H, Lazareva-Ulitsky B, Loo R, Kejariwal A, Vandergriff J, Rabkin S et al (2005) The PANTHER database of protein families, subfamilies, functions and pathways. Nucleic Acids Res 33(Database issue):D284–D288
34. Matthews L, Gopinath G, Gillespie M, Caudy M, Croft D, de Bono B et al (2009) Reactome knowledgebase of human biological pathways and processes. Nucleic Acids Res 37(Database issue):D619–D622
35. de Fougerolles AR (2008) Delivery vehicles for small interfering RNA in vivo. Hum Gene Ther 19(2):125–132
36. Jackson AL, Linsley PS (2004) Noise amidst the silence: off-target effects of siRNAs? Trends Genet 20(11):521–524
37. Dorsett Y, Tuschl T (2004) siRNAs: applications in functional genomics and potential as therapeutics. Nat Rev Drug Discov 3(4):318–329
38. Jackson AL, Burchard J, Leake D, Reynolds A, Schelter J, Guo J et al (2006) Position-specific chemical modification of siRNAs reduces "off-target" transcript silencing. RNA 12(7):1197–1205
39. Ge Q, Filip L, Bai A, Nguyen T, Eisen HN, Chen J (2004) Inhibition of influenza virus production in virus-infected mice by RNA interference. Proc Natl Acad Sci USA 101(23):8676–8681
40. Tompkins SM, Lo CY, Tumpey TM, Epstein SL (2004) Protection against lethal influenza virus challenge by RNA interference in vivo. Proc Natl Acad Sci USA 101(23):8682–8686
41. Robbins M, Judge A, Ambegia E, Choi C, Yaworski E, Palmer L et al (2008) Misinterpreting the therapeutic effects of small interfering RNA caused by immune stimulation. Hum Gene Ther 19(10):991–999
42. Behlke MA (2006) Progress towards in vivo use of siRNAs. Mol Ther 13(4):644–670
43. Arkin MR, Wells JA (2004) Small-molecule inhibitors of protein-protein interactions: progressing towards the dream. Nat Rev Drug Discov 3(4):301–317
44. Gulick RM, Lalezari J, Goodrich J, Clumeck N, DeJesus E, Horban A et al (2008) Maraviroc for previously treated patients with R5 HIV-1 infection. N Engl J Med 359(14):1429–1441
45. Liu R, Paxton WA, Choe S, Ceradini D, Martin SR, Horuk R et al (1996) Homozygous defect in HIV-1 coreceptor accounts for resistance of some multiply-exposed individuals to HIV-1 infection. Cell 86(3):367–377

Chapter 13
The Application of MicroRNAs in Cancer Diagnostics

Karina D. Sørensen, Marie S. Ostenfeld, Dennis K. Jeppesen, Helle Kristensen, Christa Haldrup, and Torben F. Ørntoft

Abstract MicroRNAs (miRNAs) play important biological roles in cancer development and progression. During the past decade, widespread use of novel high-throughput technologies for miRNA profiling (e.g., microarrays and next-generation sequencing) has revealed deregulation of miRNA expression as a common hallmark of human cancer. Furthermore, miRNAs have been found to be a new class of promising cancer biomarkers with potential to improve the accuracy of diagnosis and prognosis in several hematologic and solid malignancies, as well as to predict response to specific treatments. Recent studies have identified exosome-associated tumor-derived miRNAs in, e.g., blood samples from cancer patients, suggesting that miRNAs may be useful as circulation biomarkers for noninvasive diagnostic testing. In this chapter, we review the current state of development of miRNAs as cancer biomarkers with examples from common human malignancies and discuss remaining barriers to clinical translation. Finally, we describe new emerging classes of noncoding RNAs, including long noncoding RNAs (lncRNAs), with potential as cancer biomarkers. Conceivably, these could be used in combination with miRNAs in molecular diagnostic tests in the future.

K.D. Sørensen (✉) • M.S. Ostenfeld • D.K. Jeppesen • H. Kristensen
C. Haldrup • T.F. Ørntoft
Department of Molecular Medicine, Aarhus University Hospital,
Brendstrupgaardsvej 100, 8200 Aarhus N, Denmark
e-mail: karina.dalsgaard@ki.au.dk

K.A. Howard (ed.), *RNA Interference from Biology to Therapeutics*,
Advances in Delivery Science and Technology, DOI 10.1007/978-1-4614-4744-3_13,

13.1 Introduction to MicroRNA and Cancer

13.1.1 Biological Roles of MicroRNAs in Cancer

MicroRNAs (miRNAs) are endogenous small non-coding RNA molecules of approximately 22 nucleotides (nt) in length that control cellular processes. miRNA are initially transcribed into long hairpin-containing primary miRNAs (pri-miRNAs). These are processed to approximately 70-nt precursor miRNAs (pre-miRNAs), which are exported to the cytoplasm and cleaved to mature miRNAs that bind to perfect or near-perfect complementary target sequences in messenger RNA (mRNA), leading to mRNA degradation and/or translational inhibition [1]. Highly complex cellular and signal transduction pathways can be controlled by targeting and controlling the level of mRNAs and their translation.

Since their discovery in 1993 [2], miRNAs have been shown to play fundamental roles in a wide range of biological processes, and aberrant miRNA expression has been connected to several human pathologies, including cancer [3]. The first evidence linking miRNAs to cancer was presented in 2002 by Calin and coworkers, who reported the genomic loss of a pair of neighboring miRNAs, the miR-15a and miR-16-1 cluster, in a majority of chronic lymphocytic leukemias (CLLs) [4]. During the past decade, this finding has prompted a new and exciting field of miRNA studies in cancer research. Today, almost 5,000 publications (including ~1,000 reviews) have linked miRNAs with cancer. Deregulated miRNA expression is believed to contribute to the pathogenesis of most if not all human malignancies [1].

Several reports have shown that miRNAs directly regulate well-known oncogenes and tumor suppressor genes, including, e.g., tumor protein p53 (TP53) [5], B-cell CLL/lymphoma 2 (BCL2) [6], and v-myc myelocytomatosis viral oncogene homolog (MYC) [7]. Thus, depending on their mRNA target, miRNAs can act as tumor suppressors or oncogenes themselves [8–10]. miRNAs that are upregulated in cancer may be considered oncogenes and have been termed "oncomiRs." OncomiRs advance tumor development and/or progression by targeting and inhibiting tumor suppressor genes and genes that control, e.g., apoptosis or cell differentiation. Likewise, miRNAs with decreased expression in cancer cells may be considered potential tumor suppressor miRNAs and generally prevent tumor development and/or progression by targeting and inhibiting oncogenes, proliferation-related genes or genes that control apoptosis [11]. In addition to the direct regulation of target genes, miRNAs can affect the expression of tumor suppressors and oncogenes indirectly. This is exemplified by the observation that members of the miR-29 family directly target de novo DNA methyltransferases 3A and 3B (DNMT3A/B), causing changes in DNA methylation (and hence transcriptional activity) of multiple genes in lung cancer [12].

miRNAs have been shown to regulate all known processes involved in cancer, including apoptosis, cell cycle regulation, stress response, differentiation, survival, migration, adhesion, and invasion [13]. Thus, miRNAs are involved in both the initiation and the progression of cancer. The regulatory roles of miRNAs in cancer can be highly complex and are not covered in detail here. For example, a single miRNA may target tens or hundreds of different mRNAs, and the expression of a certain miRNA can change according to individual steps of tumorigenesis [10]. Several aspects remain to be uncovered regarding the functional roles of miRNAs in human cells and their potential importance in the fight against cancer, e.g., their potential as therapeutic targets and/or therapeutic drugs.

13.1.2 Commonly Deregulated miRNAs in Cancer

Since the first cancer-related miRNAs were discovered in 2002, the list of miRNAs that function as either tumor suppressors or oncomiRs in different cancer types has continually expanded. The expression level of several of these miRNAs has been found to be deregulated in multiple human malignancies (Table 13.1) and hence such miRNAs may have potential as new cancer biomarkers (molecular mechanisms responsible for aberrant miRNA expression in cancer cells are described below in Sect. 13.1.3). For instance, loss of miR-15a and miR-16-1 as seen in CLL patients has also been observed in patients suffering from, e.g., prostate cancer [21], pancreatic cancer [22], and multiple myeloma [13]. Likewise, overexpression of miR-155 has been found in high-risk CLL [32], acute myeloid leukemia (AML) [33], breast cancer [14], and lung cancer [25]. The fact that the expression level of some miRNAs is changed not only in one but also in many types of cancer suggests that these miRNAs may be downstream targets of commonly deregulated pathways in cancer [26].

When describing a specific miRNA as an oncomiR or tumor suppressor, it is necessary to specify the cell and tissue type in which the miRNA functions. miRNA expression patterns are highly cell specific, and a given miRNA can control different functions depending on the cellular context, and which targets and pathways that are affected in different types of cells. An example is the miR-221 and miR-222 cluster. This cluster is upregulated in CLL, glioblastoma, liver cancer, and thyroid carcinoma, and targets at least four important tumor suppressors: cyclin-dependent kinase inhibitor 1B (CDKN1B; p27/Kip1) [59], cyclin-dependent kinase inhibitor 1C (CDKN1C; p57/Kip2) [60], phosphatase and tensin homolog (PTEN), and tissue inhibitor of metalloproteinases 3 (TIMP3) [61], and hence are considered as oncomiRs in these malignancies. In contrast, the downregulation of miR-221 and miR-222 in erythroblastic leukemia, where they target the oncogene KIT (v-kit Hardy-Zuckerman 4 feline sarcoma viral oncogene homolog KIT), defines miR-221 and miR-222 as tumor suppressor miRNAs in this type of cancer [62]. Thus, it is worth noting that some miRNAs have dual tumor suppressive and oncogenic roles in cancer, that is dependent on the cell type and pattern of gene expression. Attempts to understand the complex biological roles of miRNAs in cancer are further complicated by the many ways in which a given miRNA can be deregulated.

Table 13.1 Commonly deregulated miRNAs in human malignancies

miRNA	Expression in cancer	Tumor suppressor (TS) or oncomiR (OM)	Validated targets	References
let-7 family	↓: Breast, CLL, colon, gastric, leiomyoma, lung, melanoma, ovarian, prostate	TS	CDK6, CDC25A, E2F2, HMGA2, c-MYC, RAS	[4, 7, 13–20]
miR-15a/16-1 cluster	↓: CLL, lymphoma, multiple myeloma, pancreatic, pituitary adenoma, prostate	TS	BCL2, MAGE83, WT1	[6, 13, 21–24]
miR-17-92 cluster	↑: Breast, colon, gastric, lung, lymphoma, medulloblastoma, multiple myeloma, pancreatic, prostate	OM	BIM, E2F1, PTEN	[14, 16, 18, 25–31]
miR-21	↑: AML, breast, cervical, CLL, colon, gastric, glioblastoma, head and neck, lung, myeloma pancreatic, prostate	OM	PDCD4, PTEN, TPM1	[4, 14, 16, 25, 32–39]
miR-29	↓: Breast, CLL, liver, lung, lymphoma	TS	DNMT1, DNMT3a/b, MCL1, TCL1	[12, 14, 25, 32, 40–43]
miR-34a/b/c	↓: Breast, bladder, CLL, colon, glioblastoma, kidney, liver, lymphoma, neuroblastoma, pancreatic	TS	BCL2, CDK4, CDK6, CREB, cyclin E2, E2F3, MET	[8, 13, 32, 35, 44–52]
miR-145	↓: Bladder, breast, prostate	TS	CBFB, ERG, SWAP70	[14, 53, 54]
miR-155	↑: AML, breast, CLL, colon, lung, lymphoma, pancreatic	OM	AID, SHIP1	[13, 14, 16, 25, 32–34, 55–58]
miR-221/222 cluster	↑: CLL, glioblastoma, liver, thyroid carcinoma. ↓: Erythroblastic leukemia, prostate	TS/OM	KIT, PTEN, p27, p57, TIMP3	[59–65]

AID activation induced cytidine deaminase, *AML* acute myeloid leukemia, *BCL2* B-cell leukemia/lymphoma 2, *BIM* Bcl2-interacting mediator of cell death, *CBFB* core-binding factor β subunit, *CDC25A* cell division cycle 25 homolog A, *CDK4* cyclin-dependent kinase 4, *CDK6* cyclin-dependent kinase 6, *CLL* chronic lymphocytic leukemia, *c-MYC* v-myc avian myelocytomatosis viral oncogene homolog, *CREB* cAMP responsive element binding protein, *DNMT* DNA methyltransferase, *E2F1/2/3* E2F transcription factor 1/2/3, *ERG* ETS-related gene, *HMGA2* high mobility group AT-hook 2, *KIT* tyrosine kinase receptor, *MAGE83* melanoma antigen family 83, *MCL1* myeloid cell leukemia sequence 1, *MET* met proto-oncogene (hepatocyte growth factor receptor), *PDCD4* programmed cell death 4, *PTEN* phosphatase and tensin homolog, *RAS* rat sarcoma oncogenes (protein subfamily of small GTPases), *SHIP1* SH2 domain-containing inositol-5′-phosphatase 1, *SWAP70* SWAP switching B-cell complex 70 kDa subunit, *TCL1* T-cell lymphoma breakpoint 1, *TIMP3* tissue inhibitor of metalloproteinases 3, *TPM1* tropomyosin 1, *WT1* Wilms tumor 1

13.1.3 Mechanisms of miRNA Deregulation in Cancer

Precise control of miRNA levels is essential for maintaining normal cellular homeostasis. Although the biogenesis of miRNAs is often described in a simplified linear manner, many of the steps in the processing of a given miRNA can be performed in multiple ways, since miRNAs are under precise control of developmental and/or tissue-specific signaling. Therefore, there are multiple ways by which something can go wrong [66]. This section describes major mechanisms known to be involved in miRNA deregulation in cancer.

Genetic, epigenetic, transcriptional, and posttranscriptional mechanisms have all been found to contribute to changes in miRNA expression in cancer (Fig. 13.1). Genetic mechanisms include primarily chromosomal abnormalities that can lead to genomic deletion, amplification, translocation, or mutation of miRNA genes. It has been reported that more than half of all human miRNAs are located at fragile sites or in genomic regions that are frequently involved in chromosomal alterations in multiple types of cancer [67]. Likewise, it has been shown that oncomiRs are often located in regions amplified in cancer, while tumor suppressor miRNAs are located mainly in regions of genomic loss [68]. miRNAs have a high overall conservation, consistent with the finding that miRNA genes have lower single nucleotide polymorphism (SNP) density than their neighboring regions [68]. Somatic mutations in the mature miRNA seed sequence seem rare events. Several studies have shown that so-called miRSNPs, i.e., SNPs present at or near a miRNA binding site in a functional gene, affect the interaction of miRNAs with their target mRNA, and that this mode of regulation is implicated in cancer [69–71].

Although not clearly understood, transcriptional regulation of miRNAs is known to be controlled by a variety of factors, including transcription factors [8, 72–74]. Accordingly, overexpression or downregulation of a transcription factor at an inappropriate time or in the wrong tissue can lead to upregulation or silencing of a miRNA, and hence may contribute to tumorigenesis. For example, in prostate cancer cell lines, the mutation and consequent loss of function of the tumor suppressor TP53 have been connected to downregulation of miR-145 [75]. Other mechanisms influencing the regulation of transcription are heritable changes in gene activity, which are independent of changes in the primary DNA sequence, i.e., epigenetic mechanisms. There are three main epigenetic events regulating cancer-associated genes: abnormal hypermethylation of CpG islands associated with the promoter region of tumor suppressor genes, global DNA hypomethylation, and activating/repressive posttranslational modifications of histone proteins [44, 76]. Bioinformatic analyses have shown that around half of all human miRNA genes have a promoter-associated CpG island [77, 78], potentially exposing these miRNA genes to the regulatory control of epigenetic modifications. An example of this is miR-127, which is downregulated in bladder cancer cells. The silencing of miR-127 has been shown to be mediated by epigenetic mechanisms and miR-127 is significantly induced in bladder cancer cell lines upon treatment with activating chromatin-modifying drugs [79].

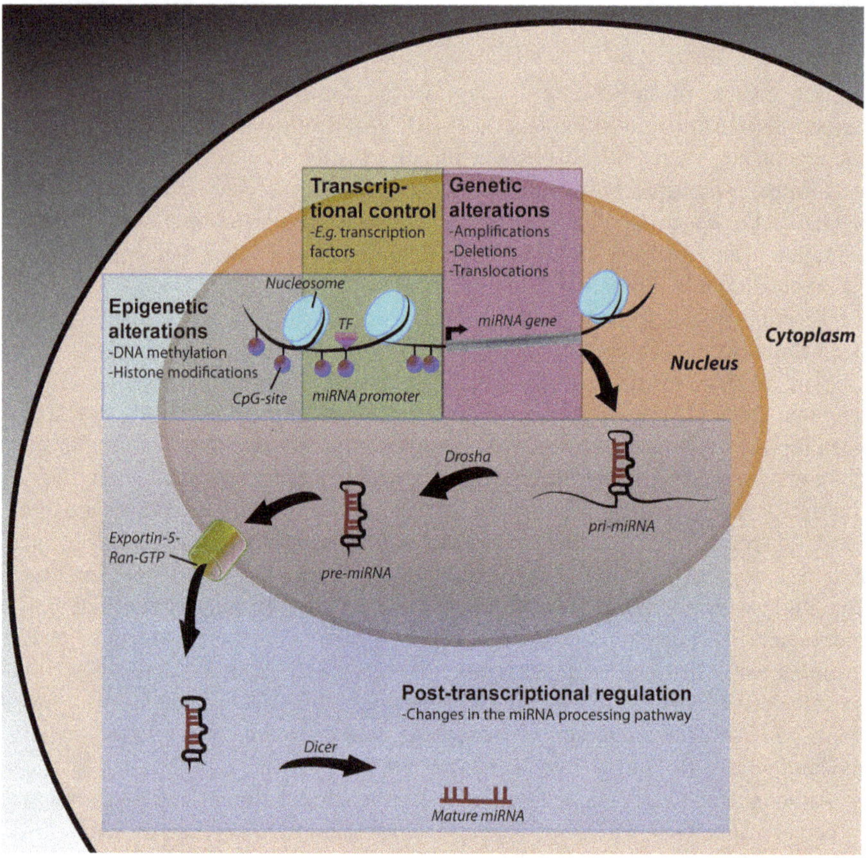

Fig. 13.1 *Mechanism of miRNA deregulation.* The regulation of miRNA expression is a tightly regulated process and many alterations might contribute to the change in expression levels of a mature miRNA in cancer. The major mechanisms of miRNA regulation can be divided into four groups. (1, *red box*): Genomic (genetic) alterations, such as deletions, amplifications, or translocations of the miRNA gene, resulting in abnormal miRNA gene copy number. (2, *green box*): Changes in the transcriptional control of the miRNA, e.g., altered levels of important transcription factors (TF). (3, *blue box*): Epigenetic alterations like DNA methylation of CpG sites and various covalent modifications of histone proteins in nucleosomes. (4, *purple box*): Post-transcriptional mechanisms such as modulations in the expression and activity of the miRNA processing enzymes, e.g., Drosha, Exportin-5, and Dicer

Changes in miRNA expression in cancer might also be attributable to posttranscriptional control, such as impairment of miRNA processing steps. Key proteins in the processing pathway of miRNAs may be deregulated or dysfunctional in cancer, and can enhance cancer development further [80]. For instance, a mutation in the exportin-5 gene (XPO5) can lead to accumulation of pre-miRNAs in the nucleus of cells [81]. Furthermore, editing of the mature miRNA has been shown to change miRNA complementarity to the target sequences [82–84].

The final expression level and sequence of the mature miRNA in a given cancer cell depend on the cooperation and net effect of all of these, and possibly other yet to be described mechanisms [85]. The disease-specific expression patterns of miRNAs as well as their key functional roles in tumorigenesis suggest that miRNAs hold great potential as cancer biomarkers and in some cases also as therapeutic targets.

13.2 MicroRNA Profiling for Discovery of Cancer Biomarkers

13.2.1 Molecular Profiling of miRNA Expression in Cancer Tissue Samples

Profiling of miRNAs expression in clinical tumor tissue samples has become a widely used approach for discovery of new cancer biomarker candidates, including diagnostic, prognostic, and predictive biomarkers [86–88]. High stability of miRNAs in freshly frozen, as well as archived formalin-fixed paraffin-embedded (FFPE), tissue samples [89] has made miRNAs particularly attractive for cancer biomarker studies. In contrast, mRNAs are quickly degraded upon formalin-fixation, making mRNA expression analysis based on FFPE samples rather difficult. Moreover, because FFPE samples are collected as part of daily clinical routines and stored long term at hospital pathology departments, such samples are generally more readily available than freshly frozen samples from patient cohorts with long-term clinical follow-up information. New miRNA markers may enable earlier detection of cancer, which is critical for a favorable outcome. Furthermore, molecular subclassification of tumors based on miRNA expression patterns may be used to guide individualized treatment of cancer, e.g., by allowing stratification of patients into subgroups with low or high risk of disease progression. Expression profiling can also facilitate the discovery of miRNAs with key regulatory roles in cancer development or progression and thereby can lead to the identification of potential new treatment targets [26].

Figure 13.2 outlines a frequently used strategy for molecular profiling of cancer. In this example, miRNA expression is compared in two sample groups, tumor vs. normal tissue, for identification of diagnostic biomarker candidates. Alternatively, miRNA expression may be analyzed in tumor samples from patients with aggressive vs. indolent (slowly growing, nonaggressive) cancer, or from patients responsive or nonresponsive to a certain anticancer drug in order to identify candidate markers with prognostic or predictive biomarker potential, respectively. Depending on the research question addressed, additional sample groups can be included. Ideally, screening for differentially expressed miRNAs between sample groups should be performed in one patient set followed by validation in at least one independent patient sample set to ensure reproducibility and selection of meaningful candidates for further biomarker validation studies.

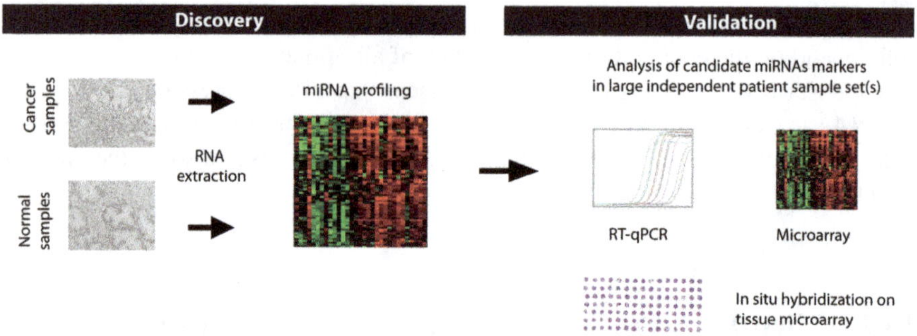

Fig. 13.2 *Strategy for identification of miRNA biomarkers in cancer*. MicroRNA expression profiling of patient samples (normal vs. cancer tissue samples shown here) is performed to identify candidate miRNA markers for a given malignancy. Microarray analysis is commonly used in the discovery phase, but other profiling methods are also available (see Fig. 13.3). Novel candidate miRNA markers identified in the discovery phase are subsequently validated in large independent patient sample set(s) by RT-qPCR or in some cases by microarray analysis or in situ hybridization using tissue microarrays. Only a subset of miRNAs will pass validation

13.2.2 Technologies for miRNA Profiling

Several technologies are available for determination of miRNA expression levels and can be divided into three overall types: hybridization-based [DNA microarrays, in situ hybridization (ISH), Northern blotting], amplification-based [reverse transcription quantitative PCR (RT-qPCR)], and sequencing-based [next-generation sequencing (NGS)] [90, 91]. Using DNA microarrays, it is possible to measure the expression of hundreds of miRNAs in parallel in a single experiment [92, 93]. Typically, fluorescently labeled RNA (or reverse-transcribed cDNA) molecules are hybridized through Watson–Crick base pairing to complementary DNA oligonucleotides (probes) immobilized on a microarray glass (or quartz) slide (Fig. 13.3, left panel). The position of each type of probe on the microarray is predefined (referred to as a "spot"), and the intensity of the fluorescent signal from a given spot after hybridization is taken as a measure for the expression level of the corresponding miRNA in the sample analyzed [94, 95]. Today, several commercial DNA microarray platforms are available for relatively cost-effective and user-friendly profiling of miRNAs [96–99].

Although intra-platform concordance is generally high, inter-platform variance can be substantial because of differences in exact microarray designs (e.g., probe type, length, and sequence) and experimental protocols [99–102]. For example, some protocols use total RNA as input (includes mature, pri-, and pre-miRNAs), while others employ an enrichment step for small RNAs (mainly mature miRNAs) prior to labeling and hybridization. These approaches will produce different results, since probes directed at the mature miRNAs sequence will also recognize the corresponding pri- and pre-miRNA. Furthermore, the short length (~22-nt) of mature miRNAs, many of which vary in sequence by only 1 or 2 nt, constrains probe design, and cross-hybridization to closely related miRNAs remains a significant challenge

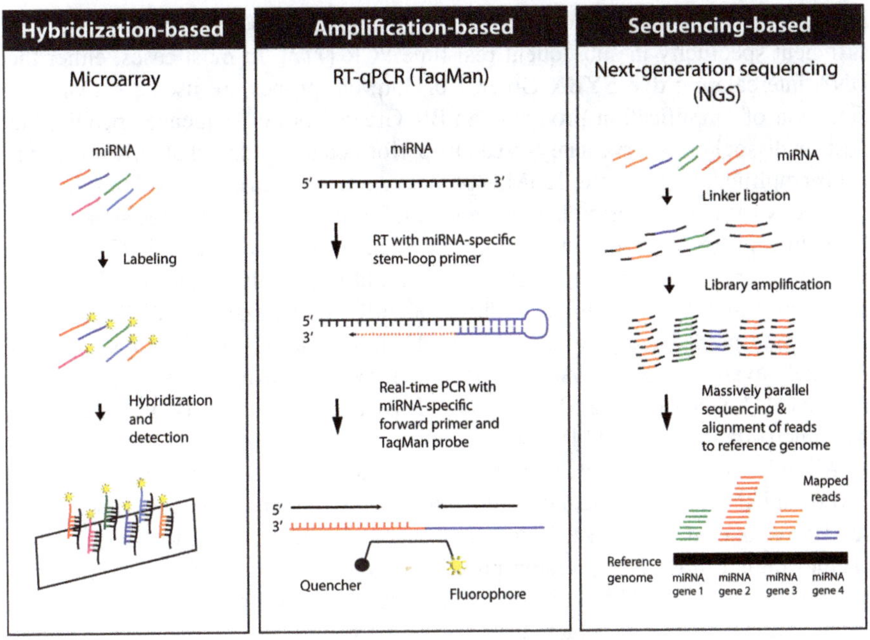

Fig. 13.3 *MicroRNA expression profiling methodologies*. Microarray analysis is shown as an example of hybridization-based miRNA expression profiling (*left panel*). TaqMan-based RT-qPCR is presented as an example of amplification-based miRNA expression analysis (*middle panel*). Next-generation sequencing (also known as deep sequencing) is a newly developed technology for sequencing-based analysis of miRNA expression (*right panel*)

for microarray-based miRNA detection. Although higher specificity can be obtained by incorporation of chemical modifications, such as locked nucleic acid (LNA), into the probes [103], it is generally recommended to validate microarray-based findings by an alternative methodology. Early studies generally used rather laborious Northern blotting [104–106], but today RT-qPCR is typically used (see below). ISH on tissue sections using labeled miRNA-specific probes (usually LNA probes) has also been used for validation, although this technology may be more suited for localization studies than differential expression analysis [107]. A main advantage of ISH is that it can be used in combination with tissue microarrays (TMAs), thus allowing the parallel analysis of a particular miRNA in clinical tumor specimens from hundreds of patients on a single TMA.

Compared to microarrays and other hybridization-based methods, RT-qPCR has increased dynamic range and can be used for profiling of very limited amounts of RNA often available from clinical samples [108–110]. The short length of mature miRNAs and sequence similarities between related miRNAs, and with pre- and pri-miRNAs, are also challenges for RT-qPCR assay. Reverse transcription is performed either directly by the use of miRNA-specific primers, including linear and stem-loop primers (Fig. 13.3, middle panel), or by universal priming after enzymatic addition of a polyA tail to template miRNAs (not shown) [111]. When using universal priming for RT, some studies suggest that it may be necessary to incorporate high affinity DNA analogs (e.g., LNA) into the miRNA-specific primers to obtain sufficient specificity in subsequent real-time PCR [112]. In most cases, either the DNA intercalating dye SYBR Green I or TaqMan probes are used for real-time detection of amplification products. SYBR Green I is not sequence specific, but postrun dissociation curve analysis can be performed to validate that only one product is amplified. In principle, TaqMan probes can provide an extra level of sequence specificity beyond that provided by the RT-PCR primers, but due to the short length of mature miRNAs, TaqMan probes will often have to overlap RT-PCR primer sequences and, therefore, in reality may not add specificity to the reaction when analyzing miRNAs. For global profiling of miRNA expression, RT-qPCR assays can be multiplexed and parallelized [109, 113]. Several RT-qPCR arrays are commercially available for analysis of tens to hundreds of miRNAs in 96- or 384-well format [111], including, e.g., TaqMan low-density microRNA cards, which are 384-well micro-fluidic cards [114].

All hybridization- and PCR-based methods suffer from the limitation that only known miRNAs can be detected due to reliance on predesigned primer/probe sequences. In contrast, NGS (*aka* massively parallel sequencing or deep sequencing) enables digital miRNA expression profiling as well as discovery of new miRNAs and miRNA variants [115, 116]. There are a number of different commercial platforms available for NGS, which differ somewhat with respect to sequencing principles, chemistries, and methods of detection, but all share the ability to sequence thousands of DNA molecules in parallel, thus producing massive amounts of data. In a typical small RNA sequencing experiment, oligonucleotide linkers are added to small RNAs extracted from a biological sample and the resulting library is amplified by PCR and sequenced (Fig. 13.3, right panel). Typically, over 10 million small RNA sequence reads are produced for each sample [117]. When sequence reads are mapped back to a reference genome, the number of reads for a particular miRNA species can be taken as a measure for its relative expression level, whereas the exact sequences of individual reads can be used to identify new miRNAs and polymorphisms [118]. Drawbacks include the risk of sequencing errors and introduction of bias during library preparation and amplification as well as high analysis costs and scarcity of standardized computational tools for data analysis [119, 120]. Nevertheless, NGS is likely to become the technology of choice in future miRNA and transcriptome profiling studies as sequencing prices are constantly reduced, experimental protocols optimized, and new computational analysis tools developed.

13.3 MicroRNAs as Cancer Biomarkers

13.3.1 Early miRNA Profiling Studies of Cancer

One of the first expression profiling studies to highlight the potential of miRNAs as cancer biomarkers was published in 2005 by Lu et al. [23]. The study used a bead-based hybridization method to measure the expression of 217 miRNAs in 334 samples representing multiple human malignancies. The authors found that miRNA expression patterns accurately reflected the developmental lineage and differentiation stage of tumors and, furthermore, showed that a miRNA expression signature was superior to an equivalent mRNA signature for determining the tissue of origin of undifferentiated tumors. Likewise, in an independent report, Rosenfeld et al. used miRNA expression profiles from microarray-based analysis of 401 clinical samples, representing 22 different tumor tissues and metastases, to construct a 48-miRNA molecular classifier that identified the tissue of origin of metastatic tumors of unknown primary site with high accuracy [89]. This miRNA-based classifier was designed as a binary decision tree and was reported as superior to previously published mRNA-based classifiers designed for the same purpose. A correct and precise diagnosis is essential for optimal clinical management of such cancers.

In another early miRNA profiling study, Volinia et al. used a custom-made microarray to measure the expression of 228 miRNAs in 540 cancer and nonmalignant samples from lung, breast, stomach, prostate, colon, and pancreas [26]. miRNA expression profiles clearly distinguished tumors from nonmalignant samples, and the authors identified a general solid cancer signature consisting of 21 miRNAs. Interestingly, the bioinformatically predicted targets for these miRNAs were statistically significantly enriched for known protein-coding cancer genes, supporting the hypothesis that key oncogenes and tumor suppressor genes are regulated by aberrant miRNA expression in human malignancies. In further support of this, the authors experimentally validated three target mRNAs in vitro and demonstrated a significant inverse correlation in miRNA abundance and target protein expression for one selected miRNA:target pair (upregulated miR-106a and downregulated retinoblastoma protein). They also identified six distinct cancer-specific miRNA signatures, one for each of the malignancies investigated.

13.3.2 Examples of Diagnostic/Prognostic miRNAs in Solid and Hematopoietic Tumors

The studies described above were among the first to demonstrate the potential of miRNA for cancer diagnostics and to highlight miRNA profiling as a fruitful strategy for biomarker discovery. MicroRNA profiles with the potential to discriminate between normal and cancer tissue have been identified for all malignancies investigated to date, and may potentially be used for early diagnosis. Cancer-specific molecular changes are often detectable before pathological and cytological abnormalities become clearly apparent. Moreover, for several cancer types, it has now been shown that miRNAs can be used for classification of tumors into clinically relevant subgroups. For example, some miRNA expression signatures distinguish histological or cytogenetic subtypes, while other signatures are associated with routine prognostic parameters such as tumor stage, grade, and metastasis status. Consistent with this, prognostic miRNA signatures have been published for numerous solid and hematopoietic malignancies. In addition, many studies have identified miRNAs that modulate responsiveness to particular anticancer drugs in vitro as well as in patients. Table 13.2 is a representative list of published miRNAs and miRNA

Table 13.2 Examples of miRNAs with cancer biomarker potential

Cancer type	MicroRNA(s)	Biomarker potential	References
Diagnosis/molecular classification			
Acute leukemia	miR-128a, miR-128b, miR-223, let-7b	Lymphoblastic vs. myeloid	[121]
Acute leukemia	miR-148, miR-151, miR-424	B-cell vs. T-cell type	[122]
Bladder cancer	miR-21, miR-145	Normal vs. cancer	[123]
Breast cancer	mir-21, mir-125b, mir-145, mir-155	Normal vs. cancer	[14]
Colon cancer	miR-142-3p, miR-144, miR-212, miR-151	Microsatellite status	[124]
Lung cancer	miR-205	Adenocarcinoma vs. squamous cell carcinoma	[125]
Prognosis			
Bladder cancer	miR-129, miR-133b miR-518c*	Progression	[123]
Colon cancer	miR-320, miR-498	Recurrence	[124]
Gastric cancer	miR-214, miR-433, let-7g	Survival	[126]
Glioblastoma	miR-196	Survival	[127]
Hepatocellular carcinoma	miR-26	Survival	[128]
Hodgkin's lymphoma	miR-135a	Relapse and survival	[129]
Lung cancer	miR-155, let-7a	Survival	[25]
Lung cancer	miR-137, miR-182*, miR-221, miR-372, let-7a	Relapse and survival	[130]
Mantle cell lymphoma	miR-29	Survival	[41]
Ovarian cancer	miR-23a, miR-27a	Recurrence and survival	[131]
Prediction of treatment response		*Drug*	
Pancreatic cancer	miR-21	Gemcitabine/5-fluorouracil	[132] [133]
Breast cancer	miR-21	Trastuzumab	[134]
CLL	miR-148a, miR-222, miR-21	Fludarabine	[135]
Ovarian cancer	mir-23a, miR-27a	Cisplatin/Carboplatin	[131]
Hepatocellular carcinoma	miR-26	Interferon alpha	[128]

signatures for diagnosis, prognosis, and prediction of treatment response for a variety of human cancers. In the remaining part of this section, we will review in more detail the current state of development for miRNAs as tissue-based cancer biomarkers using CLL and prostate cancer as examples of common hematopoietic and solid malignancies, respectively. CLL is one of the most extensively studied malignancies regarding miRNA expression profiling and biomarker potential, whereas prostate cancer is less well characterized. Hence, existing studies for these diseases exemplify different investigatory stages of miRNA biomarker development for cancer.

13.3.3 Chronic Lymphocytic Leukemia

Chronic lymphocytic leukemia (CLL) is characterized by accumulation of small, mature B cells and is the most common type of leukemia in Western countries [136]. CLL has a highly variable clinical course. Some patients present with indolent disease that requires no or little treatment, while others have an aggressive form associated with rapid disease progression and death. In a pioneering study from 2002, Calin et al. showed that miR-15a and miR-16-1 are located at 13q14, a commonly deleted chromosomal region in CLL, and that expression of these miRNAs was downregulated in more than half of all CLLs [4]. This was the first strong evidence for an association between miRNAs and human cancer. The same group of authors later showed that miR-15a and miR-16 exert tumor suppressor functions in CLL by direct targeting of the anti-apoptotic protein BCL2 (B-cell CLL/lymphoma 2) [6], suggesting that these miRNAs could have therapeutic potential in CLL. A separate (Chap. 14) in this book addresses the utility of miRNAs as cancer therapeutics.

CLL was the first human malignancy to be analyzed by genome-wide miRNA expression profiling using microarrays [106]. The initial study included 41 CLL samples from 38 patients plus 6 normal samples. The authors identified a miRNA signature that could distinguish CLL from normal CD5+ B cells as well as three distinct miRNA signatures associated with known prognostic factors for CLL, i.e., presence or absence of zeta-chain-associated protein kinase 70-kD (ZAP70) expression, immunoglobulin heavy chain variable (IgH_V) region mutations, and 13q14 deletion, respectively [106]. This early study clearly showed the potential of miRNAs for molecular classification of cancer into clinically and biologically relevant subgroups.

Consistent with this, CLL was the first malignancy for which a miRNA-based prognostic signature was reported. Using miRNA expression profiles from 94 CLL samples, Calin et al. [32] revealed an expression signature composed of 13 miRNAs that was significantly associated with known prognostic factors for CLL (ZAP70 expression and IgH_V region mutation status) and with disease progression. It is of note that miR-15a, miR-16-1, miR-29a, miR-29b, miR-29c, and miR-223 were among the 13 miRNAs in this signature. Furthermore, from the original profiling data set, nine miRNAs (including miR-29c and miR-181a) were found to be able to predict time to disease progression for patients diagnosed with CLL, although an independent patient sample set was not used to confirm the results

[32]. However, in a follow-up study miR-29b, miR-29c, miR-181a, and miR-181b were included in a miRNA-based molecular classifier that could distinguish between indolent CLL and aggressive CLL with or without 11q deletion [40]. The same report showed that miR-29b and miR-181b targets the TCL1 (T-cell leukemia/lymphoma 1A) oncogene, overexpression of which may be a causal event in the pathogenesis of CLL [40], suggesting a key role for miR-29 and miR-181 family members in CLL.

Loss of miR-181b expression was associated with therapy-refractory CLL in a separate study, which also found that low miR-181b expression and high miR-21 expression were significant independent prognostic factors for poor overall survival and for time to disease progression [137]. Another study identified 32 miRNAs that could distinguish subtypes of CLL with distinct common genomic aberrations: 11q deletion, 17p deletion, trisomy 12, 13q deletion, and normal karyotype, respectively [138]. Interestingly, this study revealed that low expression of miR-181a was associated with aggressive disease in patients with 17p deletion, but with less aggressive disease in patients with trisomy 12, demonstrating that individual miRNAs may have subtype-specific biomarker roles.

The prognostic value of miR-29 for CLL was independently confirmed in a study by Stamatopoulos et al. who showed that miR-29c and miR-223 were downregulated in advanced stage CLL and that both miRNAs significantly predicted treatment-free survival and overall survival [139]. The same authors developed a RT-qPCR-based test for risk stratification of patients diagnosed with early stage CLL, which combined four prognostic markers: miR-29c, miR-223, ZAP70, and lipoprotein lipase (LPL) [139]. Each marker was dichotomized based on expression levels and the markers were combined to generate a simplified risk score ranging from 0 (low risk for all four markers) to 4 (high risk for all four markers). The study demonstrated that increased accuracy of predictions can be obtained by combining different types of biomarkers into a molecular diagnostic test. Similarly, Rossi et al. developed a three-marker test named the 21FK score, which combined information on miR-21 expression levels, 17p deletion status, and karyotype status (normal or abnormal) to produce a three-category risk score (0, 1, or 2) that could stratify CLL patients according to overall survival times [137]. The 21FK risk score was superior to routine prognostic factors for prediction of overall survival time in two independent CLL patient cohorts [137], suggesting that such a test could have clinical utility in the future. Translation of molecular diagnostic tests to actual clinical practice, however, requires substantial independent validation in large prospective multicenter studies generally including thousands of patients.

13.3.4 Prostate Cancer

Prostate cancer (PC) is the most frequently diagnosed malignancy in men in Western countries and a major cause of cancer-associated death. The natural history of PC is highly variable. Some PCs develop slowly and do not cause significant symptoms during the remaining lifetime of the patient, while other PCs progress more rapidly, almost invariably with fatal outcome. Although localized PC can be cured by radiation therapy or surgery (radical prostatectomy), overtreatment of clinically insignificant tumors is a major problem. Accordingly, there is a strong need for improved diagnostic and prognostic biomarkers for this disease.

The development of miRNA biomarkers for PC is still at a relatively early stage [140], although several studies have shown that miRNA expression profiles can be used to distinguish PC from nonmalignant prostate as well as localized PC from advanced metastatic PC (MPC) [26, 63, 107, 141–147]. Inconsistencies have been observed between studies, which at least in part can be explained by the use of different profiling methodologies. Nevertheless, repeated findings in multiple studies show upregulation of miR-375 and miR-200c and downregulation of miR-145, miR-205, miR-221, and miR-222 in cancer vs. normal prostate. Molecular classifiers combining 2 (miR-205, miR-183), 3 (miR-143, miR-145, miR-375), or 54 different miRNAs have been shown to correctly classify more than 75% of normal vs. PC samples [63, 107, 147]. If these results are confirmed in future large-scale clinical studies, a miRNA-based test could potentially improve the accuracy of PC diagnosis, e.g., as a supplement to the currently used routine biomarker for PC detection, serum prostate-specific antigen (PSA), which has highly suboptimal sensitivity and specificity.

In addition to their diagnostic value, miRNAs have shown potential for molecular classification of PC into clinically relevant subgroups for prognostic risk stratification. For example, by analysis of miRNA expression profiles from 57 PC tumors with or without perineural invasion (PNI), Pruitt et al. identified two subgroups of patient samples each associated with a distinct miRNA expression signature [148]. One subgroup included all non-PNI tumors as well as a subset of less aggressive PNI tumors with lower stage and lower Gleason score, while the second subgroup consisted of more aggressive PNI tumors characterized by higher stage and higher Gleason score. Interestingly, clinically relevant clusters could not be generated when the same analysis was performed on matching mRNA expression data [148], again emphasizing miRNAs as promising biomarker candidates for PC.

Moreover, in a large miRNA profiling study of 102 clinical PC samples, Martens-Uzunova et al. identified 80 miRNAs that were not only differentially expressed in pairs of nonmalignant and tumor samples, but also separated clinically localized PC (LPC) samples into two subgroups with distinct prognostic characteristics [147]. The first group included 34 LPC samples clustered together with the nonmalignant samples, while the second group included 16 LPC samples and clustered together with advanced PC samples and lymph node metastases. Notably, the second group was characterized by significantly higher risk of cancer-specific death. This finding awaits confirmation in an independent patient cohort.

So far, few miRNAs have been shown to have prognostic value for PC in more than one patient sample set. Schaefer et al. found that high miR-96 expression was significantly associated with short recurrence-free survival after radical prostatectomy in two independent PC patient sets [63]. In a separate study, Spahn et al. found that low expression of miR-221 was an independent adverse prognostic factor for recurrence-free survival after radical prostatectomy based on data from 92 PC patients [145]. Consistent with this, Tong et al. identified a 16-miRNA signature, including *miR-211*, which correctly classified 75–85% of patients with and without recurrence in a distinct PC patient cohort [144]. As further confirmation of the prognostic potential of miR-211, Martens-Uzunova et al. identified a 25-miRNA classifier (including miR-211) for prediction of cancer-specific survival in a cohort of 50 PC patients [147].

13.3.5 Predictive miRNAs

Several studies have identified miRNA expression profiles associated with anticancer drug responsiveness, indicating that miRNAs have potential as predictive biomarkers. Furthermore, for a growing number of miRNAs, there is experimental evidence for direct functional roles as cellular modulators of drug resistance, suggesting that some miRNAs may also have potential as new therapeutic targets in this context [149, 150]. Multiple mechanisms are known to be involved in development of drug resistance, including, e.g., inhibition of apoptosis, blocking of drug entry, active efflux of the drug, and altered expression of the drug target [151]. Conceivably, all of these cellular mechanisms may be regulated by miRNAs. Selected examples of predictive miRNAs are described below and in Table 13.2.

Pancreatic cancer is associated with poor prognosis and surgical treatment is often followed by adjuvant chemotherapy. At least two independent reports found high expression of miR-21 to be a significant predictor of poor cancer-specific survival for pancreatic cancer patients receiving adjuvant gemcitabine-based chemotherapy [132, 133]. It was also shown that inhibition of miR-21 increased sensitivity to gemcitabine, while overexpression of miR-21 reduced chemosensitivity of pancreatic cancer cells in vitro, suggesting that miR-21 could be a new therapeutic target. High expression of miR-21 has also been associated with drug resistance in other cancer types. For breast cancer, it was reported that upregulation of miR-21-mediated resistance to Trastuzumab, a monoclonal antibody directed against human epidermal growth factor receptor 2 (HER2), which is commonly overexpressed and linked to poor clinical outcome [134]. Functional studies suggested that miR-21 induced drug resistance at least in part by targeting tumor suppressor PTEN (phosphatase and tensin homolog) in breast and pancreatic cancer cells [133, 134]. In addition, miR-21 was included in a miRNA signature that could distinguish fludarabine-refractory and -sensitive cases of CLL [135]. Finally, a large study including a total of 455 patients with hepatocellular carcinoma showed that patients with low expression of miR-26 in the tumor had a favorable response to interferon alfa [128].

13.4 Circulating miRNAs as Diagnostic Cancer Biomarkers

13.4.1 Discovery of Circulating Tumor-Specific miRNAs

The recent discovery of miRNAs circulating in body fluids has initiated intense investigation to determine if their detection may be advantageous for cancer diagnostics as an alternative to profiling of solid tumors. At present, detection of cancer relapse is often carried out by invasive testing (e.g., frequent cystoscopies for bladder cancer). Ideally, such procedures could be replaced by noninvasive testing based on monitoring of circulatory miRNAs in biological fluids. miRNAs have been detected in all body fluids examined so far, including urine, plasma, serum, semen, saliva, and cerebrospinal fluid. The miRNAs are present in nanovesicles such as exosomes, as well as in association with free protein complexes. The possible biological role of circulating miRNAs is presently the focus of numerous studies. According to available data, circulating miRNAs appear to possess a regulatory function beyond cells from which they originate and in recipient cells that take up the miRNA complexes. Exosome-mediated transfer of miRNAs has been demonstrated in cell culture models. Furthermore, in vivo models suggest that cancer cell-derived exosomes may modulate the microenvironment at distant sites into pre-metastatic niches and, thereby, promote metastasis through yet unknown mechanisms [152]. Whether present in exosomes or in free protein complexes, the detection of these miRNAs serve as a potential diagnostic tool for cancer evaluation.

Circulating tumor-specific miRNAs were first discovered in 2008, where increased levels of miR-210, miR-155, and miR-21 were detected in serum samples from patients with diffuse B-cell lymphoma [153, 154]. The same year, Mitchell and coworkers reported the detection of miRNAs in plasma samples. Plasma-derived miRNAs showed remarkable stability and were stable upon repeated freeze/thawing, exposure to low/high pH, and extended storage. They were also resistant toward endogenous RNase degradation, in contrast to synthetic spike-in miRNAs [155, 156]. Highly expressed miRNAs, such as miR-16, miR-15b, and miR-24, were detected at similar levels in both plasma and serum. Furthermore, human tumor cell-derived miRNAs, with no murine orthologs, could be detected in the bloodstream of xenografted mice, and increased serum levels of miR-141 were seen in patients with prostate cancer compared to healthy controls [155]. A study by Chen et al. described 63 miRNAs in serum from lung cancer patients that were not detected in healthy individuals [156]. Remarkably, a comparative analysis revealed highly similar miRNA profiles in blood cells and serum in normal individuals (90% identical miRNAs in the two fractions), but not in lung cancer patients (~50% of detectable miRNAs were only present in serum). Accordingly, disease-specific miRNAs present in serum seem to derive directly from cancer cells rather than reflecting an altered miRNA profile in blood cells. This reiterates the potential of circulating miRNA profiles for cancer diagnosis.

13.4.2 Circulating miRNAs Are Highly Stable and Present in Protein Complexes and Vesicles

Early work suggested that circulating miRNAs were not cell-associated [155]. It is now known that miRNAs are encapsulated in small vesicles, such as exosomes, microvesicles, and microparticles, as well as in free association with proteins such as Argonaute 2 (AGO2), Nucleophosmin 1 (NPM1), and high-density lipoprotein

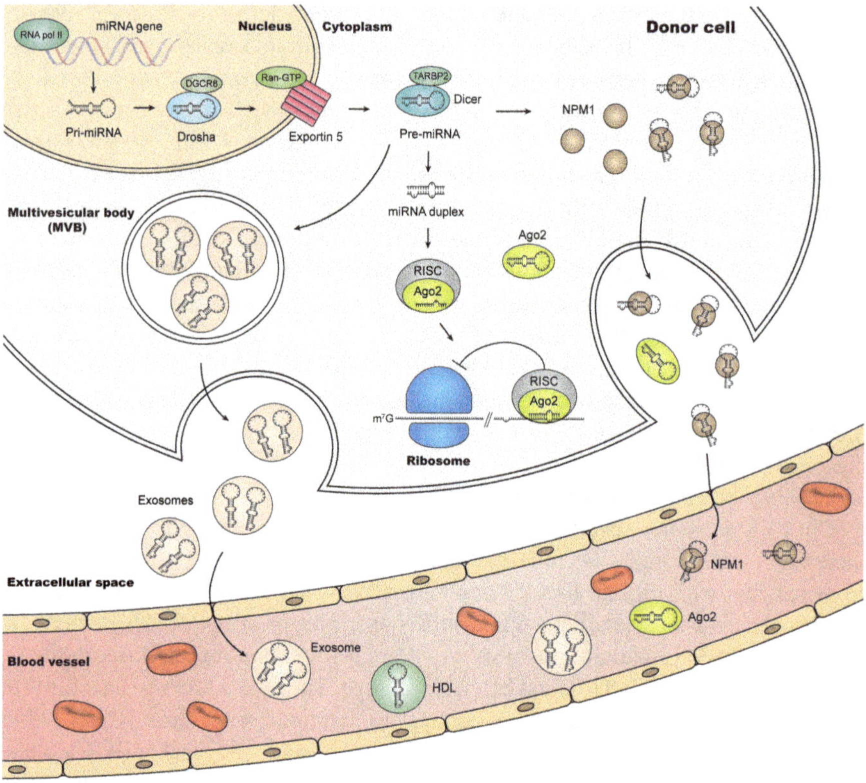

Fig. 13.4 *Biogenesis and origin of circulating miRNAs.* Primary miRNA (pri-miRNA) transcripts are generated in the nucleus and processed to precursor miRNAs (pre-miRNAs) by the Drosha enzyme and DGCR8. After export to the cytoplasm, the pre-miRNA is recognized and processed to a small double-stranded RNA (dsRNA) structure of about 22 nucleotides by Dicer, acting together with TRBP. The dsRNA structure is recognized and unwound by the RISC complex, including the AGO2 protein. Subsequently, the mature single-stranded miRNA is carried to complementary miRNA target sites within mRNAs to regulate gene expression. Alternatively, pre-miRNAs can be transported in body fluids as circulating miRNAs. Inward budding of endosomal membranes can generate intracellular multivesicular bodies (MVBs) that release their content as exosomes with encapsulated miRNAs. NPM1, AGO2, and high-density lipoprotein (HDL) have been found to transport circulating pre-miRNAs. After entering the bloodstream, circulating miRNAs can be taken up and processed by recipient cells

(HDL) [157–160] (Fig. 13.4). So far, the vast majority of studies have focused on the function of exosomes and how their constituents may impact tumor growth and cancer disease. Exosomes are small vesicles (40–100 nm) that are formed by local invaginations of endosomal membranes into multivesicular bodies (MVB) and secreted upon fusion of the MVBs with the plasma membrane for release into the extracellular space and body fluids.

miRNAs in complex with AGO2 have been found to represent the predominant form of freely circulating miRNAs in plasma and cell culture media and is stable over a 2-month period at room temperature [157, 158]. It has been hypothesized that these complexes represent by-products of dead cells secreted into the extracellular space, and that they are unlikely to possess paracrine cell signaling functions. This is in contrast to studies of miRNA-containing exosomes, for which a biological function involving paracrine delivery of miRNA, mRNA, DNA, and protein is well documented.

13.4.3 Exosome Biogenesis and Secretion

The mechanism for sorting of exosomal cargo into the exocytic MVB is largely unknown. It is speculated to involve ubiquitination and the endosomal sorting complex required for transport (ESCRT) sorting system, since ubiquitinated proteins and ESCRT components are enriched in purified exosomes from some cell types [161, 162]. Furthermore, phosphatidylinositol-(4,5)-bisphosphate (PIP2) and phosphatidylinositol-(3,4,5)-trisphosphate (PIP3)-binding domains can induce targeting of highly oligomeric cytoplasmic proteins to exosomes [163, 164]. Studies have shown that the miRNA content of exosomes resembles overall that of the donor cell [165, 166]. The proportion, however, of the most and least abundant miRNAs are often shifted, suggesting a specific enrichment or depletion of certain miRNAs in the vesicles [166, 167]. The mechanism of this differential packing into exosomes is presently unknown. Notably, miRNA target transcripts appear to be underrepresented in exosomes, while housekeeping mRNAs (less prone to miRNA-mediated repression) seem overrepresented [165]. To explain this, it was hypothesized that miRNAs and their corresponding target mRNAs are enriched in GW182 (glycine-tryptophan protein of 182-kDa) and AGO2-associated membrane fractions and thereby excluded from exosome-like vesicles.

The events leading to fusion of the MVBs to the plasma membrane and exosome release are not fully understood, but involves the action of RAB (member RAS oncogene family) proteins, specifically RAB27A, RAB27B, and RAB35 [168, 169]. The inhibition of ceramide production, via neutral sphingomyelinase (nSMAse) by small molecule antagonist GW4869, reduces exosome-mediated secretion of miRNAs [170]. In addition, intracellular rise in calcium levels and induction of TP53 upon DNA-damage have been shown to enhance exosome secretion [171, 172]. In contrast, release is inhibited by the induction of autophagy, a normal self-catabolic process involving degradation and recycling of cellular components through the lysosomal machinery [173]. For some cell types (e.g., T cells and resting B cells), activation of cell surface receptors is required for secretion [174, 175].

13.4.4 Biological Function of Circulating Exosome-Encapsulated miRNAs

Two opposing hypotheses currently exist to explain the biological functions of exosomes: (1) that vesicles are secreted from cells to carry out an active paracrine cell signaling role by the transfer to recipient cells of exosome content, such as miRNAs. By doing so, cancer cells may enhance local tumor growth and invasion, or prime distant sites for establishment of tumor metastases and (2) that vesicles are used as an alternative route to cellular degradation of unwanted material, which is then packed into exosomes and exocytosed. No data exists yet that can clearly discriminate between the two proposed models. It is currently unknown whether freely circulating AGO2-miRNA complexes can interact specifically with recipient cells to actively target mRNA downregulation [157, 158]. In contrast, complexes of HDL and exogenous miR-223 have been shown to increase the level of miR-223 by 250-fold in recipient Huh-7 hepatocarcinoma cells and to decrease the level of the miR-223 targets Ras homolog gene family, member B (RHOB) and Ephrin A1 (EFNA1) [160].

Several studies indicate that exosomes represent vehicles for intercellular communication and for active paracrine cell signaling. In 2007, Valadi and coworkers demonstrated that mRNA and miRNAs could be transferred to recipient cells, and the transferred mRNA was actively transcribed [176]. Uptake of mast cell-derived exosomal RNA was observed in other mast cell lines, but not in CD4+ T cells. Fluorescence-labeled vesicle monitoring has been used to show that ovarian cancer exosomes are transferred to natural killer cell lines, but not Jurkat T cells [177]. These studies demonstrate selective vesicle uptake that could occur either by specific endocytosis, receptor-mediated uptake, or membrane fusion. Transfer of mRNA molecules linked to cell cycle regulation and angiogenesis, and enriched in exosomes, has been demonstrated for glioblastoma and colorectal cancer cells. Furthermore, vesicular transfer stimulated the proliferation of U87 malignant glioma cells and endothelial cells [178, 179]. Microvesicle secretion from EGFRvIII + glioma tumor cells has been shown to result in transfer of this oncogenic receptor to EGFRvIII-negative cancer cells, leading to activation of oncogenic pathways, triggered production of vascular endothelial growth factor (VEGF), and anchorage-independent growth of the recipient cells [180]. Exosomes may, however, also have an anti-tumorigenic function, as has been observed for pancreatic cancer cell-derived exosomes [181].

The ability of exosomes to promote tumor metastasis was addressed by Hood et al. in a murine model of metastatic melanoma [152]. In this study, prior injection with melanoma exosomes showed sentinel lymph node homing with subsequent recruitment of melanoma cells, extracellular matrix deposition, and vascular proliferation in the lymph nodes. This suggests that exosome-mediated cargo delivery may prime a particular organ microenvironment or niche as the "soil" for metastatic cells (the "seeds") and tumor establishment at remote sites. In addition, exosomes may also exert a tumorigenic effect through modulation of the immune system. Cancer cell-derived exosomes can induce apoptosis of activated tumor-specific T cells, thereby impairing the ability of effector lymphocytes to exert cytolytic activity against tumor targets [182].

13.4.5 Methods for Exosome Isolation and Content Analysis

Exosomes differ from microvesicles (100–1,000 nm) and apoptotic bodies (50–500 nm) based on size, morphology, lipid composition, protein, RNA and DNA content, and mechanism of release from the cell. The vesicles are commonly isolated either by differential centrifugation with sedimentation at $100{,}000 \times g$ or at a specific floatation density in gradient ultracentrifugation, or by size-exclusion chromatography. Some studies have also used filtration, precipitation, and immunoaffinity capture for successful isolation (for further details on methodology, see [183]). Vesicle size and morphology are usually examined by electron microscopy, dynamic light scattering, or nanoparticle tracking analysis. The contents of exosomes (proteins, RNA, and DNA) are listed in the online database "Exocarta" [166, 178, 184, 185]. Although the protein composition of exosomes reflects the parental cell, some proteins are common to most exosomes. These include the transmembrane tetraspanin proteins, particularly CD63, CD9, CD81, and CD82 [162, 186], and the cytosolic heat-shock proteins HSP70 and HSP90 [187]. Proteins associated with the biogenesis of MVBs, such as ALG-2 interacting protein X (ALIX) and tumor susceptibility gene 101 (TSG101) as well as cytoskeleton proteins such as actin, are also prominent. Lipidomic analysis has shown an increase in phosphatidyl-ethanolamines and -cholines, as well as a possible increase in cholesterol content (reviewed in [167]).

Profiling of miRNAs incorporated within exosomes has been performed using microarray analysis [188], RT-qPCR profiling [189, 190], and deep sequencing [165]. Examination of the RNA integrity, e.g., on an Agilent bioanalyzer prior to actual miRNA profiling has in most studies revealed the complete absence of 18S/28S rRNA [166, 176, 191]; however, a few reports show intact 18S/28S rRNA [178]. Methods to identify significantly altered miRNAs using reference miRNAs for normalization are difficult as no genes are known to be expressed at the same copy number in vesicle and donor cell samples. A combination of miR-22*, miR-221, and miR-26a served as successful normalizing miRNAs compared to patients with hepatitis B in serum samples from healthy individuals [192]. However, the potential use of these miRNAs for normalization in a cancer context is unknown. Normally expression is quantified by miRNA ranking based on signal intensity or number of clones sequenced or, alternatively, by comparative CT methods involving $2^{-\Delta\Delta CT}$ calculation. When profiling circulating miRNAs from minimal sample material, qPCR-based detection platforms have shown superiority in sensitivity compared to microarray platforms [193].

13.4.6 Increased Secretion of Exosomes in Cancer

First discovered during reticulocyte maturation decades ago [194], exosome vesicles are now widely recognized as being secreted also by all other cell types [195] and to be more actively secreted from malignant cells. For example, medullablastoma cell lines were found to secrete 13,400–25,300 vesicles per cell in 48 h, compared to 3,800–6,200 vesicles secreted from normal fibroblasts [185]. Interestingly, the level of RNA in the vesicles was 210- to 310-fold higher in the medullablastoma cell-derived vesicles compared to those from normal fibroblasts. Increased endosomal release and uptake by cancer cells were exhibited at low pH indicative of the tumor microenvironment in metastatic melanoma cells [154]. In a xenograft mouse model of melanoma, the level of exosomes detected in plasma correlated with the tumor burden [196]. Accordingly, an increase in exosome concentration has been observed in the bloodstream of cancer patients with lung cancer or melanoma [196, 197]. Specific cancer-associated markers detected on a subset of the vesicles demonstrated that these vesicles were indeed tumor derived, as opposed to the result of a general host response. Several studies have now shown that exosome constituents are altered in plasma- and urine-derived exosomes in a variety of cancers that promotes the possibility of using these constituents in noninvasive tests [155, 166, 196–198].

Epithelial cell adhesion molecule (EpCAM)-positive vesicles were found in the serum of ovarian cancer patients but absent in healthy individuals and increased with the stage of the disease. Furthermore, although EpCAM-positive vesicles, isolated using an immune-affinity capture technique, were present in both benign and malignant disease, the miRNA profiles from these vesicles were distinct for patients with benign ovarian disease and patients with ovarian cancer [166]. Of note, proteolytic cleavage by metalloproteinases of EpCAM on exosomes has been observed in a subset of serum samples from breast cancer patients [199]. This emphasizes that tumor-derived exosomes are heterogeneous in composition and that isolation of cancer-derived exosomes using an EpCAM-bead subfractioning method should be considered carefully.

13.4.7 Circulating miRNAs as Diagnostic Cancer Biomarkers

Circulating miRNAs have now been associated with several cancer types (Table 13.3). As indicated in the table, three of the listed studies have identified exosome-derived

Table 13.3 Circulating miRNAs

Tumor type	MicroRNAs	Significance	References
Biliary tract cancer	miR-9	Diagnosis; stable in bile	[200]
Bladder cancer	miR-126 and miR-182	Upregulated in urine; diagnosis	[201]
	miR-96 and miR-183	Diagnosis (urine)	[202]
Breast cancer	miR-195 and let-7a; miRNA-21, 106a, 126, 155, 199a and 335	Correlated with nodal status and estrogen receptor status, and decreased postoperatively	[203, 204]
Colorectal carcinoma	miR-92; miR-29and miR-92a	Diagnosis with 70% specificity and 89% sensitivity; early detection of CRC	[190, 205]
Diffuse large B-cell lymphoma	miR-21	Relapse-free survival	[153]
Glioblastoma	miR-21	Upregulated in serum microvesicles	[179]
Gastric cancer	miR-17-5p, miR-21, miR-106a, miR-106b, let-7a; miR-1, miR-20a, miR-27a, miR-34 and miR-423-5p	Up- or downregulated; correlated to tumor stage	[206, 207]
Hepatocellular carcinoma	miR-92a; miR-500; miR-25, miR-375, and let-7f	Downregulated, upregulated after surgical treatment, and diagnosis	[208–210]
Leukemia	miR-92a/miR-638; miR-29a, -181a, and -221	Diagnosis and treatment	[211, 212]
Melanoma	Gene profiles	21 miRNAs downregulated/30 miRNAs upregulated	[213]
Multiple melanoma	miR-193b-365	Upregulated	[214]
Non-small cell lung carcinoma	miR-25 and miR-223; exosomal miRNA; miR-486, miR-30d, miR-1, and miR-499	Upregulated; screening test; overall survival	[156, 197, 215]
Oral cancer	miR-31; miR-184	Upregulated, and downregulated after surgical resection; diagnosis	[216, 217]
Ovarian cancer	miRNAs-21, 92, 93, 126, and 29a; miRNAs-155, 127, and 99b	Upregulated, and related with pre-operative CA-125; downregulated; exosomes	[166, 218, 219]
Pancreatic cancer	miR-21, miR-210, miR-155, and miR-196a	Diagnosis	[220–223]
Prostate cancer	miR-141	Diagnosis	[155]
Esophageal squamous cell carcinoma	miR-10a, miR-22, miR-100, miR-148b, miR-223, miR-133a, and miR-127-3p	Upregulated	[224]
Rhabdomyosarcoma	miR-206	Upregulated	[225]

The table is modified from [226]

miRNAs, while the origin of miRNA signal in the remaining studies was not examined. In the pioneering study by Mitchell et al., miR-141 levels were increased in serum from 25 patients with metastatic prostate cancer compared to 25 healthy controls. In a subsequent study, miR-141 was found to predict the clinical outcome of prostate cancer patients with a hazard ratio of 8.3 as well as to correlate with PSA levels under longitudinal evaluations [227]. More recently, circulating miR-141 has been reported as an independent adverse prognostic factor for patients with metastatic colon cancer [228]. The combination of miR-141 and the widely used colorectal marker carcinoembryogenic antigen (CEA) further improved the accuracy of detection. In the study by Chen et al. in 2008, miR-25 and miR-223 were found to be deregulated in serum samples from lung cancer patients, and the two miRNAs were validated as noninvasive diagnostic markers among additional 152 cases [156]. Several of the circulating miRNAs identified by Chen et al. for colorectal cancer overlapped with those identified for lung cancer. This suggests that specific circulating miRNAs may exist in cancer disease in general. This would be in agreement with data reported from profiling of solid tumor tissue, showing a general upregulation of oncogenic miR-21 and downregulation of tumor suppressor miR-145 [123, 229].

Several studies indicate that circulating miRNAs have potential as diagnostic markers for lung cancer (reviewed in [230]). In one study, a fivefold difference in expression was detected for 11 serum miRNAs between long- and short-term survivors, and the levels of four miRNAs (miR-486, miR-30d, miR-1, and miR-499) were significantly associated with overall survival [215]. Early stages of lung cancer have also been associated with changes in serum miRNAs [231]. Concordantly, a diagnostic test of 34 circulating miRNAs has been developed for detection of early non-small cell lung cancer (NSCLC) [232]. This test was able to identify NSCLC cases among asymptomatic high-risk individuals with 80% accuracy and could discriminate between benign and malignant lesions. Rabinowits et al. examined epithelial exosome concentration by size-exclusion chromatography and magnetic activated cell sorting using an anti-EpCAM antibody [197]. Here, an increase in exosome concentration from 0.77 mg/ml for the control group to 2.85 mg/ml for the lung adenocarcinoma group was observed. Small RNA extraction from the epithelial-derived exosomes (=carcinoma-derived) was profiled for miRNAs using microarray hybridization. Twelve specific miRNAs elevated in NSCLC tumor tissue were mirrored in the circulating exosomes. For the control samples, the total levels of miRNAs were low, and the level of the 12 miRNAs was below the detection limit. The results were recently supported by the finding that a combination of miR-21, miR-210, and miR-486-5p detection was shown to identify lung cancer patients with malignant solitary pulmonary nodules (SPN), compared to benign SPN or healthy controls, with a sensitivity of 75% and specificity of 85% [233]. This holds promise for improvement of the pre-operative diagnosis based on SPN status presently conducted.

Several studies have addressed whether changes in circulating miRNAs is directly related to the presence of a tumor. The upregulated plasma concentrations of miR-17-5p, miR-21, miR-106a, and miR-106b in gastric cancer [206], miR-184 in squamous cell carcinoma [155], and miR-195 and let-7a in breast cancer patients [203] were significantly reduced after surgical removal of the tumor tissues. For hepatocellular carcinoma, a decreased level of miR-92a relative to miR-638 was observed among cancer patients [208]. After surgical removal of the tumor, the miR-92a/miR-638 level was restored with borderline significance. Future studies may show whether sustained or subsequent rise of a cancer-specific circulating miRNA can point to disease relapse. Downregulation of circulating miR-92 in plasma has been observed in patients with acute leukemia [211]. The cellular origin of the normal levels of miR-92a is unknown and it remains speculative why this miRNA is downregulated in cancer. In addition to profiling of serum and plasma, a recent study utilized bile for the identification of miR-9 as a potential circulating biomarker for biliary tract cancer [200]. In urine, four miRNAs (miR-96, miR-183, miR-126, and miR-182) have been associated with bladder cancer and disease stage and progression [201, 202]. Future studies are likely to incorporate additional body fluids for the identification of circulating miRNAs with diagnostic and prognostic potential.

13.5 Long Non-coding RNAs with Potential in Cancer Diagnostics

13.5.1 Long Non-coding RNAs are Deregulated in Cancer and Modulate Gene Expression

Although miRNAs constitute the largest group of non-coding RNAs (ncRNAs) known to be involved in cancer development and progression, other classes of ncRNAs with great potential for cancer prognosis and diagnosis including various long non-coding RNAs (lncRNAs) are currently being discovered. lncRNAs constitute a large group of RNA transcripts, which are more than 200-nt long, not translated into protein, and may or may not be poly-adenylated and spliced. A consensus annotation system for lncRNAs does not yet exist, as new molecules are continuously discovered. Genes for lncRNAs can be found in intergenic regions, within introns of protein coding genes (intragenic) as well as overlapping with exons of protein coding genes. lncRNAs are expressed in all human cell types and can be purified using standard mRNA purification protocols. The remaining part of this section describes lncRNAs in relation to cancer in more detail.

Although the biological functions are not fully understood, some long ncRNAs are known to play important roles in regulation of gene expression. For example, a number of newly discovered large intergenic non-coding RNAs (lincRNAs) have been shown to repress gene transcription by targeting polycomb repressor complex 1 (PRC1) or 2 (PRC2) to specific genes (reviewed in [234, 235]). PRCs regulate gene expression via chromatin modifications and are known to be involved in maintaining pluripotency of embryonic stem cells by repressing developmental genes required for differentiation [236, 237]. A central theory of carcinogenesis is the dedifferentiation of cancer cells into a phenotype resembling stem cells, possibly through mechanisms involving the PRCs (reviewed in [238, 239]). The close association of deregulated lincRNAs with PRCs, therefore, suggests a pivotal role for such lincRNAs in cancer.

Hox antisense intergenic RNA (HOTAIR) is one of the most publicized examples of a cancer-associated lincRNA capable of interacting with PRC2. HOTAIR resides in the *HOXC* gene locus and is frequently overexpressed in breast cancer, colorectal cancer, and hepatocellular carcinoma [240–242]. Overexpression of HOTAIR in breast cancer-derived cell lines modulates the pattern of PRC2 occupancy into resembling that of embryonic fibroblasts, and many of the genes targeted by HOTAIR/PRC2 are associated with breast cancer. Furthermore, HOTAIR transfected cells have a greater invasive potential in matrix invasion assays in vitro as well as when injected into mice [240]. These observations likely explain the finding of high HOTAIR expression in breast cancer, colorectal cancer, and hepatocellular carcinoma as a significant predictor of metastasis and death [240–242]. Thus, potentially, HOTAIR may be useful as a future biomarker for aggressive cancer types.

Prostate cancer-associated ncRNA transcript 1 (PCAT-1) is also known to be overexpressed in prostate cancer [243]. PCAT-1 (like HOTAIR) interacts with PRC2 in order to regulate expression of downstream genes. However, PCAT-1 itself is also regulated by PRC2. Accordingly, expression of PCAT-1 and the PRC2 subunit EZH2 are almost mutually exclusive [243]. The implications of this relationship for cancer development are still unresolved, but it appears that PCAT-1 is upregulated in a subset of metastatic and high-grade tumors, which express low levels of EZH2. Overexpression of PCAT-1 in prostate cancer cell lines results in higher levels of proliferation, providing a likely explanation for the advantage of overexpression of PCAT-1 in cancer cells. In summary, HOTAIR and PCAT-1 are two examples of novel lincRNAs which may be used as diagnostic and/or prognostic markers for different cancer types. Many more are likely to be discovered in the near future.

Long ncRNAs, including lincRNAs, are also emerging as possible enhancers of gene transcription (reviewed in [234]). The mechanism(s) employed by activating lncRNAs are unclear, but they seem capable of activating neighboring genes as well as genes located several kilobases and more away [244]. Transcribed ultraconserved regions (T-UCRs) constitute a likely subgroup of novel activating lncRNAs. The corresponding UCRs are defined as genomic regions more than 200 base pairs long that are 100% identical in mouse, rat, and human genomes [245]. UCRs are strongly depleted in genomic regions containing germline copy number variations [246, 247]. The extreme evolutionary conservation and the apparent selection against alterations in copy number of these sequences suggest that they perform essential cellular functions.

Based on the criteria listed above, there are 481 UCRs in the human genome (excluding rRNA genes), most of which are transcribed [248]. UCRs are found within protein coding genes as well as in introns and in intergenic regions [245].

Since the UCRs are merely conserved genomic sequences, they are unlikely to be all involved in the same cellular functions. However, many of the T-UCRs have been shown to harbor enhancer functions [249]. T-UCR expression is deregulated in cancer, and different types of cancer are associated with specific expression signatures of the T-UCRs [247, 248]. The mechanisms behind deregulation of T-UCRs in cancer may involve epigenetic aberrations [250], genomic rearrangements [247], or miRNA-mediated degradation [247]. Thus, T-UCR expression as well as changes in DNA methylation and/or larger scale DNA alterations may have potential as cancer biomarkers. This is illustrated in a number of malignancies. In neuroblastoma, high expression of 28 T-UCRs, which correlate negatively with specific miRNAs, is associated with a good prognosis [251]. Moreover, in hepatocellular carcinoma, expression levels of 56 T-UCRs are deregulated [252]. In colorectal cancer and lymphocytic leukemia, cancer-associated sequence abnormalities in UCRs have been reported [253]. Furthermore, in a collection of different types of primary tumors, three specific T-UCRs (Uc.160+, Uc.283+A, and Uc.346+) are downregulated by DNA methylation, which is associated with the formation of lymph node metastases [250]. Although these observations need further validation, they all seem to be of potential use in future cancer diagnostics.

13.5.2 PCA3 and Examples of Other lncRNAs as Potential Cancer Diagnostics

Prostate Cancer Antigen 3 (PCA3) is a non-coding RNA that can be found in urine from prostate cancer patients. It has been developed as a diagnostic biomarker for prostate cancer and is now commercially available (reviewed in [254]). The cellular function of PCA3 has not been fully clarified, but several reports suggest that PCA3 regulates the transcript level of its host gene *BMCC1/PRUNE2* (Prune homolog 2) (reviewed in [255]). Numerous studies have evaluated the sensitivity and specificity of PCA3 for prostate cancer detection in urine (reviewed in [254]). Together, results from these studies indicate that PCA3 may not be the ideal biomarker in urine, primarily, because it does not clearly reflect the abundance of PCA3 in the tumor where PCA3 is frequently overexpressed [256–258]. Nevertheless, in combination with the routinely used serum biomarker PSA, a urine test for PCA3 may be added to increase the specificity of diagnosis.

Vault non-coding RNAs (vRNAs) are components of vault ribonucleoprotein complexes, which confer drug resistance in cancer cells. Three vRNAs, highly similar in sequence and secondary structure, are known to be expressed in mammalian cells. Overexpression of vRNAs in cancer cell lines confers chemoresistance, while knock-down of vRNAs renders cancer cells more sensitive to chemotherapeutics [259]. Hence, the expression status of vault RNAs in cancer cells may aid in prediction of resistance to chemotherapeutic compounds.

The non-coding RNA Growth Arrest Specific 5 (GAS5) is known to bind as a decoy to the glucocorticoid receptor (GR), thereby preventing its interaction with the glucocorticoid response elements in genomic DNA. GR target genes are involved in repression of apoptosis and are normally upregulated upon GR binding. Thus, interaction of GAS5 with the GR may allow or even promote apoptosis (reviewed in [235]). In agreement with this model, GAS5 transcript levels have been found to be significantly downregulated in breast cancer cells compared to normal breast epithelium [260]. Thus, GAS5 may have potential as a future cancer biomarker or even as an anticancer therapeutic agent.

13.6 Future Perspectives

Although a large number of molecular profiling studies have identified several miRNAs with potential as future cancer biomarkers, no miRNA-based diagnostic test has yet been developed and approved for use in routine clinical practice. Lack of sufficient independent validation and inconsistencies between results from different miRNA profiling studies of cancer tissue samples remain major limitations for the clinical translation of novel miRNA biomarker candidates. Such inconsistencies may be explained, at least in part, by the use of different microarray platforms and RT-qPCR methodologies in studies of the same tumor type [140]. Standardized procedures for sample preparation and miRNA quantification could accelerate the discovery and validation of new miRNA cancer biomarkers, a prerequisite for their translation into clinical usage. Improved and more detailed clinical annotation of patient samples included in miRNA expression profiling studies may also help push this field forward.

Furthermore, miRNA profiling studies conducted so far have often been underpowered, i.e., insufficient sample numbers in the discovery and/or validation phase. Consequently, the results obtained from these studies may not be representative for a given cancer type and, hence, less likely to lead to the development of clinically useful biomarkers. Recent studies focusing on long non-coding RNAs tend to suffer from similar limitations in sample numbers. As an additional complicating factor, most tumors are characterized by marked histologic and genetic heterogeneity [261, 262]. Accordingly, the exact cancer tissue specimen(s) used for molecular profiling analyses may not be representative for the entire tumor. To address this issue, some studies have used laser-microdissection for isolation of specific cell fractions from tumor tissue samples prior to miRNA profiling. Although this is a laborious process not compatible with routine daily clinical procedures, such studies can contribute invaluable new and more detailed insights into tumor biology and mechanisms of carcinogenesis [263].

Identification of circulating tumor biomarkers could be an alternative solution to this problem. Indeed, the expression of specific circulating miRNAs seems to be a good surrogate of tumor miRNA expression and, thus, has initiated a new paradigm useful for noninvasive testing for early diagnosis, prognosis, and/or therapeutic decision making. Challenges, however, remain for the identification of circulating miRNA biomarkers. First, the preparation of samples (serum, plasma, urine, etc.) requires very strict standardization in order to generalize findings from different patients, research groups, and laboratories. One possible reason for variation is that thrombocytes present in the plasma fraction are activated upon incubation below 20°C. This will lead to release of small miRNA-containing granula not removed at 20,000 × *g* centrifugation typically used. Accordingly, any variation in incubation times across samples before complete processing can lead to variations in the miRNA expression detected. Second, the quantification of miRNAs is not straightforward. Further investigations should clarify the biological half-life of miRNAs, since these may vary individually, as well as uncover suitable normalization strategies involving identification of ideal "housekeeping" small RNAs.

Moreover, the protocol for isolating exosomes is not standardized. By using specific exosome surface markers such as EpCAM for affinity purification, it may be possible to avoid the problem of contaminating exosomes from blood cells. However, the amount of circulating tumor exosomes is presumably low in the bloodstream and likely display a heterogenous expression of surface markers. Thus, using surface markers for exosome isolation will reduce any miRNA signal considerably and introduce bias in the subset of vesicles isolated. It is currently unknown whether miRNA profiling of crude samples (without discrimination between vesicular or protein complex origin) is better at distinguishing between patient groups, compared to profiling after specific purification steps. Using a nonspecific procedure (a simple 15,000–20,000 × *g* spin prior to RNA extraction) will remove large cellular debris and protein aggregates, but may not be adequate for detection of small amounts of tumor-derived material. Finally, hemolysis of plasma samples as well as activation of platelets during sample processing will influence the results greatly and demands for precise handling of samples to avoid this. If future studies can successfully overcome these challenges, the distinct expression profiles of miRNAs in cancer tissues together with the remarkable stability of circulating miRNAs should make these new easily accessible candidate biomarkers ideal for translation into clinical use. Because cancer disease is often diagnosed at a late stage with a concomitant poor prognosis, development of minimally invasive tests for detection and monitoring of common solid tumors may significantly help to reduce the worldwide health burden caused by cancer.

During the past decade, expression profiling studies of microRNAs, and more recently also of long ncRNAs, have made important contributions to basic, clinical, and translational cancer research. In order to transfer these findings into actual clinical utility, it will be essential to conduct large multicenter clinical studies for prospective validation of the many candidate biomarkers already identified. Such studies should also define in which exact clinical setting and in which type of patients a given biomarker (panel) has true potential for, e.g., diagnosis, prognosis, prediction of treatment response, or disease monitoring.

The recent availability of new advanced technologies for sequencing of the human genome, NGS, has already revolutionized biological research and likely will have similar effects on molecular diagnostics in the very near future. NGS technologies can be used for discovery of new disease markers with unsurpassed efficiency and speed. Also, new systems biology approaches which integrate genome-wide sequencing data obtained at different molecular levels (e.g., genome, transcriptome, and epigenome) for the clinical sample may help to reveal the complex molecular mechanisms that regulate cancer development and progression in more depth. An improved understanding of the fundamental molecular and biological processes occurring during carcinogenesis could not only provide the basis for development of new biomarkers, but may also contribute to the future development of new pathway-targeted therapies.

References

1. Croce CM (2009) Causes and consequences of microRNA dysregulation in cancer. Nat Rev Genet 10(10):704–714
2. Lee RC, Feinbaum RL, Ambros V (1993) The *C. elegans* heterochronic gene lin-4 encodes small RNAs with antisense complementarity to lin-14. Cell 75(5):843–854
3. Kim VN, Han J, Siomi MC (2009) Biogenesis of small RNAs in animals. Nat Rev Mol Cell Biol 10(2):126–139
4. Calin GA et al (2002) Frequent deletions and down-regulation of micro-RNA genes miR15 and miR16 at 13q14 in chronic lymphocytic leukemia. Proc Natl Acad Sci USA 99(24):15524–15529
5. Hu W et al (2010) Negative regulation of tumor suppressor p53 by microRNA miR-504. Mol Cell 38(5):689–699
6. Cimmino A et al (2005) miR-15 and miR-16 induce apoptosis by targeting BCL2. Proc Natl Acad Sci USA 102(39):13944–13949
7. Sampson VB et al (2007) MicroRNA let-7a down-regulates MYC and reverts MYC-induced growth in Burkitt lymphoma cells. Cancer Res 67(20):9762–9770
8. He L et al (2007) A microRNA component of the p53 tumour suppressor network. Nature 447(7148):1130–1134
9. Voorhoeve PM et al (2006) A genetic screen implicates miRNA-372 and miRNA-373 as oncogenes in testicular germ cell tumors. Cell 124(6):1169–1181
10. Di Leva G, Croce CM (2010) Roles of small RNAs in tumor formation. Trends Mol Med 16(6):257–267
11. Allis CD et al (2007) New nomenclature for chromatin-modifying enzymes. Cell 131(4): 633–636
12. Fabbri M et al (2007) MicroRNA-29 family reverts aberrant methylation in lung cancer by targeting DNA methyltransferases 3A and 3B. Proc Natl Acad Sci USA 104(40): 15805–15810
13. Garofalo M, Croce CM (2011) microRNAs: master regulators as potential therapeutics in cancer. Annu Rev Pharmacol Toxicol 51:25–43
14. Iorio MV et al (2005) MicroRNA gene expression deregulation in human breast cancer. Cancer Res 65(16):7065–7070
15. Johnson SM et al (2005) RAS is regulated by the let-7 microRNA family. Cell 120(5): 635–647
16. Schetter AJ et al (2008) MicroRNA expression profiles associated with prognosis and therapeutic outcome in colon adenocarcinoma. JAMA 299(4):425–436
17. Yang N et al (2008) MicroRNA microarray identifies Let-7i as a novel biomarker and therapeutic target in human epithelial ovarian cancer. Cancer Res 68(24):10307–10314
18. Petrocca F et al (2008) E2F1-regulated microRNAs impair TGFbeta-dependent cell-cycle arrest and apoptosis in gastric cancer. Cancer Cell 13(3):272–286

19. Peter ME (2009) Let-7 and miR-200 microRNAs: guardians against pluripotency and cancer progression. Cell Cycle 8(6):843–852
20. Trang P et al (2010) Regression of murine lung tumors by the let-7 microRNA. Oncogene 29(11):1580–1587
21. Bonci D et al (2008) The miR-15a-miR-16-1 cluster controls prostate cancer by targeting multiple oncogenic activities. Nat Med 14(11):1271–1277
22. Finnerty JR et al (2010) The miR-15/107 group of microRNA genes: evolutionary biology, cellular functions, and roles in human diseases. J Mol Biol 402(3):491–509
23. Lu J et al (2005) MicroRNA expression profiles classify human cancers. Nature 435(7043): 834–838
24. Bottoni A et al (2005) miR-15a and miR-16-1 down-regulation in pituitary adenomas. J Cell Physiol 204(1):280–285
25. Yanaihara N et al (2006) Unique microRNA molecular profiles in lung cancer diagnosis and prognosis. Cancer Cell 9(3):189–198
26. Volinia S et al (2006) A microRNA expression signature of human solid tumors defines cancer gene targets. Proc Natl Acad Sci USA 103(7):2257–2261
27. Pichiorri F et al (2008) MicroRNAs regulate critical genes associated with multiple myeloma pathogenesis. Proc Natl Acad Sci USA 105(35):12885–12890
28. Poliseno L et al (2010) Identification of the miR-106b 25 microRNA cluster as a proto-oncogenic PTEN-targeting intron that cooperates with its host gene MCM7 in transformation. Sci Signal 3(117):ra29
29. Uziel T et al (2009) The miR-17 92 cluster collaborates with the Sonic Hedgehog pathway in medulloblastoma. Proc Natl Acad Sci USA 106(8):2812–2817
30. Mendell JT (2008) miRiad roles for the miR-17-92 cluster in development and disease. Cell 133(2):217–222
31. Xiao C et al (2008) Lymphoproliferative disease and autoimmunity in mice with increased miR-17-92 expression in lymphocytes. Nat Immunol 9(4):405–414
32. Calin GA et al (2005) A MicroRNA signature associated with prognosis and progression in chronic lymphocytic leukemia. N Engl J Med 353(17):1793–1801
33. Garzon R et al (2008) MicroRNA signatures associated with cytogenetics and prognosis in acute myeloid leukemia. Blood 111(6):3183–3189
34. Garzon R et al (2008) Distinctive microRNA signature of acute myeloid leukemia bearing cytoplasmic mutated nucleophosmin. Proc Natl Acad Sci USA 105(10):3945–3950
35. Ciafre SA et al (2005) Extensive modulation of a set of microRNAs in primary glioblastoma. Biochem Biophys Res Commun 334(4):1351–1358
36. Jazbutyte V, Thum T (2010) MicroRNA-21: from cancer to cardiovascular disease. Curr Drug Targets 11(8):926–935
37. Meng F et al (2007) MicroRNA-21 regulates expression of the PTEN tumor suppressor gene in human hepatocellular cancer. Gastroenterology 133(2):647–658
38. Frankel LB et al (2008) Programmed cell death 4 (PDCD4) is an important functional target of the microRNA miR-21 in breast cancer cells. J Biol Chem 283(2):1026–1033
39. Zhu S et al (2007) MicroRNA-21 targets the tumor suppressor gene tropomyosin 1 (TPM1). J Biol Chem 282(19):14328–14336
40. Pekarsky Y et al (2006) Tcl1 expression in chronic lymphocytic leukemia is regulated by miR-29 and miR-181. Cancer Res 66(24):11590–11593
41. Zhao JJ et al (2010) microRNA expression profile and identification of miR-29 as a prognostic marker and pathogenetic factor by targeting CDK6 in mantle cell lymphoma. Blood 115(13):2630–2639
42. Xiong Y et al (2010) Effects of microRNA-29 on apoptosis, tumorigenicity, and prognosis of hepatocellular carcinoma. Hepatology 51(3):836–845
43. Mott JL et al (2007) mir-29 regulates Mcl-1 protein expression and apoptosis. Oncogene 26(42):6133–6140
44. Lujambio A et al (2008) A microRNA DNA methylation signature for human cancer metastasis. Proc Natl Acad Sci USA 105(36):13556–13561

45. Cole KA et al (2008) A functional screen identifies miR-34a as a candidate neuroblastoma tumor suppressor gene. Mol Cancer Res 6(5):735–742
46. Li N et al (2009) miR-34a inhibits migration and invasion by down-regulation of c-Met expression in human hepatocellular carcinoma cells. Cancer Lett 275(1):44–53
47. Raver-Shapira N et al (2007) Transcriptional activation of miR-34a contributes to p53-mediated apoptosis. Mol Cell 26(5):731–743
48. Chang TC et al (2007) Transactivation of miR-34a by p53 broadly influences gene expression and promotes apoptosis. Mol Cell 26(5):745–752
49. Toyota M et al (2008) Epigenetic silencing of microRNA-34b/c and B-cell translocation gene 4 is associated with CpG island methylation in colorectal cancer. Cancer Res 68(11): 4123–4132
50. Sun F et al (2008) Downregulation of CCND1 and CDK6 by miR-34a induces cell cycle arrest. FEBS Lett 582(10):1564–1568
51. Yamakuchi M, Ferlito M, Lowenstein CJ (2008) miR-34a repression of SIRT1 regulates apoptosis. Proc Natl Acad Sci USA 105(36):13421–13426
52. Pigazzi M et al (2009) miR-34b targets cyclic AMP-responsive element binding protein in acute myeloid leukemia. Cancer Res 69(6):2471–2478
53. Ostenfeld MS et al (2010) miR-145 induces caspase-dependent and -independent cell death in urothelial cancer cell lines with targeting of an expression signature present in Ta bladder tumors. Oncogene 29(7):1073–1084
54. Chiyomaru T et al (2011) SWAP70, actin-binding protein, function as an oncogene targeting tumor-suppressive miR-145 in prostate cancer. Prostate doi:10.1002/pros.21372. [Epub ahead of print]
55. Metzler M et al (2004) High expression of precursor microRNA-155/BIC RNA in children with Burkitt lymphoma. Genes Chromosomes Cancer 39(2):167–169
56. Kluiver J et al (2005) BIC and miR-155 are highly expressed in Hodgkin, primary mediastinal and diffuse large B cell lymphomas. J Pathol 207(2):243–249
57. Borchert GM, Holton NW, Larson ED (2011) Repression of human activation induced cytidine deaminase by miR-93 and miR-155. BMC Cancer 11:347
58. Pedersen IM et al (2009) Onco-miR-155 targets SHIP1 to promote TNFalpha-dependent growth of B cell lymphomas. EMBO Mol Med 1(5):288–295
59. le Sage C et al (2007) Regulation of the p27(Kip1) tumor suppressor by miR-221 and miR-222 promotes cancer cell proliferation. EMBO J 26(15):3699–3708
60. Medina R et al (2008) MicroRNAs 221 and 222 bypass quiescence and compromise cell survival. Cancer Res 68(8):2773–2780
61. Garofalo M et al (2009) miR-221&222 regulate TRAIL resistance and enhance tumorigenicity through PTEN and TIMP3 downregulation. Cancer Cell 16(6):498–509
62. Felli N et al (2005) MicroRNAs 221 and 222 inhibit normal erythropoiesis and erythroleukemic cell growth via kit receptor down-modulation. Proc Natl Acad Sci USA 102(50): 18081–18086
63. Schaefer A et al (2010) Diagnostic and prognostic implications of microRNA profiling in prostate carcinoma. Int J Cancer 126(5):1166–1176
64. Fornari F et al (2008) MiR-221 controls CDKN1C/p57 and CDKN1B/p27 expression in human hepatocellular carcinoma. Oncogene 27(43):5651–5661
65. Quintavalle C et al (2011) miR-221/222 overexpession in human glioblastoma increases invasiveness by targeting the protein phosphate PTPmu. Oncogene 31(7):858–868
66. Kim YK et al (2009) Functional links between clustered microRNAs: suppression of cell-cycle inhibitors by microRNA clusters in gastric cancer. Nucleic Acids Res 37(5): 1672–1681
67. Calin GA et al (2004) Human microRNA genes are frequently located at fragile sites and genomic regions involved in cancers. Proc Natl Acad Sci USA 101(9):2999–3004
68. Wang D et al (2010) Human microRNA oncogenes and tumor suppressors show significantly different biological patterns: from functions to targets. PLoS One 5(9) pii:e13067

69. Mishra PJ, Banerjee D, Bertino JR (2008) MiRSNPs or MiR-polymorphisms, new players in microRNA mediated regulation of the cell: introducing microRNA pharmacogenomics. Cell Cycle 7(7):853–858
70. Adams BD, Furneaux H, White BA (2007) The micro-ribonucleic acid (miRNA) miR-206 targets the human estrogen receptor-alpha (ERalpha) and represses ERalpha messenger RNA and protein expression in breast cancer cell lines. Mol Endocrinol 21(5):1132–1147
71. He H et al (2005) The role of microRNA genes in papillary thyroid carcinoma. Proc Natl Acad Sci USA 102(52):19075–19080
72. Gao P et al (2009) c-Myc suppression of miR-23a/b enhances mitochondrial glutaminase expression and glutamine metabolism. Nature 458(7239):762–765
73. Velu CS, Baktula AM, Grimes HL (2009) Gfi1 regulates miR-21 and miR-196b to control myelopoiesis. Blood 113(19):4720–4728
74. Yan HL et al (2009) Repression of the miR-17-92 cluster by p53 has an important function in hypoxia-induced apoptosis. EMBO J 28(18):2719–2732
75. Suh SO et al (2011) MicroRNA-145 is regulated by DNA methylation and p53 gene mutation in prostate cancer. Carcinogenesis 32(5):772–778
76. Guil S, Esteller M (2009) DNA methylomes, histone codes and miRNAs: tying it all together. Int J Biochem Cell Biol 41(1):87–95
77. Weber B et al (2007) Methylation of human microRNA genes in normal and neoplastic cells. Cell Cycle 6(9):1001–1005
78. Lehmann U et al (2008) Epigenetic inactivation of microRNA gene hsa-mir-9-1 in human breast cancer. J Pathol 214(1):17–24
79. Saito Y et al (2006) Specific activation of microRNA-127 with downregulation of the proto-oncogene BCL6 by chromatin-modifying drugs in human cancer cells. Cancer Cell 9(6):435–443
80. Thomson JM et al (2006) Extensive post-transcriptional regulation of microRNAs and its implications for cancer. Genes Dev 20(16):2202–2207
81. Melo SA et al (2010) A genetic defect in exportin-5 traps precursor microRNAs in the nucleus of cancer cells. Cancer Cell 18(4):303–315
82. Yang W et al (2006) Modulation of microRNA processing and expression through RNA editing by ADAR deaminases. Nat Struct Mol Biol 13(1):13–21
83. Kawahara Y et al (2007) RNA editing of the microRNA-151 precursor blocks cleavage by the Dicer-TRBP complex. EMBO Rep 8(8):763–769
84. Kawahara Y et al (2007) Redirection of silencing targets by adenosine-to-inosine editing of miRNAs. Science 315(5815):1137–1140
85. Deng S et al (2008) Mechanisms of microRNA deregulation in human cancer. Cell Cycle 7(17):2643–2646
86. Fabbri M (2010) miRNAs as molecular biomarkers of cancer. Expert Rev Mol Diagn 10(4):435–444
87. Ferracin M, Veronese A, Negrini M (2010) Micromarkers: miRNAs in cancer diagnosis and prognosis. Expert Rev Mol Diagn 10(3):297–308
88. Munker R, Calin GA (2011) MicroRNA profiling in cancer. Clin Sci (Lond) 121(4):141–158
89. Rosenfeld N et al (2008) MicroRNAs accurately identify cancer tissue origin. Nat Biotechnol 26(4):462–469
90. Ferdin J, Kunej T, Calin GA (2010) Non-coding RNAs: identification of cancer-associated microRNAs by gene profiling. Technol Cancer Res Treat 9(2):123–138
91. Kong W et al (2009) Strategies for profiling microRNA expression. J Cell Physiol 218(1):22–25
92. Liu CG et al (2004) An oligonucleotide microchip for genome-wide microRNA profiling in human and mouse tissues. Proc Natl Acad Sci USA 101(26):9740–9744
93. Liu CG et al (2008) Expression profiling of microRNA using oligo DNA arrays. Methods 44(1):22–30
94. Yin JQ, Zhao RC, Morris KV (2008) Profiling microRNA expression with microarrays. Trends Biotechnol 26(2):70–76

95. Li W, Ruan K (2009) MicroRNA detection by microarray. Anal Bioanal Chem 394(4):1117–1124
96. Chen J et al (2008) Highly sensitive and specific microRNA expression profiling using BeadArray technology. Nucleic Acids Res 36(14):e87
97. Wang H, Ach RA, Curry B (2007) Direct and sensitive miRNA profiling from low-input total RNA. RNA 13(1):151–159
98. Castoldi M et al (2006) A sensitive array for microRNA expression profiling (miChip) based on locked nucleic acids (LNA). RNA 12(5):913–920
99. Git A et al (2010) Systematic comparison of microarray profiling, real-time PCR, and next-generation sequencing technologies for measuring differential microRNA expression. RNA 16(5):991–1006
100. Pradervand S et al (2010) Concordance among digital gene expression, microarrays, and qPCR when measuring differential expression of microRNAs. Biotechniques 48(3):219–222
101. Sato F et al (2009) Intra-platform repeatability and inter-platform comparability of microRNA microarray technology. PLoS One 4(5):e5540
102. Yauk CL et al (2010) Cross-platform analysis of global microRNA expression technologies. BMC Genomics 11:330
103. Castoldi M et al (2008) miChip: an array-based method for microRNA expression profiling using locked nucleic acid capture probes. Nat Protoc 3(2):321–329
104. Sempere LF et al (2004) Expression profiling of mammalian microRNAs uncovers a subset of brain-expressed microRNAs with possible roles in murine and human neuronal differentiation. Genome Biol 5(3):R13
105. Cummins JM et al (2006) The colorectal microRNAome. Proc Natl Acad Sci USA 103(10):3687–3692
106. Calin GA et al (2004) MicroRNA profiling reveals distinct signatures in B cell chronic lymphocytic leukemias. Proc Natl Acad Sci USA 101(32):11755–11760
107. Wach S et al (2012) MicroRNA profiles of prostate carcinoma detected by multiplatform microRNA screening. Int J Cancer 130(3):611–621
108. Chen C et al (2005) Real-time quantification of microRNAs by stem-loop RT-PCR. Nucleic Acids Res 33(20):e179
109. Mestdagh P et al (2008) High-throughput stem-loop RT-qPCR miRNA expression profiling using minute amounts of input RNA. Nucleic Acids Res 36(21):e143
110. Schmittgen TD et al (2008) Real-time PCR quantification of precursor and mature microRNA. Methods 44(1):31–38
111. Benes V, Castoldi M (2010) Expression profiling of microRNA using real-time quantitative PCR, how to use it and what is available. Methods 50(4):244–249
112. Jacobsen N, Andreasen D, Mouritzen P (2011) Profiling microRNAs by real-time PCR. Methods Mol Biol 732:39–54
113. Schmittgen TD, Lee EJ, Jiang J (2008) High-throughput real-time PCR. Methods Mol Biol 429:89–98
114. Tombol Z et al (2010) MicroRNA expression profiling in benign (sporadic and hereditary) and recurring adrenal pheochromocytomas. Mod Pathol 23(12):1583–1595
115. Landgraf P et al (2007) A mammalian microRNA expression atlas based on small RNA library sequencing. Cell 129(7):1401–1414
116. Newman MA, Mani V, Hammond SM (2011) Deep sequencing of microRNA precursors reveals extensive 3' end modification. RNA 17(10):1795–1803
117. Jima DD et al (2010) Deep sequencing of the small RNA transcriptome of normal and malignant human B cells identifies hundreds of novel microRNAs. Blood 116(23):e118–e127
118. Creighton CJ, Reid JG, Gunaratne PH (2009) Expression profiling of microRNAs by deep sequencing. Brief Bioinform 10(5):490–497
119. Linsen SE et al (2009) Limitations and possibilities of small RNA digital gene expression profiling. Nat Methods 6(7):474–476

120. Willenbrock H et al (2009) Quantitative miRNA expression analysis: comparing microarrays with next-generation sequencing. RNA 15(11):2028–2034
121. Mi S et al (2007) MicroRNA expression signatures accurately discriminate acute lymphoblastic leukemia from acute myeloid leukemia. Proc Natl Acad Sci USA 104(50):19971–19976
122. Fulci V et al (2009) Characterization of B- and T-lineage acute lymphoblastic leukemia by integrated analysis of MicroRNA and mRNA expression profiles. Genes Chromosomes Cancer 48(12):1069–1082
123. Dyrskjot L et al (2009) Genomic profiling of microRNAs in bladder cancer: miR-129 is associated with poor outcome and promotes cell death in vitro. Cancer Res 69(11):4851–4860
124. Schepeler T et al (2008) Diagnostic and prognostic microRNAs in stage II colon cancer. Cancer Res 68(15):6416–6424
125. Lebanony D et al (2009) Diagnostic assay based on hsa-miR-205 expression distinguishes squamous from nonsquamous non-small-cell lung carcinoma. J Clin Oncol 27(12): 2030–2037
126. Ueda T et al (2010) Relation between microRNA expression and progression and prognosis of gastric cancer: a microRNA expression analysis. Lancet Oncol 11(2):136–146
127. Guan Y et al (2010) MiRNA-196 is upregulated in glioblastoma but not in anaplastic astrocytoma and has prognostic significance. Clin Cancer Res 16(16):4289–4297
128. Ji J et al (2009) MicroRNA expression, survival, and response to interferon in liver cancer. N Engl J Med 361(15):1437–1447
129. Navarro A et al (2009) Regulation of JAK2 by miR-135a: prognostic impact in classic Hodgkin lymphoma. Blood 114(14):2945–2951
130. Yu SL et al (2008) MicroRNA signature predicts survival and relapse in lung cancer. Cancer Cell 13(1):48–57
131. Eitan R et al (2009) Tumor microRNA expression patterns associated with resistance to platinum based chemotherapy and survival in ovarian cancer patients. Gynecol Oncol 114(2):253–259
132. Hwang JH et al (2010) Identification of microRNA-21 as a biomarker for chemoresistance and clinical outcome following adjuvant therapy in resectable pancreatic cancer. PLoS One 5(5):e10630
133. Giovannetti E et al (2010) MicroRNA-21 in pancreatic cancer: correlation with clinical outcome and pharmacologic aspects underlying its role in the modulation of gemcitabine activity. Cancer Res 70(11):4528–4538
134. Gong C et al (2011) Up-regulation of miR-21 mediates resistance to trastuzumab therapy for breast cancer. J Biol Chem 286(21):19127–19137
135. Ferracin M et al (2010) MicroRNAs involvement in fludarabine refractory chronic lymphocytic leukemia. Mol Cancer 9:123
136. Chiorazzi N, Rai KR, Ferrarini M (2005) Chronic lymphocytic leukemia. N Engl J Med 352(8):804–815
137. Rossi S et al (2010) microRNA fingerprinting of CLL patients with chromosome 17p deletion identify a miR-21 score that stratifies early survival. Blood 116(6):945–952
138. Visone R et al (2009) Karyotype-specific microRNA signature in chronic lymphocytic leukemia. Blood 114(18):3872–3879
139. Stamatopoulos B et al (2009) microRNA-29c and microRNA-223 down-regulation has in vivo significance in chronic lymphocytic leukemia and improves disease risk stratification. Blood 113(21):5237–5245
140. Sørensen KD, Ørntoft TF (2010) Discovery of prostate cancer biomarkers by microarray gene expression profiling. Expert Rev Mol Diagn 10(1):49–64
141. Porkka KP et al (2007) MicroRNA expression profiling in prostate cancer. Cancer Res 67(13):6130–6135
142. Ozen M et al (2008) Widespread deregulation of microRNA expression in human prostate cancer. Oncogene 27(12):1788–1793
143. Ambs S et al (2008) Genomic profiling of microRNA and messenger RNA reveals deregulated microRNA expression in prostate cancer. Cancer Res 68(15):6162–6170

144. Tong AW et al (2009) MicroRNA profile analysis of human prostate cancers. Cancer Gene Ther 16(3):206–216
145. Spahn M et al (2010) Expression of microRNA-221 is progressively reduced in aggressive prostate cancer and metastasis and predicts clinical recurrence. Int J Cancer 127(2):394–403
146. Szczyrba J et al (2010) The microRNA profile of prostate carcinoma obtained by deep sequencing. Mol Cancer Res 8(4):529–538
147. Martens-Uzunova ES et al (2012) Diagnostic and prognostic signatures from the small non-coding RNA transcriptome in prostate cancer. Oncogene 31(8):978–991
148. Prueitt RL et al (2008) Expression of microRNAs and protein-coding genes associated with perineural invasion in prostate cancer. Prostate 68(11):1152–1164
149. Gandellini P et al (2011) MicroRNAs as new therapeutic targets and tools in cancer. Expert Opin Ther Targets 15(3):265–279
150. Zheng T et al (2010) Role of microRNA in anticancer drug resistance. Int J Cancer 126(1):2–10
151. Allen KE, Weiss GJ (2010) Resistance may not be futile: microRNA biomarkers for chemoresistance and potential therapeutics. Mol Cancer Ther 9(12):3126–3136
152. Hood JL, San Roman S, Wickline SA (2011) Exosomes released by melanoma cells prepare sentinel lymph nodes for tumor metastasis. Cancer Res 71(11):3792–3801
153. Lawrie CH et al (2008) Detection of elevated levels of tumour-associated microRNAs in serum of patients with diffuse large B-cell lymphoma. Br J Haematol 141(5):672–675
154. Parolini I et al (2009) Microenvironmental pH is a key factor for exosome traffic in tumor cells. J Biol Chem 284(49):34211–34222
155. Mitchell PS et al (2008) Circulating microRNAs as stable blood-based markers for cancer detection. Proc Natl Acad Sci USA 105(30):10513–10518
156. Chen X et al (2008) Characterization of microRNAs in serum: a novel class of biomarkers for diagnosis of cancer and other diseases. Cell Res 18(10):997–1006
157. Turchinovich A et al (2011) Characterization of extracellular circulating microRNA. Nucleic Acids Res 39(16):7223–7233
158. Arroyo JD et al (2011) Argonaute2 complexes carry a population of circulating microRNAs independent of vesicles in human plasma. Proc Natl Acad Sci USA 108(12):5003–5008
159. Wang K et al (2010) Export of microRNAs and microRNA-protective protein by mammalian cells. Nucleic Acids Res 38(20):7248–7259
160. Vickers KC et al (2011) MicroRNAs are transported in plasma and delivered to recipient cells by high-density lipoproteins. Nat Cell Biol 13(4):423–433
161. Buschow SI et al (2005) Exosomes contain ubiquitinated proteins. Blood Cells Mol Dis 35(3):398–403
162. Mathivanan S et al (2010) Proteomics analysis of A33 immunoaffinity-purified exosomes released from the human colon tumor cell line LIM1215 reveals a tissue-specific protein signature. Mol Cell Proteomics 9(2):197–208
163. Shen B et al (2011) Protein targeting to exosomes/microvesicles by plasma membrane anchors. J Biol Chem 286(16):14383–14395
164. Fang Y et al (2007) Higher-order oligomerization targets plasma membrane proteins and HIV gag to exosomes. PLoS Biol 5(6):e158
165. Gibbings DJ et al (2009) Multivesicular bodies associate with components of miRNA effector complexes and modulate miRNA activity. Nat Cell Biol 11(9):1143–1149
166. Taylor DD, Gercel-Taylor C (2008) MicroRNA signatures of tumor-derived exosomes as diagnostic biomarkers of ovarian cancer. Gynecol Oncol 110(1):13–21
167. Subra C et al (2007) Exosome lipidomics unravels lipid sorting at the level of multivesicular bodies. Biochimie 89(2):205–212
168. Ostrowski M et al (2010) Rab27a and Rab27b control different steps of the exosome secretion pathway. Nat Cell Biol 12(1):19–30, sup pp. 1–13
169. Hsu C et al (2010) Regulation of exosome secretion by Rab35 and its GTPase-activating proteins TBC1D10A-C. J Cell Biol 189(2):223–232

170. Kosaka N et al (2010) Secretory mechanisms and intercellular transfer of microRNAs in living cells. J Biol Chem 285(23):17442–17452
171. Savina A et al (2003) Exosome release is regulated by a calcium-dependent mechanism in K562 cells. J Biol Chem 278(22):20083–20090
172. Lehmann BD et al (2008) Senescence-associated exosome release from human prostate cancer cells. Cancer Res 68(19):7864–7871
173. Fader CM et al (2008) Induction of autophagy promotes fusion of multivesicular bodies with autophagic vacuoles in k562 cells. Traffic 9(2):230–250
174. Rialland P et al (2006) BCR-bound antigen is targeted to exosomes in human follicular lymphoma B-cells. Biol Cell 98(8):491–501
175. Blanchard N et al (2002) TCR activation of human T cells induces the production of exosomes bearing the TCR/CD3/zeta complex. J Immunol 168(7):3235–3241
176. Valadi H et al (2007) Exosome-mediated transfer of mRNAs and microRNAs is a novel mechanism of genetic exchange between cells. Nat Cell Biol 9(6):654–659
177. Keller S et al (2009) Systemic presence and tumor-growth promoting effect of ovarian carcinoma released exosomes. Cancer Lett 278(1):73–81
178. Hong BS et al (2009) Colorectal cancer cell-derived microvesicles are enriched in cell cycle-related mRNAs that promote proliferation of endothelial cells. BMC Genomics 10:556
179. Skog J et al (2008) Glioblastoma microvesicles transport RNA and proteins that promote tumour growth and provide diagnostic biomarkers. Nat Cell Biol 10(12):1470–1476
180. Al-Nedawi K et al (2008) Intercellular transfer of the oncogenic receptor EGFRvIII by microvesicles derived from tumour cells. Nat Cell Biol 10(5):619–624
181. Ristorcelli E et al (2008) Human tumor nanoparticles induce apoptosis of pancreatic cancer cells. FASEB J 22(9):3358–3369
182. Huber V et al (2005) Human colorectal cancer cells induce T-cell death through release of proapoptotic microvesicles: role in immune escape. Gastroenterology 128(7):1796–1804
183. Taylor DD, Zacharias W, Gercel-Taylor C (2011) Exosome isolation for proteomic analyses and RNA profiling. Methods Mol Biol 728:235–246
184. Gonzales PA et al (2009) Large-scale proteomics and phosphoproteomics of urinary exosomes. J Am Soc Nephrol 20(2):363–379
185. Balaj L et al (2011) Tumour microvesicles contain retrotransposon elements and amplified oncogene sequences. Nat Commun 2:180
186. Escola JM et al (1998) Selective enrichment of tetraspan proteins on the internal vesicles of multivesicular endosomes and on exosomes secreted by human B-lymphocytes. J Biol Chem 273(32):20121–20127
187. Thery C, Zitvogel L, Amigorena S (2002) Exosomes: composition, biogenesis and function. Nat Rev Immunol 2(8):569–579
188. Lodes MJ et al (2009) Detection of cancer with serum miRNAs on an oligonucleotide microarray. PLoS One 4(7):e6229
189. Kogure T et al (2011) Intercellular nanovesicle-mediated microRNA transfer: a mechanism of environmental modulation of hepatocellular cancer cell growth. Hepatology 54(4):1237–1248
190. Ng EK et al (2009) Differential expression of microRNAs in plasma of colorectal cancer patients: a potential marker for colorectal cancer screening. Gut 58(10):1375–1381
191. Ohshima K et al (2010) Let-7 microRNA family is selectively secreted into the extracellular environment via exosomes in a metastatic gastric cancer cell line. PLoS One 5(10):e13247
192. Zhu HT et al (2012) Identification of suitable reference genes for qRT-PCR analysis of circulating microRNAs in hepatitis B virus-infected patients. Mol Biotechnol 50(1):49–56
193. Jensen SG et al (2011) Evaluation of two commercial global miRNA expression profiling platforms for detection of less abundant miRNAs. BMC Genomics 12(1):435
194. Johnstone RM et al (1987) Vesicle formation during reticulocyte maturation. Association of plasma membrane activities with released vesicles (exosomes). J Biol Chem 262(19):9412–9420
195. Cocucci E, Racchetti G, Meldolesi J (2009) Shedding microvesicles: artefacts no more. Trends Cell Biol 19(2):43–51

196. Logozzi M et al (2009) High levels of exosomes expressing CD63 and caveolin-1 in plasma of melanoma patients. PLoS One 4(4):e5219
197. Rabinowits G et al (2009) Exosomal microRNA: a diagnostic marker for lung cancer. Clin Lung Cancer 10(1):42–46
198. Smalley DM et al (2008) Isolation and identification of potential urinary microparticle biomarkers of bladder cancer. J Proteome Res 7(5):2088–2096
199. Rupp AK et al (2011) Loss of EpCAM expression in breast cancer derived serum exosomes: role of proteolytic cleavage. Gynecol Oncol 122(2):437–446
200. Shigehara K et al (2011) Real-time PCR-based analysis of the human bile microRNAome identifies miR-9 as a potential diagnostic biomarker for biliary tract cancer. PLoS One 6(8):e23584
201. Hanke M et al (2010) A robust methodology to study urine microRNA as tumor marker: microRNA-126 and microRNA-182 are related to urinary bladder cancer. Urol Oncol 28(6):655–661
202. Yamada Y et al (2011) MiR-96 and miR-183 detection in urine serve as potential tumor markers of urothelial carcinoma: correlation with stage and grade, and comparison with urinary cytology. Cancer Sci 102(3):522–529
203. Heneghan HM et al (2010) Circulating microRNAs as novel minimally invasive biomarkers for breast cancer. Ann Surg 251(3):499–505
204. Wang F et al (2010) Correlation and quantitation of microRNA aberrant expression in tissues and sera from patients with breast tumor. Gynecol Oncol 119(3):586–593
205. Huang Z et al (2010) Plasma microRNAs are promising novel biomarkers for early detection of colorectal cancer. Int J Cancer 127(1):118–126
206. Tsujiura M et al (2010) Circulating microRNAs in plasma of patients with gastric cancers. Br J Cancer 102(7):1174–1179
207. Liu R et al (2011) A five-microRNA signature identified from genome-wide serum microRNA expression profiling serves as a fingerprint for gastric cancer diagnosis. Eur J Cancer 47(5):784–791
208. Shigoka M et al (2010) Deregulation of miR-92a expression is implicated in hepatocellular carcinoma development. Pathol Int 60(5):351–357
209. Yamamoto Y et al (2009) MicroRNA-500 as a potential diagnostic marker for hepatocellular carcinoma. Biomarkers 14(7):529–538
210. Li LM et al (2010) Serum microRNA profiles serve as novel biomarkers for HBV infection and diagnosis of HBV-positive hepatocarcinoma. Cancer Res 70(23):9798–9807
211. Tanaka M et al (2009) Down-regulation of miR-92 in human plasma is a novel marker for acute leukemia patients. PLoS One 4(5):e5532
212. Moussay E et al (2010) Determination of genes and microRNAs involved in the resistance to fludarabine in vivo in chronic lymphocytic leukemia. Mol Cancer 9:115
213. Leidinger P et al (2010) High-throughput miRNA profiling of human melanoma blood samples. BMC Cancer 10:262
214. Unno K et al (2009) Identification of a novel microRNA cluster miR-193b-365 in multiple myeloma. Leuk Lymphoma 50(11):1865–1871
215. Hu Z et al (2010) Serum microRNA signatures identified in a genome-wide serum microRNA expression profiling predict survival of non-small-cell lung cancer. J Clin Oncol 28(10):1721–1726
216. Wong TS et al (2008) Mature miR-184 as Potential Oncogenic microRNA of Squamous Cell Carcinoma of Tongue. Clin Cancer Res 14(9):2588–2592
217. Liu CJ et al (2010) Increase of microRNA miR-31 level in plasma could be a potential marker of oral cancer. Oral Dis 16(4):360–364

218. Resnick KE et al (2009) The detection of differentially expressed microRNAs from the serum of ovarian cancer patients using a novel real-time PCR platform. Gynecol Oncol 112(1):55–59
219. Hausler SF et al (2010) Whole blood-derived miRNA profiles as potential new tools for ovarian cancer screening. Br J Cancer 103(5):693–700
220. Li A et al (2010) Pancreatic cancers epigenetically silence SIP1 and hypomethylate and overexpress miR-200a/200b in association with elevated circulating miR-200a and miR-200b levels. Cancer Res 70(13):5226–5237
221. Wang J et al (2009) MicroRNAs in plasma of pancreatic ductal adenocarcinoma patients as novel blood-based biomarkers of disease. Cancer Prev Res 2(9):807–813
222. Ali S et al (2010) Differentially expressed miRNAs in the plasma may provide a molecular signature for aggressive pancreatic cancer. Am J Transl Res 3(1):28–47
223. Ho AS et al (2010) Circulating miR-210 as a Novel Hypoxia Marker in Pancreatic Cancer. Transl Oncol 3(2):109–113
224. Zhang C et al (2010) Expression profile of microRNAs in serum: a fingerprint for esophageal squamous cell carcinoma. Clin Chem 56(12):1871–1879
225. Miyachi M et al (2010) Circulating muscle-specific microRNA, miR-206, as a potential diagnostic marker for rhabdomyosarcoma. Biochem Biophys Res Commun 400(1):89–93
226. Yu DC et al (2011) Circulating MicroRNAs: potential biomarkers for cancer. Int J Mol Sci 12(3):2055–2063
227. Gonzales JC et al (2011) Comparison of circulating MicroRNA 141 to circulating tumor cells, lactate dehydrogenase, and prostate-specific antigen for determining treatment response in patients with metastatic prostate cancer. Clin Genitourin Cancer 9(1):39–45
228. Cheng H et al (2011) Circulating plasma MiR-141 is a novel biomarker for metastatic colon cancer and predicts poor prognosis. PLoS One 6(3):e17745
229. Nana-Sinkam SP, Croce CM (2011) MicroRNAs as therapeutic targets in cancer. Transl Res 157(4):216–225
230. Markou A, Liang Y, Lianidou E (2011) Review: prognostic, therapeutic and diagnostic potential of microRNAs in non-small cell lung cancer. Clin Chem Lab Med 49(10):1591–1603
231. Foss KM et al (2011) miR-1254 and miR-574-5p: serum-based microRNA biomarkers for early-stage non-small cell lung cancer. J Thorac Oncol 6(3):482–488
232. Bianchi F et al (2011) A serum circulating miRNA diagnostic test to identify asymptomatic high-risk individuals with early stage lung cancer. EMBO Mol Med 3(8):495–503
233. Shen J et al (2011) Diagnosis of lung cancer in individuals with solitary pulmonary nodules by plasma microRNA biomarkers. BMC Cancer 11(1):374
234. Orom UA, Shiekhattar R (2011) Noncoding RNAs and enhancers: complications of a long-distance relationship. Trends Genet 27(10):433–439
235. Wapinski O, Chang HY (2011) Long noncoding RNAs and human disease. Trends Cell Biol 21(6):354–361
236. Lee TI et al (2006) Control of developmental regulators by Polycomb in human embryonic stem cells. Cell 125(2):301–313
237. Boyer LA et al (2006) Polycomb complexes repress developmental regulators in murine embryonic stem cells. Nature 441(7091):349–353
238. Monteiro J, Fodde R (2010) Cancer stemness and metastasis: therapeutic consequences and perspectives. Eur J Cancer 46(7):1198–1203
239. Gieni RS, Hendzel MJ (2009) Polycomb group protein gene silencing, non-coding RNA, stem cells, and cancer. Biochem Cell Biol 87(5):711–746
240. Gupta RA et al (2010) Long non-coding RNA HOTAIR reprograms chromatin state to promote cancer metastasis. Nature 464(7291):1071–1076
241. Kogo R et al (2011) Long non-coding RNA HOTAIR regulates Polycomb-dependent chromatin modification and is associated with poor prognosis in colorectal cancers. Cancer Res 71(20):6320–6326

242. Yang Z et al (2011) Overexpression of long non-coding RNA HOTAIR predicts tumor recurrence in hepatocellular carcinoma patients following liver transplantation. Ann Surg Oncol 18(5):1243–1250
243. Prensner JR et al (2011) Transcriptome sequencing across a prostate cancer cohort identifies PCAT-1, an unannotated lincRNA implicated in disease progression. Nat Biotechnol 29(8):742–749
244. Orom UA et al (2010) Long noncoding RNAs with enhancer-like function in human cells. Cell 143(1):46–58
245. Bejerano G et al (2004) Ultraconserved elements in the human genome. Science 304(5675):1321–1325
246. Derti A et al (2006) Mammalian ultraconserved elements are strongly depleted among segmental duplications and copy number variants. Nat Genet 38(10):1216–1220
247. Calin GA et al (2007) Ultraconserved regions encoding ncRNAs are altered in human leukemias and carcinomas. Cancer Cell 12(3):215–229
248. Licastro D et al (2010) Promiscuity of enhancer, coding and non-coding transcription functions in ultraconserved elements. BMC Genomics 11:151
249. Pennacchio LA et al (2006) In vivo enhancer analysis of human conserved non-coding sequences. Nature 444(7118):499–502
250. Lujambio A et al (2010) CpG island hypermethylation-associated silencing of non-coding RNAs transcribed from ultraconserved regions in human cancer. Oncogene 29(48): 6390–6401
251. Scaruffi P et al (2009) Transcribed-Ultra Conserved Region expression is associated with outcome in high-risk neuroblastoma. BMC Cancer 9:441
252. Braconi C et al (2011) Expression and functional role of a transcribed noncoding RNA with an ultraconserved element in hepatocellular carcinoma. Proc Natl Acad Sci USA 108(2): 786–791
253. Mestdagh P et al (2010) An integrative genomics screen uncovers ncRNA T-UCR functions in neuroblastoma tumours. Oncogene 29(24):3583–3592
254. Roobol MJ, Haese A, Bjartell A (2011) Tumour markers in prostate cancer III: biomarkers in urine. Acta Oncol 50(Suppl 1):85–89
255. Day JR et al (2011) PCA3: from basic molecular science to the clinical lab. Cancer Lett 301(1):1–6
256. Bussemakers MJ et al (1999) DD3: a new prostate-specific gene, highly overexpressed in prostate cancer. Cancer Res 59(23):5975–5979
257. de Kok JB et al (2002) DD3(PCA3), a very sensitive and specific marker to detect prostate tumors. Cancer Res 62(9):2695–2698
258. Hessels D et al (2003) DD3(PCA3)-based molecular urine analysis for the diagnosis of prostate cancer. Eur Urol 44(1):8–15, discussion 15–16
259. Gopinath SC, Wadhwa R, Kumar PK (2010) Expression of noncoding vault RNA in human malignant cells and its importance in mitoxantrone resistance. Mol Cancer Res 8(11):1536–1546
260. Mourtada-Maarabouni M et al (2009) GAS5, a non-protein-coding RNA, controls apoptosis and is downregulated in breast cancer. Oncogene 28(2):195–208
261. Arora R et al (2004) Heterogeneity of Gleason grade in multifocal adenocarcinoma of the prostate. Cancer 100(11):2362–2366
262. Navin N et al (2011) Tumour evolution inferred by single-cell sequencing. Nature 472(7341):90–94
263. Tomlins SA et al (2007) Integrative molecular concept modeling of prostate cancer progression. Nat Genet 39(1):41–51

Chapter 14
Therapeutic Application of MicroRNAs in Cancer

Nobuyoshi Kosaka, Fumitaka Takeshita, Yusuke Yoshioka, and Takahiro Ochiya

Abstract MicroRNAs (miRNAs) are small noncoding RNAs that regulate gene expression thereby controlling many biological and cellular processes, including development, organogenesis, and homeostasis. Due to miRNAs ability to target multiple mRNAs, if miRNA expression is altered, diseases such as cancer can occur as a consequence of the misregulation of target gene networks. Deregulation of miRNA expression in cancer cells is caused by a variety of mechanisms such as genetic alterations, epigenetic regulation, or altered expression of transcription factors, which target miRNAs. Many recent studies have focused on the development of novel diagnostic tools and therapeutics in the field of oncology. In this chapter, we summarize the latest and most significant discoveries for the use of miRNA-based therapy in various physiological and pathological conditions with particular focus on cancer. In addition, we discuss a new method for the delivery of miRNA to a desired site using biologically significant exosomes.

14.1 Introduction

In 1993, Ambros and colleagues discovered a gene, lin-4, whose product was a small nonprotein coding RNA, microRNA (miRNA) that affected development in *Caenorhabditis elegans* [1]. Currently, there are 1,527 mature human miRNA sequences listed in the miRNA registry (Sanger miRBase release 18; http://www.mirbase.org/). To yield the functional mature miRNA, biogenesis requires several posttranscriptional processing steps [2]. The primary miRNA transcript (pri-miRNA) is generated in the form of long polycistronic RNA transcripts by RNA

N. Kosaka (✉) • F. Takeshita • Y. Yoshioka • T. Ochiya
Division of Molecular and Cellular Medicine, National Cancer Center Research Institute, 5-1-1, Tsukiji, Chuo-ku, Tokyo 104-0045, Japan
e-mail: nkosaka@ncc.go.jp

K.A. Howard (ed.), *RNA Interference from Biology to Therapeutics*, Advances in Delivery Science and Technology, DOI 10.1007/978-1-4614-4744-3_14,

polymerase II. The pri-miRNAs are processed in the nucleus by the RNase III enzyme Drosha into a 60–110-bp fragment called precursor miRNA (pre-miRNA). Exportin-5 transports the pre-miRNAs into the cytoplasm, and another RNase III enzyme, Dicer, processes the pre-miRNAs into double-stranded 18–24-bp mature miRNAs. The double-strand mature miRNA is composed of two complementary single-stranded molecules, known as the guide strand and the antiguide or passenger strand. The mature miRNA can bind to a complementary sequence, typically in the 3′ untranslated region (3′ UTR), of target mRNAs as part of the RNA-induced silencing complex (RISC). This base pairing subsequently causes degradation of the mRNA and/or inhibition of protein translation. miRNAs are predicted to regulate the expression of ~90% of all human genes. The expression of miRNAs is highly specific for tissues and developmental stages, and essential miRNA functions have been observed in several fundamental biological processes, such as development, organogenesis, and homeostasis.

Recently, dysregulation of miRNA expression was found to contribute to the initiation and progression of cancer [3]. The function and expression patterns of miRNAs have been intensively investigated in various human cancers. The deregulation of miRNA expression has been shown to contribute to cancer development through various kinds of mechanisms, including deletions, amplifications, or mutations involving miRNA loci, epigenetic silencing, the dysregulation of transcription factors that regulate specific miRNAs, or by inhibition of miRNA processing [4]. miRNA expression profiling is becoming increasingly important as a useful diagnostic and prognostic tool, and many studies indicate that miRNAs can act as oncogenes and/or tumor suppressor genes.

14.1.1 Tumor-Suppressive miRNAs Attenuate Cancer Cell Malignancy

Among the first and most intensively studied tumor-suppressive miRNAs are let-7 and miR-16. It has been reported that the expression of let-7 is frequently reduced in lung cancers and that reduced let-7 expression is significantly associated with shorter patient survival [5]. In addition, overexpression of let-7 in the lung adenocarcinoma cell line, A549, inhibited cellular growth in vitro. The known targets of let-7 include representative oncogenes such as RAS and HMGA2 [6–8]. Expression of let-7 is lower in lung tumors than in normal lung tissue, while RAS protein is significantly higher in lung tumors, providing a possible mechanism for let-7 in cancer. Furthermore, the high-mobility group A2, HMGA2, is oncogenic in a variety of tumors, including benign mesenchymal tumors and lung cancers. Ectopic expression of let-7 has shown to reduce HMGA2 expression and cell proliferation in lung cancer cells.

Emerging evidence suggests that cancer stem cells (CSCs) are responsible for tumor formation, maintenance, and progression [9]. These cancer-initiating cells are rare tumor cells characterized by their strong tumorigenic properties and the ability to self-renew. The CSCs also contribute to gaining resistance to chemotherapy and radiotherapy. If a cancer treatment fails to eliminate all self-renewing cancer stem cells, residual surviving cancer stem cells are able to repopulate the tumor, causing relapse. However, the molecular mechanisms underlying self-renewal, multipotent differentiation, and tumorigenicity remain obscure. Interestingly, let-7 miRNA levels are markedly reduced in breast cancer tumor-initiating cells but increase with differentiation [10]. Infection of breast cancer tumor-initiating cells with let-7-lentivirus reduced cell proliferation, mammosphere formation, the proportion of undifferentiated cells in vitro, as well as tumor formation and metastasis in NOD/SCID mice [10]. In addition, blocking let-7 using antisense oligonucleotides (Antagomirs) resulted in enhanced in vitro self-renewal capacity of nontumor initiating cells [10]. Of note, overexpression of let-7 reduced the expression of the known let-7 targets RAS and HMGA2. Therefore, let-7 seems to regulate the stem cell-like properties of multiple breast cancer tumor-initiating cells by silencing more than one target [10]. Interestingly, RNA-binding proteins, LIN28 and LIN28B, which are important factors for maintaining pluripotency in stem cells, bind to the terminal loops of let-7 family precursor miRNAs and block their processing into mature miRNAs [11]. The contribution of LIN28 was reported, not only in pluripotency but also in oncogenesis. LIN28 and LIN28B are overexpressed in primary human tumors and human cancer cell lines. Although downregulation of the let-7 gene by LIN28 can lead to carcinogenesis, their interaction is essential for development.

miR-15a and miR-16-1 are deleted or downregulated in the majority of chronic lymphocytic leukemia (CLL) [12] [13], which is the most common human leukemia. It is characterized predominantly by nondividing malignant B cells overexpressing the anti-apoptotic B cell lymphoma 2 (BCL2) protein. These miRNAs negatively regulate BCL2 at the posttranscriptional level [14]. Furthermore, in cancer cells of advanced prostate tumors, the miR-15a and miR-16 levels are significantly decreased, whereas the expressions of BCL2, CCND1, and WNT3A are inversely upregulated [15]. Delivery of miR-15a- and miR-16-specific antagomirs into normal mouse prostate results in marked hyperplasia. Likewise, knockdown of miR-15a and miR-16 promotes survival, proliferation, and invasiveness of untransformed prostate cells, which become tumorigenic in immunodeficient NOD-SCID mice.

14.1.2 Oncogenic miRNAs Promote Cancer Progression

One of the most widely investigated oncogenic miRNAs is miR-21. The miR-21 has been identified as the only miRNA commonly overexpressed in solid tumors of the lung, breast, stomach, prostate, colon, brain, head and neck, esophagus, and pancreas [16]. A known target mRNA for miR-21 is programmed cell death 4 (PDCD4) [17–19]. PDCD4 protein levels are reduced by miR-21 in cancer cells, such as breast and colorectal cancer cells. Furthermore, overexpression of miR-21 in a human breast cancer cell line promoted soft agar colony formation by the downregulation of Pdcd4 protein levels. Moreover, miR-21 was highly overexpressed in hepatocellular carcinoma (HCC) and cell lines [20]. Inhibition of miR-21 in cultured HCC cells increased expression of the phosphatase and tensin homolog (PTEN) tumor suppressor and decreased tumor cell proliferation, migration, and invasion. In contrast, enhanced miR-21 expression by transfection with precursor miR-21 increased tumor cell proliferation, migration, and invasion. Moreover, an increase in cell migration was observed in normal human hepatocytes transfected with precursor miR-21. PTEN was shown to be a direct target of miR-21 and to contribute to miR-21 effects on cell invasion. Aberrant expression of miR-21 can contribute to HCC growth and metastatic spread by modulating PTEN expression and PTEN-dependent pathways involved in mediating the phenotypic characteristics of cancer cells, such as cell growth, migration, and invasion.

The definition of oncogenic miRNAs and tumor-suppressive miRNAs is determined by their expression and function in cancer cells; however, designation to a particular subset varies dependent on the cell type or the microenvironment. For instance, let-7 induces translation and upregulation of target mRNAs upon cell cycle arrest, yet represses translation in proliferating cells [21]. Knowing the precise mechanisms of miRNA regulation and the cellular function of miRNAs is critical for the design and choice of miRNA-mediated therapy.

14.2 Therapeutic Control of miRNA

14.2.1 Overproduction of miRNA Can Be Suppressed by Antisense Oligonucleotides

Single-strand antisense oligonucleotides (ASOs) are an attractive approach to inhibit the function of miRNAs (Fig. 14.1). Previously, ASOs were used as inhibitors of mRNA expression; however, after the discovery of RNA interference (RNAi), small interfering RNA (siRNA) has become the preferred method for gene silencing. ASOs against miRNAs are referred to as either anti-miRs, antagomirs, or anti-miRNA oligonucleotides (AMOs). There are several types of antisense nucleotide inhibiting miRNAs in the cells. The difference in these molecules depends on the chemical modification or the structure of nucleic acid constructs.

Antagomirs are 2′-*O*-methyl oligonucleotides with a cholesterol terminus and a few phosphorothioates complementary to the miRNA [22]. In addition, 2′-*O*-methyl oligonucleotide is used with a full phosphorothioate backbone [23]. The 2′-*O*-methyl-group (OMe) is one of the most classical and most often used modifications to oligonucleotides [24]. The methyl group contributes a limited amount of nuclease resistance and improves binding affinity to RNA compared to unmodified oligonucleotides. Furthermore, 2′-*O*-methoxyethyl (MOE)-modified oligonucleotides have higher affinity and specificity to RNA than their OMe analogs [24]. Locked nucleic acid (LNA)-modified oligonucleotides are distinctive 2′-*O*-modified RNA in which

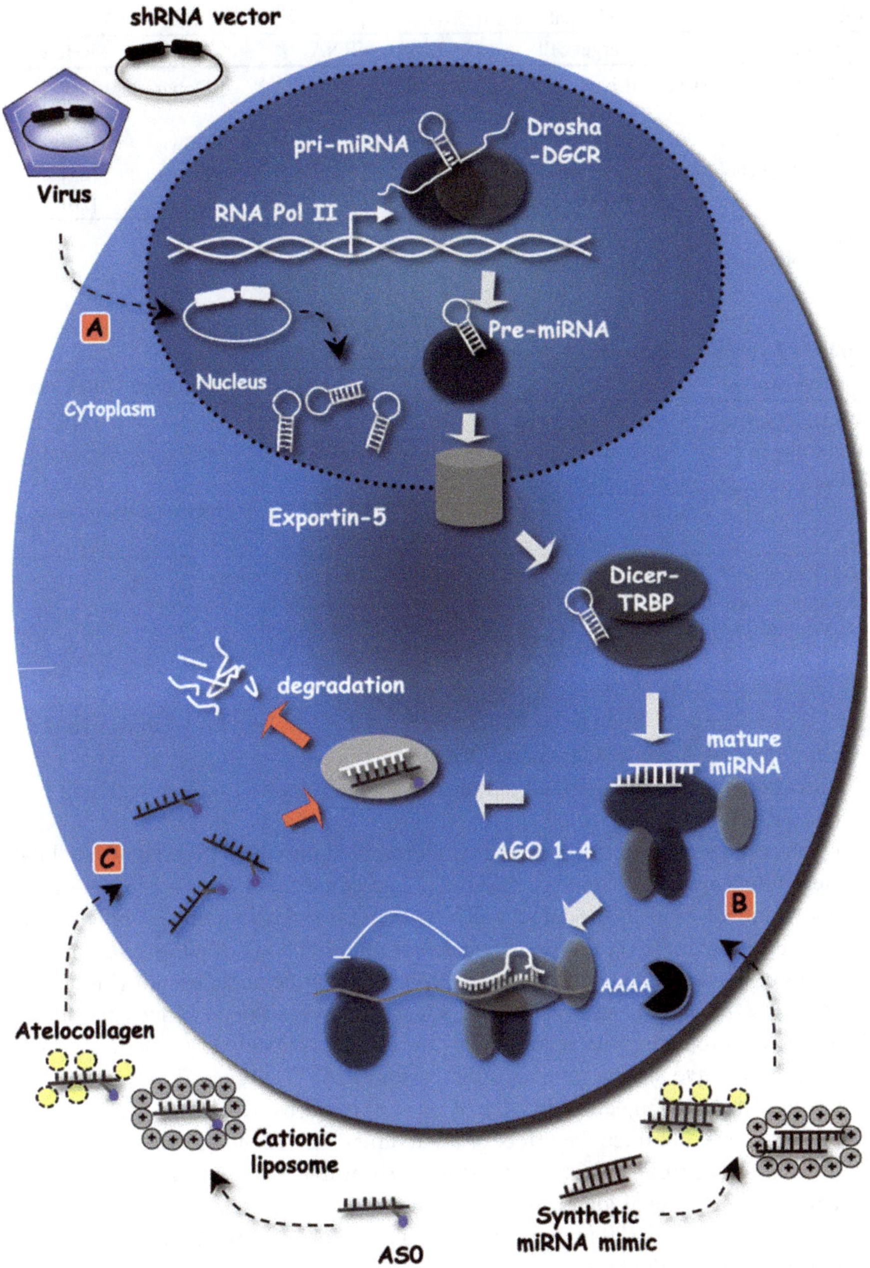

Fig. 14.1 Schematic representation of different delivery systems for miRNAs. Viral vector delivers the miRNA expression vector into the nucleus, then the expression of miRNA is increased after correct transcription and processing of miRNAs (A). Upregulation of miRNA can also be performed by synthetic miRNAs with carriers, such as atelocollagen, polymeric nanoparticles, or cationic liposomes (B). Alternatively, ASO can also be delivered to the cells in a similar way to synthetic miRNAs (C)

Table 14.1 Animal studies with antisense miRNAs

Types of nuclestide	Target cell	Target miRNA	Outcome[a]	Reference
2-*O*′-methyl cholesterol	Liver	miR-19, miR-122, miR-192, miR-194	Down	[19]
2-*O*′-MOE[b]	Liver	miR-122	Down	[20]
2-*O*′-methyl cholesterol	Liver, brain	miR-122	Down	[25]
2-*O*′-methyl cholesterol	Breast cancer cell	miR-10b	Down	[26]

[a]Outcome means the result of target miRNA expression in target cells
[b]2′-*O*-methoxyethyl phosphorothioate

Table 14.2 Animal studies with miRNA mimics

Delivery method	Target cell	Target miRNA	Outcome[a]	Reference
Atelocollagen	Prostate cancer cell	miR-16	Up	[37]
Atelocollagen	Osteosarcoma cell	miR-143	Up	[36]
JetPEI	Breast cancer cell	miR-22	Up	[38]
Lentivirus	Lung cancer (mouse model)[b]	let-7	Up	[40]
AAV	Liver cancer (mouse model)[b]	miR-26a	Up	[39]

let-7: Kras$^{LSL\text{-}G12D}$, Trp-53$^{flox/flox}$ mice; miR-26a: tet-o-*MYC*, LAP-tTA mice
[a]Outcome means the result of target miRNA expression in target cells
[b]Mouse model refers to naturally arising tumors induced by transgenes

the 2′-*O*-oxygen is bridged to the 40-position via a methylene linker to form a rigid bicycle, locked into a C3′-endo(RNA) sugar conformation [24]. The LNA modification leads to the thermodynamically stable duplex formation with complementary RNA.

Upon cellular transfection, AMO are incorporated into the RISC. They then bind to the guide sequence of mature miRNAs. The consequent AMO–miRNA–RISC complex cannot bind to the target mRNA; therefore, the function of miRNAs is impaired. There are several types of chemical modifications, such as 2′-*O*-methyl, phosphorothioate, locked nucleic acid (LNA), and peptide nucleic acids (PNA) that can enhance the binding strength to miRNA as well as resistance to nuclease degradation in vivo reducing the requirement for a protective delivery system. Indeed, recent reports have shown miRNA suppression of ASO in vivo (Table 14.1). The first study of AMO injection in vivo was performed by Krutzfeldt [22]. In this report, they tested in mice the effect of antagomir-122 and antagomir-16 to reduce the expression of miR-122 in liver and miR-16 in liver, kidney, lung, heart, skin, skeletal muscle, adrenal gland, small intestine, colon, fat, brain, and bone marrow after tail vein injection (80 mg/kg in 0.2 ml). They found that miR-16 was repressed in all tissues except the brain.

The second trial of AMO was focused on inhibition of miR-122 [27]. miR-122 inhibition was mediated by a 2′-*O*-methoxyethyl phosphorothioate ASO after intraperitoneal injection (12.5–75 mg/kg ASO twice weekly for 4 weeks) in normal mice. Reduced plasma cholesterol levels and increased hepatic fatty acid oxidation were observed. miR-122 inhibition in a diet-induced obesity mouse model resulted in decreased plasma cholesterol levels and a significant improvement in liver steatosis that was accompanied by reductions in several lipogenic genes (subcutaneous injection with 12.5 mg/kg miR-122 or control ASO twice weekly for 5 1/2 weeks). These results suggest that miR-122 may be an attractive therapeutic target for metabolic disease. Furthermore, the simple systemic delivery of a nonconjugated LNA-antimiR in phosphate-buffered saline effectively blocked the liver-specific miR-122 in nonhuman primates [27]. Intravenous acute administration of LNA-antimiR-122 (1, 3, and 10 mg/kg) to African green monkeys resulted in depletion of mature miR-122 and dose-dependent lowering of plasma cholesterol. In addition to its role in metabolism, miR-122 binds two closely spaced target sites in the 5′ noncoding region of the hepatitis C virus (HCV) genome that are essential for HCV RNA accumulation in cultured liver cells and required for modulation of HCV RNA abundance [28]. Treatment of chronically infected chimpanzees with an LNA-modified antimiR-122 (SPC3649) led to long-lasting suppression of HCV viremia, and there was no evidence of viral resistance or side effects in the treated animals. As shown by the lack of rebound in viremia during the 12-week treatment of SPC3649 and the lack of adaptive mutations in the two miR-122 seed sites of HCV 5′ noncoding region, the prolonged virological response to SPC3649 treatment without HCV rebound holds promise of a new antiviral therapy with a high barrier to the development of viral resistance mutations.

Previously, Ma et al. reported that miR-10b is highly expressed in metastatic cancer cells propagated in culture as well as in metastatic breast tumors from patients [29]. miR-10b is also upregulated in highly metastatic and/or invasive cancers, such as human pancreatic adenocarcinomas [26] and glioblastomas [30]. Furthermore, miR-10b is upregulated to a higher extent in metastatic than in nonmetastatic hepatocellular carcinomas [31]. To show the therapeutic potential of anti-miRNA in cancer treatment, the effects of miR-10b silencing in a highly metastatic mouse mammary carcinoma cell model were examined [32]. Both in vitro and in vivo, silencing of miR-10b with antagomirs significantly decreased miR-10b levels and increased the levels of its target gene, Hoxd10. Intravenous administration of 50 mg/kg miR-10b antagomirs (twice weekly for 3 weeks) to mice bearing highly metastatic cells did not reduce primary mammary tumor growth but markedly suppressed the formation of lung metastases in a sequence-specific manner.

As shown above, synthesized ASO can be taken into cells. The precise mechanism of ASO incorporation inside cells, however, has not been clarified yet. One possible candidate molecule that helps ASO incorporation is the mammalian homolog of SID-1 in *C. elegans* [33, 34]. SID-1 is a multispan transmembrane protein that sensitizes cells to soaking RNAi molecules with a potency that is dependent on double-stranded RNA (dsRNA) length [34]. RNA interference (RNAi) spreads systemically in plants and nematodes to silence gene expression distant from the site of initiation. Previous reports showed that SID-1 mediate siRNA uptake in mammalian cells [35]. Studying the precise molecular mechanism of ASO uptake may answer several aspects of ASO therapy, such as dosage of ASO.

14.2.2 Upregulation of miRNA Provided by miRNA Mimics

Certain miRNAs are known to be downregulated in cancer cells. Therefore, recovering the expression of these miRNAs by the introduction of miRNA mimics into the cell may reduce cancer progression (Fig. 14.1). The structure of miRNA mimic is small, chemically modified double-stranded RNAs that mimic endogenous miRNAs and enable miRNA functional analysis of miRNA by upregulation of miRNA activity. Delivery, however, is a major barrier to the clinical application of nucleic acid-based drugs.

14.2.2.1 Delivery Systems for miRNA Mimics

Atelocollagen is a cationic biomaterial used to overcome the delivery requirements of miRNA. It is a highly purified type I collagen derived from calf dermis with pepsin treatment [36] (Table 14.2). Atelocollagen is low in immunogenicity due to the absence of telopeptides. It has been used in the clinic for a wide range of purposes, including wound healing, vessel prosthesis, and as a bone cartilage substitute and haemostatic agent. We have previously demonstrated the efficacy of atelocollagen for nucleotide delivery including plasmid DNA, antisense oligonucleotides, and siRNA both in vitro and in vivo [37, 38]. Atelocollagen complexed with siRNA confers nuclease resistance and facilitates cellular entry and prolonged gene silencing [38]. Furthermore, we have demonstrated that systemic administration of Atelocollagen/siRNA complexes, in addition to intratumor injection against oncogenes (such as fibroblast growth factor 4 or EZH2), inhibited tumor growth in orthotopic xenograft models as well as bone metastasis mouse models of human nonseminomatous germ cell tumors and metastatic prostate cancer [39, 40]. Importantly, no toxic side effects were exhibited. In these experiments, siRNA/atelocollagen complexes showed greater selective accumulation in tumor tissues, compared with normal tissues, possibly due to the enhanced permeability and retention (EPR) effect. The EPR effect facilitates extravasation of polymeric drugs more selectively at tumor tissues, and this selective targeting to solid tumor tissues may lead to superior therapeutic benefits with fewer systemic adverse effects. This EPR effect is attributed to anatomical and pathophysiological alterations such as increased vascular density due to neoangiogenesis, impaired lymphatic recovery, and lack of smooth muscle layer in solid tumor vessels.

We have utilized the atelocollagen system for delivery of tumor-suppressive miRNAs downregulated in cancer cells as a strategy to prevent cancer cell metastasis [41]. Takeshita et al. reported that transient transfection with synthetic miR-16 expressed at lower levels in prostate cancer cells, significantly reduced cell proliferation of prostate cancer cell lines in vitro [41]. In addition, mouse tail vein injection of 50 μg of miR-16 complexed with atelocollagen in a 200-μl volume significantly inhibited the growth of prostate tumors in bone in a therapeutic bone metastasis model. Furthermore, Osaki et al. revealed that miR-143 was the most downregulated miRNA in metastatic human osteosarcoma cell lines relative to their parental cell line [40]. Likewise, transfection of miR-143 into metastatic human osteosarcoma cell lines significantly decreased invasiveness but not cell proliferation in an in vitro study [40]. In addition, intravenous injection of 50 μg of miR-143 with atelocollagen once in 3 days for 3 weeks to mouse significantly suppressed lung metastasis of metastatic human osteosarcoma cell lines. Moreover, cells positive for MMP13, a target of miR-143 in osteosarcoma cells, were found in lung metastasis-positive cases but not, in at least three cases, in the nonmetastasis group showing higher miR-143 expression.

Cellular senescence is a barrier to cancer progression. Therefore, senescence induction is a novel approach to cancer treatment. Xu et al. reported that the expression of miR-22, which is upregulated in human senescent fibroblasts and epithelial cells but downregulated in various cancer cell lines, induces growth suppression and acquisition of a senescent phenotype in human normal and cancer cells through the suppression of CDK6, SIRT1, and Sp1 genes involved in the senescence program [42]. Significantly, injection of miR-22 complexed with the cationic polymer, polyethylenimine (PEI), led to reduced lung tumor growth and metastasis suppression in the mouse model. This suggests that miR-22 plays an important role in tumor suppression.

An adeno-associated virus (AAV) vector was used for delivery of miR-26a that exhibits high expression in normal adult liver but low expression in both human and murine liver tumors. miR-26a directly downregulates cyclins D2 and E2 and induces a G_1 arrest in human liver cancer cells in vitro [43]. Systemic administration of miR-26a in a mouse model of HCC resulted in the inhibition of cancer cell proliferation, induction of tumor-specific apoptosis, and dramatic protection from disease progression without toxicity. In addition, overexpression of let-7 that has been proposed to function as a tumor suppressor, and whose expression is commonly downregulated in nonsmall cell lung cancer, showed significant tumor growth reduction of nonsmall cell lung tumors in an autochthonous model of NSCLC in mouse using a lentiviral vector [44]. In these studies, the authors employed primary miRNA sequences for miRNA expression by the virus vector. Precise sequences of primary miRNAs are needed for the correct processing of miRNAs. Knowing the sequences of primary transcripts enables the therapeutic application of vector-based delivery systems.

These findings suggest that delivery of miRNAs that are highly expressed in normal tissues but lost in disease cells may provide a general strategy for miRNA replacement therapies and that systemic delivery of tumor-suppressive miRNAs could be used to treat patients with advanced cancers (Fig. 14.1).

14.2.3 *Exosome Delivery for miRNA Therapeutics*

Recently, the group of Lotvall reported the surprising discovery that miRNAs are contained inside exosomes, which are lipoprotein complexes, including small membrane vesicles of endocytic origin (30–100 nm) [45]. Exosomes can be formed through inward budding of endosomal membranes, giving rise to intracellular [46] multivesicular bodies (MVBs) that later fuse with the plasma membrane, releasing the exosomes to the extracellular space [46]. Exosomes are released by many types of cells and are able to mediate communication between cells. The work by Lotvall demonstrated that mouse and human mast cell-derived exosomes contain RNA and miRNA and that RNA from mast cell exosomes is transferable to other mouse and human mast cells. After the transfer of mouse exosomal RNA to human mast cells, new mouse proteins were found in the recipient human cells, indicating that transferred exosomal mRNA can be translated after entering another cell. Observations from this report indicated the important fact that miRNA could exist outside of the cell and remain functional in an RNase-abundant environment.

Following the discovery of miRNA in exosomes, three reports showed transfer and functionality of miRNAs contained within secretory exosomes. Pegtel et al. demonstrated that mature EBV-encoded miRNAs are secreted in exosomes by EBV-infected B cells and that these miRNAs repress the EBV target immunoregulatory genes in primary EBV-associated lymphomas by in vitro study [47]. Zhang et al. reported that miR-150 expressed in a human monocyte/macrophage cell line is contained within exosomes and that such exosomes have the ability to deliver miR-150 into human microvascular endothelial cells, thereby inducing the downregulation of c-Myb to enhance cell migration [48]. Our group has demonstrated that secreted tumor-suppressive miR-146a, downregulated in prostate cancer, can be transported to prostate cancer cells and exert gene silencing in the recipient prostate cancer cells through the suppression of its target gene, ROCK1 protein, resulting in cell growth inhibition [49]. In this study, we employed the approach to overexpress miR-146a vector in donor cells, enabling us to obtain miR-146a highly enriched exosome. We avoided the use of synthetic analogs of mature miRNAs for the overexpression experiment because they might persist in the medium and interfere with accurate quantification of extracellular miRNAs. Our study raises the possibility that secreted miRNA could function as a cell–cell communication tool between the cancer cells and cells in their microenvironment, such as endothelial cells, immune cells, and fibroblast. These three reports demonstrated important biological functions of exosomal miRNAs in various physiological and pathological conditions, including virus infection, vascular disease, and cancer.

Recently, the application of exosomes for targeted siRNA delivery to the brain in mice was performed using exosomes [50]. In this paper, self-derived dendritic cells for exosome production were used to reduce the immunogenicity and loaded with *GAPDH* siRNA by an electroporation method. Targeting was achieved by engineering the dendritic cells to express Lamp2b, an exosomal membrane protein, fused to the neuron-specific RVG peptide3. Intravenously injected RVG-targeted exosomes showed specific gene knockdown in neurons, microglia, and oligodendrocytes in the brain. The application of this system for delivery of ASO or miRNA could be used for miRNA-based therapies.

Exploitation of natural carriers of miRNAs is not limited to exosomes. Vickers et al. revealed that high-density lipoprotein (HDL) transports endogenous miRNAs and delivers them to recipient cells with functional targeting capabilities [51].

Several other groups found that circulating miRNA complexes in human plasma and serum are associated with RNA-binding proteins, such as Argonaute 2 or nucleophosmin 1 [52–54]. It is tempting to postulate that natural processes can be used for delivery of miRNA to reduce the side effects or increase target specificity. However, it is essential to understand the molecular mechanisms of the uptake of these carriers in recipient cells. Nevertheless, these natural carriers might represent a breakthrough in the complex field of drug delivery systems.

14.3 Future Perspectives

A particular miRNA regulates several mRNAs and might influence multiple types of signaling pathways. Therefore, undesirable side effects may occur upon administration of miRNA mimics or antagomirs (Fig. 14.2). It has been reported that many of the mice administered with the miRNA treatment showed liver impairment and that some of them ultimately died after intravenous infusion of shRNA using AAV [55]. Morbidity was associated with downregulation of liver-derived miRNAs, indicating possible competition of the latter with shRNAs for limiting cellular factors required for the processing of various small RNAs. This report raised the possibility that the introduction of small RNA itself induced the unwanted side effect. Indeed, Khan et al. showed that targets of endogenous miRNAs are expressed at significantly higher levels after transfection, consistently with impaired effectiveness of endogenous miRNA repression [56]. This effect exhibited concentration and temporal dependence. Notably, the profile of endogenous miRNAs can be largely inferred by correlating miRNA sites with gene expression changes after transfection. These results suggest that this upregulation results from a saturation effect, which is a competition among the transfected small RNAs and the endogenous pool of miRNAs for the intracellular machinery that processes small RNAs.

It is also essential to consider the degradation mechanism of miRNA. Recently, Suzuki et al. identified mammalian immune regulator MCPIP1 ribonuclease as a broad suppressor of miRNA activity and biogenesis [57]. MCPIP1 suppresses miRNA biosynthesis via cleavage of the terminal loops of pre-miRNAs. Furthermore, they clearly showed that elevated MCPIP1 expression is accompanied with poor survival in lung adenocarcinoma patients and lung squamous cell carcinoma patients. On the contrary, it has been reported that low expression of Dicer is associated with poor prognosis of lung cancer. Suzuki et al. also found a regulatory role of MCPIP1 in the signaling axis comprising miR-155 and its target c-Maf. In addition to MCPIP1, degradation of mature miRNAs in *C. elegans*, mediated by the $5' \rightarrow 3'$ exoribonuclease XRN-2, affects functional miRNA homeostasis in vivo [58]. Although release and degradation can both be blocked by the addition of miRNA target RNA, after the release of the miRNA, exposing it to degradation by XRN-2. These reports demonstrate the need to address the type and dosage of miRNAs. Overdose of shRNA leads to the saturation of small RNA in the cells and then to unwanted effects in nontargeting cells. Moreover, if the expression of MCPIP1 is

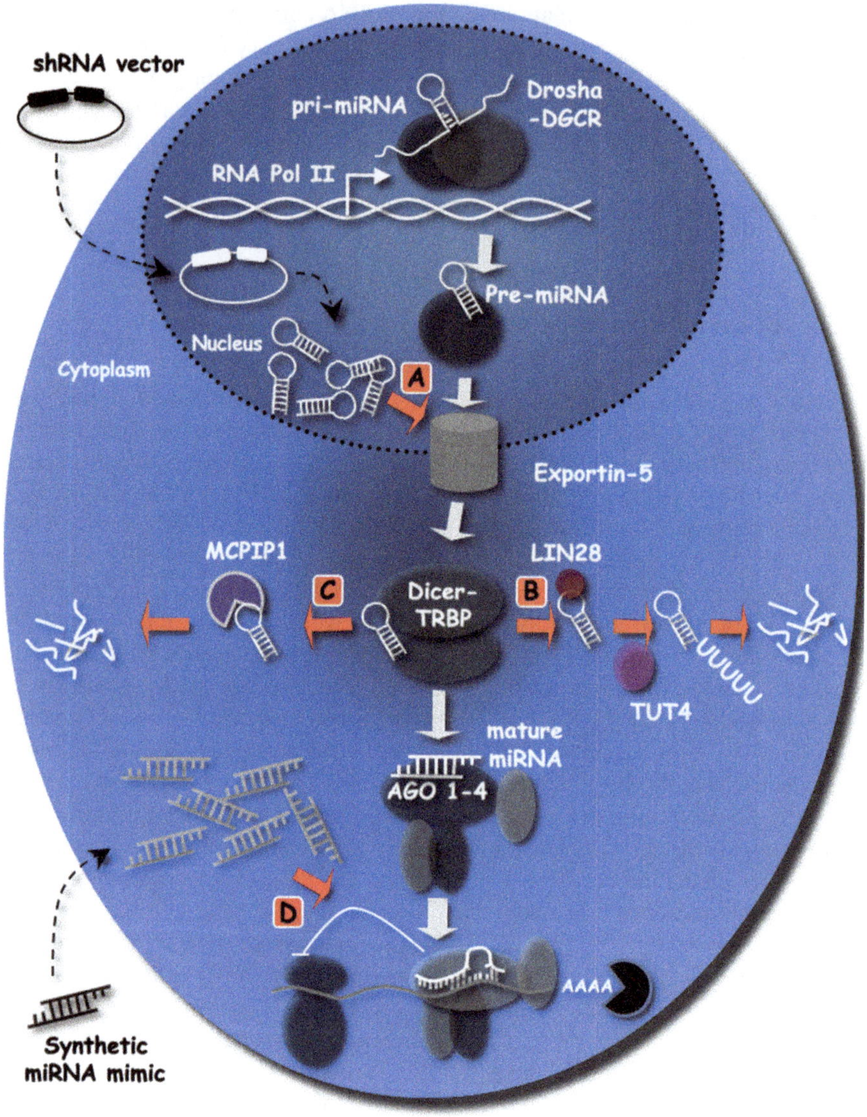

Fig. 14.2 Predicted obstacles to the therapeutic application of miRNA. miRNAs are processed by various types of proteins. If the introduction of plasmid-based delivery or virus-based miRNA vectors is extremely large, an excessive number of pre-miRNAs are produced, that may disrupt the transportation of endogenous pre-miRNAs from the nucleus to the cytoplasm mediated by Exportin-5, resulting in the downregulation of endogenous miRNAs (A). In addition, there are several inhibitory pathways for the processing of pre-miRNAs to mature miRNAs. For instance, LIN28, which is upregulated in several kinds of human cancer, specifically binds to the precursor of let-7 and induces its degradation (B). As another example, MCPIP1 binds to many kinds of pre-miRNAs and is a nuclease itself that degrades multiple miRNAs by directly cleaving the terminal loop (C). Administration of a miRNA mimic greatly increases the amounts of miRNAs in the cells, resulting in the competition between the miRNA mimic and the endogenous miRNAs for the intracellular machinery that processes miRNAs, leading to the possibility that endogenous miRNAs might lose their function (D)

extremely high in cancer cells, shRNA cannot be available in cancer cells. The study of MCPIP1 indicates that the expression of MCPIP1 may be correlated with the stage of various cancers. Similarly, if the delivered miRNA cannot enter the cells without binding to RNA-binding protein, Ago2, and target mRNA immediately, it might be degraded by the XRN-2 mammalian homolog. Taken together, the findings indicate that the basic mechanisms of miRNA homeostasis in the cells, along with the development of miRNA delivery methods, should be studied in greater depth.

Although there are many obstacles for the therapeutic use of miRNAs to control disease status, the great potential, however, is motivation for taking this field forward.

Acknowledgment This work was supported in part by a Grant-in-Aid for the Third Term Comprehensive 10-Year Strategy for Cancer Control, a Grant-in-Aid for Scientific Research on Priority Areas Cancer from the Ministry of Education, Culture, Sports, Science, and Technology, and the Program for Promotion of Fundamental Studies in Health Sciences of the National Institute of Biomedical Innovation (NiBio), and the Japan Society for the Promotion of Science (JSPS) through the "Funding Program for World-Leading Innovative R&D on Science and Technology (FIRST Program)" initiated by the Council for Science and Technology Policy (CSTP).

References

1. Lee RC, Feinbaum RL, Ambros V (1993) The *C. elegans* heterochronic gene lin-4 encodes small RNAs with antisense complementarity to lin-14. Cell 75(5):843–854
2. Kim VN, Han J, Siomi MC (2009) Biogenesis of small RNAs in animals. Nat Rev Mol Cell Biol 10(2):126–139
3. Croce CM (2009) Causes and consequences of microRNA dysregulation in cancer. Nat Rev Genet 10(10):704–714
4. Suzuki HI, Miyazono K (2011) Emerging complexity of microRNA generation cascades. J Biochem 149(1):15–25
5. Takamizawa J, Konishi H, Yanagisawa K, Tomida S, Osada H, Endoh H et al (2004) Reduced expression of the let-7 microRNAs in human lung cancers in association with shortened post-operative survival. Cancer Res 64(11):3753–3756
6. Johnson SM, Grosshans H, Shingara J, Byrom M, Jarvis R, Cheng A et al (2005) RAS is regulated by the let-7 microRNA family. Cell 120(5):635–647
7. Lee YS, Dutta A (2007) The tumor suppressor microRNA let-7 represses the HMGA2 oncogene. Genes Dev 21(9):1025–1030
8. Mayr C, Hemann MT, Bartel DP (2007) Disrupting the pairing between let-7 and Hmga2 enhances oncogenic transformation. Science 315(5818):1576–1579
9. Zhou BB, Zhang H, Damelin M, Geles KG, Grindley JC, Dirks PB (2009) Tumour-initiating cells: challenges and opportunities for anticancer drug discovery. Nat Rev Drug Discov 8(10):806–823
10. Yu F, Yao H, Zhu P, Zhang X, Pan Q, Gong C et al (2007) let-7 regulates self renewal and tumorigenicity of breast cancer cells. Cell 131(6):1109–1123
11. Viswanathan SR, Daley GQ (2010) Lin28: A microRNA regulator with a macro role. Cell 140(4):445–449
12. Calin GA, Dumitru CD, Shimizu M, Bichi R, Zupo S, Noch E et al (2002) Frequent deletions and down-regulation of micro-RNA genes miR15 and miR16 at 13q14 in chronic lymphocytic leukemia. Proc Natl Acad Sci USA 99(24):15524–15529

13. Calin GA, Ferracin M, Cimmino A, Di Leva G, Shimizu M, Wojcik SE et al (2005) A microRNA signature associated with prognosis and progression in chronic lymphocytic leukemia. N Engl J Med 353(17):1793–1801
14. Cimmino A, Calin GA, Fabbri M, Iorio MV, Ferracin M, Shimizu M et al (2005) miR-15 and miR-16 induce apoptosis by targeting BCL2. Proc Natl Acad Sci USA 102(39):13944–13949
15. Bonci D, Coppola V, Musumeci M, Addario A, Giuffrida R, Memeo L et al (2008) The miR-15a-miR-16-1 cluster controls prostate cancer by targeting multiple oncogenic activities. Nat Med 14(11):1271–1277
16. Volinia S, Calin GA, Liu CG, Ambs S, Cimmino A, Petrocca F et al (2006) A microRNA expression signature of human solid tumors defines cancer gene targets. Proc Natl Acad Sci USA 103(7):2257–2261
17. Asangani IA, Rasheed SA, Nikolova DA, Leupold JH, Colburn NH, Post S et al (2008) MicroRNA-21 (miR-21) post-transcriptionally downregulates tumor suppressor Pdcd4 and stimulates invasion, intravasation and metastasis in colorectal cancer. Oncogene 27(15): 2128–2136
18. Frankel LB, Christoffersen NR, Jacobsen A, Lindow M, Krogh A, Lund AH (2008) Programmed cell death 4 (PDCD4) is an important functional target of the microRNA miR-21 in breast cancer cells. J Biol Chem 283(2):1026–1033
19. Lu Z, Liu M, Stribinskis V, Klinge CM, Ramos KS, Colburn NH et al (2008) MicroRNA-21 promotes cell transformation by targeting the programmed cell death 4 gene. Oncogene 27(31):4373–4379
20. Meng F, Henson R, Wehbe-Janek H, Ghoshal K, Jacob ST, Patel T (2007) MicroRNA-21 regulates expression of the PTEN tumor suppressor gene in human hepatocellular cancer. Gastroenterology 133(2):647–658
21. Vasudevan S, Tong Y, Steitz JA (2007) Switching from repression to activation: microRNAs can up-regulate translation. Science 318(5858):1931–1934
22. Krutzfeldt J, Rajewsky N, Braich R, Rajeev KG, Tuschl T, Manoharan M et al (2005) Silencing of microRNAs in vivo with 'antagomirs'. Nature 438(7068):685–689
23. Esau C, Davis S, Murray SF, Yu XX, Pandey SK, Pear M et al (2006) miR-122 regulation of lipid metabolism revealed by in vivo antisense targeting. Cell Metab 3(2):87–98
24. Eckstein F (2007) The versatility of oligonucleotides as potential therapeutics. Expert Opin Biol Ther 7(7):1021–1034
25. Krutzfeldt J, Kuwajima S, Braich R, Rajeev KG, Pena J, Tuschl T et al (2007) Specificity, duplex degradation and subcellular localization of antagomirs. Nucleic Acids Res 35(9):2885–2892
26. Preis M, Gardner TB, Gordon SR, Pipas JM, Mackenzie TA, Klein EE et al (2011) MicroRNA-10b expression correlates with response to neoadjuvant therapy and survival in pancreatic ductal adenocarcinoma. Clin Cancer Res 17(17):5812–5821
27. Elmen J, Lindow M, Schutz S, Lawrence M, Petri A, Obad S et al (2008) LNA-mediated microRNA silencing in non-human primates. Nature 452(7189):896–899
28. Jopling CL, Yi M, Lancaster AM, Lemon SM, Sarnow P (2005) Modulation of hepatitis C virus RNA abundance by a liver-specific MicroRNA. Science 309(5740):1577–1581
29. Ma L, Teruya-Feldstein J, Weinberg RA (2007) Tumour invasion and metastasis initiated by microRNA-10b in breast cancer. Nature 449(7163):682–688
30. Sasayama T, Nishihara M, Kondoh T, Hosoda K, Kohmura E (2009) MicroRNA-10b is overexpressed in malignant glioma and associated with tumor invasive factors, uPAR and RhoC. Int J Cancer 125(6):1407–1413
31. Ladeiro Y, Couchy G, Balabaud C, Bioulac-Sage P, Pelletier L, Rebouissou S et al (2008) MicroRNA profiling in hepatocellular tumors is associated with clinical features and oncogene/tumor suppressor gene mutations. Hepatology 47(6):1955–1963
32. Rojas-Fernandez CH, Lapane KL, MacKnight C, Howard KA (2002) Undertreatment of osteoporosis in residents of nursing homes: population-based study with use of the Systematic Assessment of Geriatric Drug Use via Epidemiology (SAGE) database. Endocr Pract 8(5):335–342

33. Winston WM, Molodowitch C, Hunter CP (2002) Systemic RNAi in C. elegans requires the putative transmembrane protein SID-1. Science 295(5564):2456–2459
34. Feinberg EH, Hunter CP (2003) Transport of dsRNA into cells by the transmembrane protein SID-1. Science 301(5639):1545–1547
35. Wolfrum C, Shi S, Jayaprakash KN, Jayaraman M, Wang G, Pandey RK et al (2007) Mechanisms and optimization of in vivo delivery of lipophilic siRNAs. Nat Biotechnol 25(10):1149–1157
36. Ochiya T, Takahama Y, Nagahara S, Sumita Y, Hisada A, Itoh H et al (1999) New delivery system for plasmid DNA in vivo using atelocollagen as a carrier material: the Minipellet. Nat Med 5(6):707–710
37. Honma K, Ochiya T, Nagahara S, Sano A, Yamamoto H, Hirai K et al (2001) Atelocollagen-based gene transfer in cells allows high-throughput screening of gene functions. Biochem Biophys Res Commun 289(5):1075–1081
38. Minakuchi Y, Takeshita F, Kosaka N, Sasaki H, Yamamoto Y, Kouno M et al (2004) Atelocollagen-mediated synthetic small interfering RNA delivery for effective gene silencing in vitro and in vivo. Nucleic Acids Res 32(13):e109
39. Takeshita F, Minakuchi Y, Nagahara S, Honma K, Sasaki H, Hirai K et al (2005) Efficient delivery of small interfering RNA to bone-metastatic tumors by using atelocollagen in vivo. Proc Natl Acad Sci USA 102(34):12177–12182
40. Osaki M, Takeshita F, Sugimoto Y, Kosaka N, Yamamoto Y, Yoshioka Y et al (2011) MicroRNA-143 regulates human osteosarcoma metastasis by regulating matrix metalloprotease-13 expression. Mol Ther 19(6):1123–1130
41. Takeshita F, Patrawala L, Osaki M, Takahashi RU, Yamamoto Y, Kosaka N et al (2010) Systemic delivery of synthetic microRNA-16 inhibits the growth of metastatic prostate tumors via downregulation of multiple cell-cycle genes. Mol Ther 18(1):181–187
42. Xu D, Takeshita F, Hino Y, Fukunaga S, Kudo Y, Tamaki A et al (2011) miR-22 represses cancer progression by inducing cellular senescence. J Cell Biol 193(2):409–424
43. Kota J, Chivukula RR, O'Donnell KA, Wentzel EA, Montgomery CL, Hwang HW et al (2009) Therapeutic microRNA delivery suppresses tumorigenesis in a murine liver cancer model. Cell 137(6):1005–1017
44. Kumar MS, Erkeland SJ, Pester RE, Chen CY, Ebert MS, Sharp PA et al (2008) Suppression of non-small cell lung tumor development by the let-7 microRNA family. Proc Natl Acad Sci USA 105(10):3903–3908
45. Valadi H, Ekstrom K, Bossios A, Sjostrand M, Lee JJ, Lotvall JO (2007) Exosome-mediated transfer of mRNAs and microRNAs is a novel mechanism of genetic exchange between cells. Nat Cell Biol 9(6):654–659
46. Simons M, Raposo G (2009) Exosomes–vesicular carriers for intercellular communication. Curr Opin Cell Biol 21(4):575–581
47. Pegtel DM, Cosmopoulos K, Thorley-Lawson DA, van Eijndhoven MA, Hopmans ES, Lindenberg JL et al (2010) Functional delivery of viral miRNAs via exosomes. Proc Natl Acad Sci USA 107(14):6328–6333
48. Zhang Y, Liu D, Chen X, Li J, Li L, Bian Z et al (2010) Secreted monocytic miR-150 enhances targeted endothelial cell migration. Mol Cell 39(1):133–144
49. Kosaka N, Iguchi H, Yoshioka Y, Takeshita F, Matsuki Y, Ochiya T (2010) Secretory mechanisms and intercellular transfer of microRNAs in living cells. J Biol Chem 285(23):17442–17452
50. Alvarez-Erviti L, Seow Y, Yin H, Betts C, Lakhal S, Wood MJ (2011) Delivery of siRNA to the mouse brain by systemic injection of targeted exosomes. Nat Biotechnol 29(4):341–345
51. Vickers KC, Palmisano BT, Shoucri BM, Shamburek RD, Remaley AT (2011) MicroRNAs are transported in plasma and delivered to recipient cells by high-density lipoproteins. Nat Cell Biol 13(4):423–433
52. Arroyo JD, Chevillet JR, Kroh EM, Ruf IK, Pritchard CC, Gibson DF et al (2011) Argonaute2 complexes carry a population of circulating microRNAs independent of vesicles in human plasma. Proc Natl Acad Sci USA 108(12):5003–5008

53. Turchinovich A, Weiz L, Langheinz A, Burwinkel B (2011) Characterization of extracellular circulating microRNA. Nucleic Acids Res 39(16):7223–7233
54. Wang K, Zhang S, Weber J, Baxter D, Galas DJ (2010) Export of microRNAs and microRNA-protective protein by mammalian cells. Nucleic Acids Res 38(20):7248–7259
55. Grimm D, Streetz KL, Jopling CL, Storm TA, Pandey K, Davis CR et al (2006) Fatality in mice due to oversaturation of cellular microRNA/short hairpin RNA pathways. Nature 441(7092):537–541
56. Khan AA, Betel D, Miller ML, Sander C, Leslie CS, Marks DS (2009) Transfection of small RNAs globally perturbs gene regulation by endogenous microRNAs. Nat Biotechnol 27(6):549–555
57. Suzuki HI, Arase M, Matsuyama H, Choi YL, Ueno T, Mano H et al (2011) MCPIP1 ribonuclease antagonizes dicer and terminates microRNA biogenesis through precursor microRNA degradation. Mol Cell 44(3):424–436
58. Chatterjee S, Grosshans H (2009) Active turnover modulates mature microRNA activity in Caenorhabditis elegans. Nature 461(7263):546–549

Chapter 15
Preclinical and Clinical Studies Employing RNA Interference as a Therapeutic for Respiratory Syncytial Virus (RSV) Infection in the Lung

Amy R. Simon, Verena Karsten, Rachel Meyers, Jeff Cehelsky, Jared Gollob, Akshay Vaishnaw, and Sara Nochur

Abstract Here, we describe the rational design, preclinical, and clinical development of a small interfering RNA (siRNA) directed against human respiratory syncytial virus (RSV). Infection with RSV can cause significant morbidity in young children and in elderly or immunosuppressed adults. In recipients of lung transplants, RSV infection has been associated with early onset of bronchiolitis obliterans syndrome (BOS), which leads to significant morbidity and mortality. Human studies included a Phase II study in adults experimentally infected with RSV. In this study, ALN-RSV01 significantly reduced the rate of RSV infection and resulted in lower overall viral loads and a shorter duration of viral shedding. In two subsequent Phase II studies, ALN-RSV01 was administered as a nebulized aerosol to lung transplant patients naturally infected with RSV. Safety and significant reduction in BOS was first demonstrated in 16 patients. Results from a larger study in 87 patients are currently pending. The approach for using a nebulized siRNA to treat a local lung pathogen is exciting to the field, given the specificity that can be employed by RNA interference technology, the minimal systemic exposure, and the potential to silence endogenous and invading targets that impact lung function and lung diseases.

A.R. Simon (✉) • V. Karsten • R. Meyers • J. Cehelsky • J. Gollob • A. Vaishnaw • S. Nochur
Alnylam Pharmaceuticals, 300 3rd Street, Cambridge, MA 02142, USA
e-mail: asimon@alnylam.com

K.A. Howard (ed.), *RNA Interference from Biology to Therapeutics*, Advances in Delivery Science and Technology, DOI 10.1007/978-1-4614-4744-3_15,

15.1 Introduction

15.1.1 RSV Epidemiology and Diagnosis

RSV is the most commonly identified viral cause of moderate-to-severe acute upper and lower respiratory tract infection (LRTI) in infancy. The global burden of RSV infection is high, with an estimated 34 million episodes of LRTI occurring in children under 5 years old, resulting in an estimated death rate from 66,000 to 200,000 per year [1]. Healthy adults also are infected with RSV repetitively over their lifetime, but typically have a milder course that is confined to the upper respiratory tract. However, in both elderly and immunocompromised adults, RSV can cause severe and life-threatening pneumonitis as well as lead to a deterioration in underlying cardiac or pulmonary disease [2–4]. Importantly, elderly patients requiring hospitalization for RSV infection have similar lengths of stay, rates of intensive care use, and mortality (~11,000 deaths per year) as elderly patients with influenza A infection [2].

RSV is a single-stranded RNA virus that is a member of the Paramyxoviridae family. Two major human RSV subtypes, A and B, have pathogenic roles in lower respiratory tract infections, with the A subtype reported to cause a more severe disease [5]. Surveillance data from the Centers for Disease Control indicate that the most widespread RSV activity occurs between November and May in the Northern Hemisphere (http://www.cdc.gov/surveillance/nrevss/rsv/natl-trend.html). Transmission of RSV occurs by inoculation of the nasopharyngeal or ocular mucosa after contact with virus-containing secretions. While the mechanism by which RSV infects a cell has not been fully elucidated, RSV attaches to the host cell via its surface glycoprotein (G) and enters the cell via binding of its fusion protein (F) to the host epithelial cell membrane [6]. Once inside the cell, RSV replication is facilitated by the nucleocapsid (N) protein. After replicating in the epithelial cells of nasopharynx, RSV can spread to the bronchiolar epithelium of the lower respiratory tract via cell-to-cell spread or aspiration of secretions. A monoclonal antibody (Palivizumab, Synagis®) that targets the fusion protein of RSV and prevents F-protein-mediated cell entry has been approved in 45 countries worldwide for the prevention of RSV in high-risk pediatric patients [5, 6]. Drugs that block RSV attachment, fusion, or replication have been developed, but to date, only two are in Phase II clinical trials, and both of these drugs target the N protein [7]. The diagnosis of RSV in respiratory secretions (nasal swab/wash, oropharyngeal wash, bronchoalveolar lavage) can be made by various different assays including viral culture, detection of viral antigens, and detection of viral RNA (PCR-based assays).

15.1.2 Current RSV Treatment Strategy

The treatment of a LRTI secondary to RSV in children is mainly supportive, with oxygen typically given to those patients requiring hospitalization. The need for mechanical ventilation is dependent on the subject's age and underlying health status: for example, it is required in 14% of hospitalized infants versus in 30% of children who have chronic lung disease [8]. Strict infection control measures are also employed to prevent hospital outbreaks. Pharmacotherapy may include bronchodilators and corticosteroids, depending on the age and severity of the infection. Ribavirin, a nucleoside analogue with in vitro activity against RSV, is currently the only drug that is approved for use in RSV infection by the US Food and Drug Administration (FDA) and is typically administered by inhalation, although IV and oral administration are also used. The routine use of ribavirin, however, is not recommended by the American Academy of Pediatrics and is reserved for immunosuppressed patients with severe infection [9]. In mainly retrospective studies, the use of ribavirin in bone marrow patients infected with RSV resulted in a lower rate of progression of disease to the LRT as well as decreased mortality [10]. Ribavirin is also administered to patients with solid organ transplants with RSV infection; however, the efficacy of this is less clear [11].

Standard immunoglobulins (IVIG) and RSV-specific immunoglobulins (RSV-IVIG) have also been used for prophylaxis and treatment in high-risk children. RSV-IVIG was approved by the FDA in 1996 for prophylaxis in high-risk children, but is no longer available. Palivizumab (PVZ), a humanized monoclonal antibody against the RSV fusion protein, has also been in use since 1998 for RSV infection prophylaxis in high-risk children, including premature infants, and children with bronchopulmonary dysplasia or congenital heart disease [12]. While still controversial, immune modulators (e.g., PVZ and IVIG) are also used alone or in combination with ribavirin to treat adults with severe RSV infection. Combining ribavirin with immune modulators has also shown a trend toward improved outcome with regard to progression to LRTI or death, in adult hematopoietic stem cell transplant patients over ribavirin alone [10]. Given the unclear efficacy of ribavirin, along with its significant side effect profile (e.g., teratogenicity of the inhaled product) and ability to induce viral resistance, safer and more effective agents are needed for both RSV treatment and prophylaxis in pediatric as well as adult populations [13, 14].

15.1.3 Scientific Rationale for Targeting the RSV Nucleocapsid Gene

RSV has a negative single-strand RNA genome-encoding nonstructural proteins (NS2, NS1) and several other proteins such as nucleocapsid (N), phosphoprotein (P), and RNA-dependent RNA polymerase (L). The N, P, and L proteins are contained within the nucleocapsid of the virion and are required for steps in the RSV replication cycle. Consistent with their absence from the outer virus surface, and their crucial role in viral replication, the RNAs encoding the N, P, and L proteins are among the most highly conserved regions of the RSV genome [15]. Therefore, siRNAs targeting these mRNAs would be expected to result in potent inhibition of viral replication across a wide range of RSV strains. Recent studies have demonstrated the efficacy of siRNAs in inhibiting several viruses, in vitro and in vivo, such as hepatitis C [16, 17], hepatitis B [18], and influenza [19]. For RSV, several groups have demonstrated in vitro and in vivo inhibition by targeting the P or NS1 protein, confirming the feasibility of an siRNA targeting strategy for this virus [20, 21]. However, the P siRNA described is limited by its specificity for only a subset of RSV strains, while targeting NS1 may result in an antiviral effect through an immune modulatory mechanism rather than direct targeting of viral RNA through RNAi. A major consideration when developing an RSV therapeutic is that the molecule should provide activity across a wide spectrum of viral genotypes, which can vary in different seasons and regions of the world.

15.2 ALN-RSV01 Preclinical Development

15.2.1 RSV siRNA Screening and Candidate Selection

Alnylam Pharmaceuticals developed a siRNA to target the RSV N gene for the treatment of RSV infection as this gene is highly conserved and is less prone to mutation than surface viral proteins (for additional preclinical pharmacology study details, see Alvarez et al.) [22]. To select appropriate siRNAs, GenBank sequences AF035006 (RSV/A2), AF013255 (RSV/B1), AY911262 (RSV/A Long), and D00736 (RSV/18537) were aligned using the Clustal W algorithm to identify conserved 19 mers among all RSV sequences analyzed. To determine the uniqueness of each 19 mer across the human genome, a Basic Local Alignment Search Tool (BLAST) analysis was performed against the Reference Sequence (RefSeq) database. Only siRNAs with homology of 16 nucleotides or fewer to any gene in the human genome were selected for further analysis to ensure that the selected siRNA would have good specificity. Seventy siRNAs targeting the RSV N, P, and L genes were analyzed after transfection into Vero cells using lipofectamine (Invitrogen) in an in vitro plaque inhibition assay. From these candidates, 19 siRNAs exhibited >80% inhibition of plaque formation versus a PBS control at siRNA concentrations of 20 nM [22]. The unmodified siRNA designated "ALN-RSV01" that targets the N gene, consistently demonstrated the highest antiviral activity, with an IC50 of 0.7 nM in the RSV plaque inhibition assay.

In order to evaluate the ability of ALN-RSV01 to provide broad-spectrum activity across RSV variants, a series of primary isolates taken from nasal washes of children with confirmed RSV disease were sequenced across the ALN-RSV01 recognition element. Of the RSV primary isolates sequenced, 94% (89/95) showed absolute conservation across the ALN-RSV01 target site. The six isolates that were not 100% conserved had single-base alterations. Twelve of these primary isolates, including those with single-base alterations in target sequence, were tested further in the in vitro viral inhibition assay to examine if there were differences in ALN-RSV01 activity. Of these, 12/12 (100%) exhibited ~70% inhibition at 80 nM ALN-RSV01 as compared to PBS control, and all had similar dose–response curves for inhibition by ALN-RSV01.

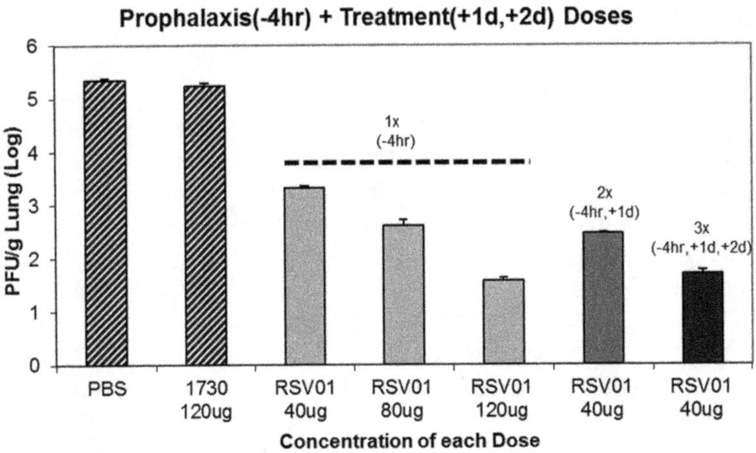

Fig. 15.1 In vivo activity of ALN-RSV01 in BALB/c mice. ALN-RSV01 dose–response curve of mice treated with single doses of 40, 80, or 120 μg ALN-RSV01 intranasally 4 h prior to infection or with multiple doses (as indicated) pre- and post-infection with 1×10^6 PFU of RSV A2. The lungs were harvested, and the virus was quantified by a standard immunostaining plaque assay. Each bar represents the mean and standard deviation of data from five animals. 1730 is a mismatch siRNA control

15.2.2 In Vivo Studies of ALN-RSV01

The BALB/c mouse is a well-established model for RSV infection and was thus chosen as the in vivo system for evaluating antiviral efficacy of ALN-RSV01. Studies were initially performed in a prophylaxis model where the ALN-RSV01 was administered intranasally (i.n.) by pipette (25 μl per nostril) to mice 4 h prior to infection with 10^6 pfu of RSV/A2. There was a dose-dependent inhibition of RSV/A2 replication in the lungs of mice, with a 120 μg dose reducing RSV titers by >3.0 $\log_{10}$ pfu/g lung as compared to either PBS or a nonspecific siRNA control (Fig. 15.1). In addition, ALN-RSV01 was equally efficacious when the same total siRNA dose was delivered over two split doses (4 h prior and 1 day after) or three split doses (4 h prior, 1 day and 2 days after) (Fig. 15.1).

To evaluate the efficacy of viral inhibition in a treatment paradigm, ALN-RSV01 was delivered i.n., in single or multiple daily doses 1–3 days post-infection. When ALN-RSV01 was administered in a treatment regimen as a single dose following viral inoculation, dose-dependent antiviral efficacy was found to decrease as a function of time of dosing post viral infection: when multiple 40 μg doses of ALN-RSV01 were delivered daily on days 1–3, potent antiviral activity was seen, and viral titers were reduced to background levels, however, by day 3 postinfection; single doses as high as 120 μg did not result in significant viral inhibition.

15.2.3 ALN-RSV01 and Cytokine Induction

Many nucleic acids have been shown to stimulate innate immune responses through a variety of RNA-binding receptors [23, 24]. While siRNA-induced immune stimulation could act synergistically with an RNAi-mediated mechanism for the treatment of a viral infection, this occurrence could also confound the interpretation of results related to a siRNA treatment strategy. Using an in vitro peripheral blood mononuclear cell (PBMC) assay [25, 26], ALN-RSV01 was evaluated for its ability to stimulate IFN-α and TNF-α. High concentrations of ALN-RSV01 were used in these assays (133 nM), greatly exceeding the IC_{50} for antiviral effect by over 100-fold. Both IFN-α and TNF-α were detected by ELISA 24 h after ALN-RSV01 treatment, with an average induction of ~147 pg/ml of IFN-α and 1,500 pg/ml of TNF-α [22].

To evaluate whether or not the antiviral activity of ALN-RSV01 was mediated by the cytokine induction seen, two non-RSV-specific siRNAs shown to more significantly induce IFN-α or TNF-α were assayed in the in vivo BALB/c mouse model. Neither of these cytokine-inducing siRNAs inhibited RSV/A2 when administered intranasally (100 μg) into mice [22]. To further evaluate the role of immune activation on the antiviral efficacy of ALN-RSV01, a chemically modified version the same siRNA sequence (containing 2′-OMe), which had no detectable cytokine induction, was also evaluated in both in vitro and in vivo studies. Treatment with this modified siRNA resulted in potent antiviral activity that was comparable to that measured for ALN-RSV01. Consistent with these findings, when ALN-RSV01 was administered to RSV-infected TLR3 and TLR7 knockout mice, robust antiviral activity was also observed. These combined data suggest that ALN-RSV01 antiviral effects are mediated via an RNAi mechanism and not via the induction of innate immunity [22].

15.2.4 In Vitro and In Vivo RACE Analysis of ALN-RSV01 Cleavage Product

To definitively confirm an RNAi-mediated mechanism of action for ALN-RSV01, a 5′ Rapid Amplification of cDNA Ends (RACE) assay was used. This assay allows for the capture and sequence analysis of the specific RNAi cleavage product mRNA intermediate following ALN-RSV01 treatment both in vitro and in vivo. Following siRNA transfection (200 nM) into Vero cells and subsequent infection with RSV/A2, a specific cleavage fragment could be detected only in the samples treated with ALN-RSV01 as compared to either PBS or a nonspecific siRNA control. In these experiments, 92% of the sequenced clones resulted from site-specific cleavage (between positions 26/27 of RSV/A2 N mRNA). When analyzed in vivo, 82% of

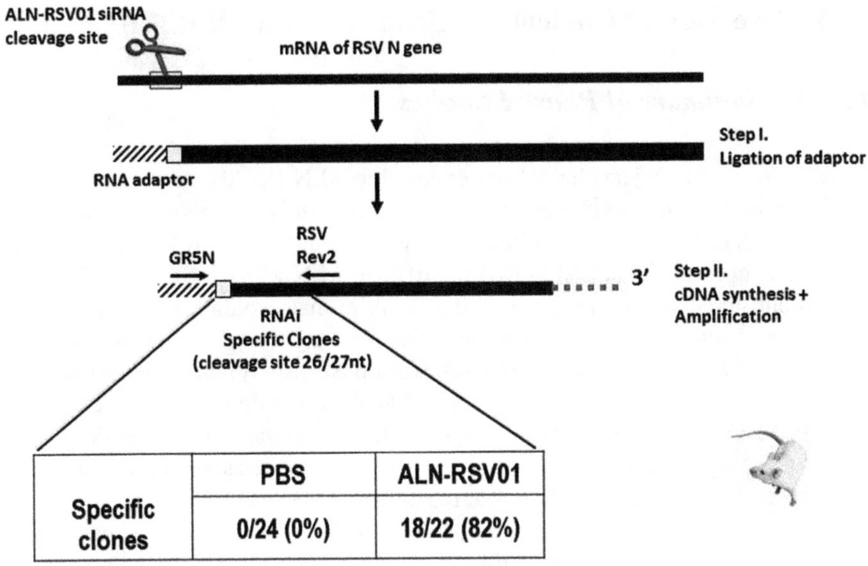

Fig. 15.2 ALN-RSV01 antiviral activity is mediated by RNAi in vivo. Shown is a schematic representation of the 5′ RACE assay used to demonstrate site-specific cleavage product. The results from the sequence analysis of individual clones from PCR amplification of cDNA generated from linker-adapted RSV N-gene mRNA isolated from an in vivo experiment are shown in *box*. In this experiment, mice were treated with ALN-RSV01 or PBS (negative control) 3 days after infection. Lungs were harvested 5 days after RSV01 treatment for RACE analysis

the clones isolated from lung tissue of ALN-RSV01 treated, RSV-infected mice demonstrated site-specific cleavage of the specific N-gene transcript, while animals treated with PBS did not (Fig. 15.2) [22].

15.2.5 Preclinical Toxicology

A comprehensive set of preclinical repeat-dose toxicology studies via intranasal (i.n.), inhalation, and intravenous (i.v.) routes was performed with ALN-RSV01 in rat and cynomolgus monkey (unpublished data). No significant drug-related toxicities were seen in either species across a range of ALN-RSV01 doses. PK data from the inhalation studies in monkeys showed short-lived, low-level plasma concentrations of ALN-RSV01 postdose, consistent with very low systemic exposure following inhalation due to rapid degradation and clearance of siRNA entering the circulation. Similarly, i.v. injection in monkeys also demonstrated no significant toxicities and a short plasma half-life, which is consistent with the short in vitro half-life in human serum (~13 min) (unpublished data).

15.3 Overview of Clinical Development for ALN-RSV01

15.3.1 Summary of Phase I Studies

To date, over 270 subjects have been exposed to ALN-RSV01 via nasal spray or inhalation in the setting of clinical studies. The safety and pharmacokinetic distribution of intranasal and inhaled ALN-RSV01 was evaluated in four Phase 1 studies. In the early clinical studies, ALN-RSV01-101 and ALN-RSV01-102, ALN-RSV01 was administered intranasally to 65 adult healthy male volunteers in single and multiple ascending doses ranging from 1.5 to 150 mg. Nasal administration was done with nasal spray devices (Becton-Dickinson Accuspray) that were filled with siRNA diluted in saline. ALN-RSV01 was found to be safe and well-tolerated at doses up to 150 mg once daily for 5 days. Systemic exposure of ALN-RSV01 was very low as only trace amounts of ALN-RSV01 were detected shortly after administration in the serum of a few patients at the highest dose level [27].

The next Phase I studies, ALN-RSV01-104 and ALN-RSV01-107, administered single and multiple doses of inhaled ALN-RSV01 ranging from 0.01 to 3.0 mg/kg using an investigational PARI eFlow® nebulizer (PARI Pharma) to a total of 107 adult male and female healthy volunteers. Administration of ALN-RSV01 by inhalation was generally safe and well-tolerated. At higher inhaled doses (1 mg/kg and above), transient flu-like symptoms were noted, including headache, chills, cough, noncardiac chest, and throat pain. Similar to the i.n. studies, little systemic exposure was observed after inhalation of ALN-RSV01. A dose of 0.6 mg/kg was well-tolerated and anticipated to be active based on allometric scaling of ALN-RSV01 doses that were effective in the mouse RSV model.

15.3.2 Experimental RSV Infection Model

An RSV experimental infection model was utilized in ALN-RSV01-105 to explore the effectiveness of ALN-RSV01 in healthy human volunteers inoculated with RSV [28]. In this double-blind placebo-controlled trial, 85 healthy male volunteers received either 75 or 150 mg ALN-RSV01 or placebo per nasal administration for 2 days. This was followed by i.n. inoculation with log3 to log5 PFU of RSV A strain (Memphis 37, GMP manufactured), as well as 3 more days of ALN-RSV01 or placebo administration. RSV infection was detected in 72% of the volunteers with a mean incubation period of 3 days. Overall, the proportion of subjects infected with RSV was significantly lower in those treated with ALN-RSV01 compared to placebo as determined by qRT-PCR (data not shown) and quantitative culture (Fig. 15.3a). The proportion of subjects that remained uninfected after RSV inoculation was also significantly greater in ALN-RSV01-treated than placebo as determined by quantitative culture (Fig. 15.3b). Measurements of viral load, viral

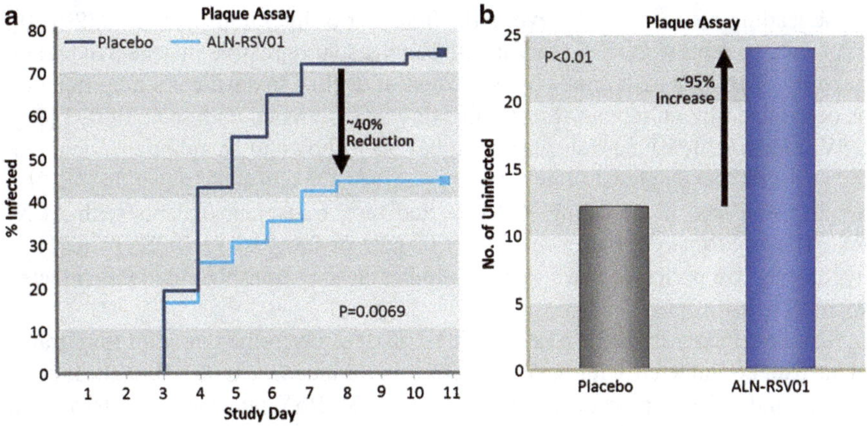

Fig. 15.3 Estimates of RSV infection. (**a**) Percent of cumulative subjects infected with RSV are shown for ALN-RSV01 (*light blue line*) and placebo (*dark blue line*) with RSV infection determined by quantitative culture (plaque assay) ($p=0.0069$), or (**b**) percent of cumulative subjects that were uninfected with RSV are shown for ALN-RSV01 (*blue bar*) and placebo (*gray bar*) with RSV infection determined by plaque assay ($p<0.01$)

shedding duration, and clinical symptom scores showed favorable trends for ALN-RSV01 but were not statistically significant due to insufficient powering of the study for these end points. The antiviral effect was shown to be independent of intranasal inflammatory cytokine concentrations, suggesting that the effect is mediated by specific inhibition of RSV and not an indirect immune-stimulatory mechanism.

Treatment with ALN-RSV01 was safe and well-tolerated. There were no severe adverse events (SAE), and the incidence of adverse events (AEs) was similar between subjects receiving ALN-RSV01 and placebo. There was no difference in the level of i.n. cytokines (TNF-alpha, IFN-alpha, G-CSF, and IL-1RA) between ALN-RSV01-treated subjects and placebo. Thus, this study determined that i.n. administration of ALN-RSV01 was both safe and effective, demonstrating proof of concept for the first time that ALN-RSV01 had antiviral activity in humans in an experimental model of RSV infection. Based on the results from our Phase I studies as well as allometric scaling from our preclinical toxicology studies, the dose of ALN-RSV01 to be used for further clinical studies was chosen to be 0.6 mg/kg.

15.3.3 Phase 2 Studies in RSV-Infected Lung Transplant Patients

Lung transplant patients have increased susceptibility to infection with community acquired viruses due to their immunosuppression and the fact that the transplanted lung is in direct communication with the environment. Chronic graft dysfunction in the transplanted lungs, as manifested by bronchiolitis obliterans syndrome (BOS), is the leading cause of death beyond the first year of lung transplantation [29, 30]. Typically, patients with BOS present with cough and progressive shortness of breath. Severity of BOS is determined by the extent of decline in a patient's lung function as determined by pulmonary function testing and is graded on a scale of 0–3 [31]. RSV infection after lung transplantation has been shown to be an independent risk factor for the development of new or progressive BOS and acute rejection [32–36]. We hypothesized that treating RSV-infected lung transplant patients with ALN-RSV01 would prevent the development of new or progressive BOS. To test this hypothesis, we performed two Phase 2 studies in lung transplant patients infected with RSV.

The first Phase 2a study, ALN-RSV01-106, was performed in adult lung transplant patients infected with RSV [37]. In this multicenter, multinational, double-blind, randomized, placebo-controlled study, 24 RSV-infected lung transplant patients (randomized 2:1, ALN-RSV01 to placebo) were administered ALN-RSV01 at 0.6 mg/kg or placebo once daily by nebulization for 3 days, in addition to the hospital's standard-of-care. Enrolled patients were stratified by ribavirin use to avoid an imbalance of this antiviral medication in two the treatment arms. The primary end point was safety and tolerability. Secondary clinical end points of the study included RSV infection characteristics (duration of viral shedding, viral clearance, and overall viral load) and patient-reported symptom scores through day 14 as well as pulmonary function parameters and BOS grade measured at day 90.

ALN-RSV01 was safe and well-tolerated. There was no difference between the treatment groups in the incidence of respiratory AEs or serum cytokine measurements. There were no drug-related SAEs, deaths, or discontinuations. Two independent clinical outcomes indicated a possible therapeutic effect of ALN-RSV01. First, the clinical symptom score (comprised of 16 symptom categories) showed improvement starting on day 1 in ALN-RSV01-treated patients, including a statistically significant difference in mean daily total symptom score from days 0 to 14 ($p=0.037$) and cumulative daily total symptom score compared to the placebo group ($p=0.035$). These effects were independent of treatment with ribavirin or high-dose steroids. Secondly, the incidence of new or progressive BOS (primary end point) was significantly decreased by 87.5% in the ALN-RSV01-treated subjects compared to placebo ($p=0.027$) (Table 15.1). Supporting data was also seen in the number of subjects that had a forced expiratory volume in 1 second (FEV1) < 80% of baseline at day 90, which trended lower in the ALN-RSV01 group compared to placebo (12.5 vs. 37.5%, *p*-value NS). No difference in any of the viral parameters could be detected, likely due to the small numbers of patient in the study as well as the imbalance in the baseline mean viral loads between the ALN-RSV01 and placebo groups. In addition, studies indicate that adults have both lower RSV titers (up to 1,000-fold lower) and a shorter duration of viral shedding than infants, which may lead to greater variability in viral data and contribute to the potential difficulty of showing antiviral efficacy in adult patient populations [38, 39].

To confirm these encouraging results, a larger Phase 2b randomized, multicenter, multinational, double-blind, placebo-controlled study in RSV-infected lung transplant patients was commenced (ALN-RSV01-109) in February 2010 at 33 transplant

Table 15.1 Day 90 clinical outcomes (intent-to-treat population)

Day 90 outcome	ALN-RSV01 $N=16$	Placebo $N=8$	p-Value
Survival	16 (100)	8 (100)	NS
Intubation	0	0	NS
Acute rejection	2 (12.5)	1 (12.5)	NS
Respiratory Infections after day 30	4 (25)	1 (12.5)	0.62
Change in BOS from baseline			
New onset	0	3 (37.5)	0.027
Progressive	1 (6.3)	1 (12.5)	1.00
Total new onset or progressive	1 (6.3)	4 (50)	0.027

Data presented as *n* (%)
p-Value are for Fisher exact test for mean score
Adapted from Zamora et al. [36]
BOS bronchiolitis obliterans syndrome, *NS* not significant

centers throughout the world. In this study, the primary end point was the effect of ALN-RSV01 on the incidence of new or progressive BOS at day 180. Secondary end points include the impact of ALN-RSV01 on BOS assessment at day 90, symptom scores, antiviral parameters, and safety. RSV-positive subjects were randomized 1:1 to receive either nebulized ALN-RSV01 (0.6 mg/kg) or placebo daily for 5 days, in addition to the hospital's standard-of-care. During randomization, subjects were stratified to the treatment arms based on two binary factors (1) time from symptom onset to treatment start and (2) preinfection BOS grade. Stratifying by onset from diagnosis to treatment start was implemented to try to ensure similar baseline viral load in the treatment groups.

Of the 3,985 subjects prescreened with respiratory symptoms at 33 centers, 218 were RSV positive (5.5% of the total), of which 87 were randomized into the study. BOS scores at days 90 and 180 were adjudicated by an independent committee made up of physicians who were lung transplant specialists not participating in the trial and who were blinded to study drug treatment assignment. In addition, RSV samples (oropharyngeal wash and nasal swabs) were analyzed by PCR at a central laboratory, and all results were reviewed by a quality oversight committee comprised of physicians who were infectious disease specialists not participating in the trial. Enrollment for this study was completed (October 2011), and subjects are now in the follow-up phase through the first half of the year. Results from this trial are expected to be reported in 2012.

15.4 Future Perspectives for RNAi in the Clinic

The unmet need for safe and effective therapies remains high as RSV-induced lower respiratory infection results in hospitalization in up to 10% of children (<5 year old) and in 16% of elderly patients [1, 2]. While a large number of anti-RSV therapeutics have been studied, such as antisense oligonucleotides, small molecule inhibitors, vaccines, and fusion protein inhibitors, no approach has generated positive clinical data leading to FDA approval of a new RSV therapeutic [7]. A positive result in the ALN-RSV01-109 trial has the potential to lead to the development of a novel therapeutic for lung transplant patients who are at risk of developing BOS and chronic graft dysfunction after RSV infection. In addition, if ALN-RSV01 is found to be clinically efficacious in lung transplant patients, it opens up the possibility that it may also be beneficial in other high-risk populations that do poorly when infected with RSV, such as bone marrow transplant patients, the elderly, and pediatric patients with prematurity or congenital heart disease. Targets like RSV that are exogenous to the lung as well as those that are endogenous to the lung (e.g., airway epithelial genes) are potentially amenable to modulation by an RNAi therapeutic delivered via nebulization, thereby limiting systemic exposure, and thus will likely continue to be a focus of research in the RNAi field and in pulmonary medicine over the next decade.

Over the past 20 years, RNAi has gone from being a Nobel Prize-worthy scientific discovery in plants and worms to a novel therapeutic approach that can be utilized to silence the expression of genes that play a role in human disease. To date, ~15 clinical programs have administered siRNA both locally (e.g., to the eye or lung) and systemically (e.g., intravenously), and these approaches have been shown to be safe and well-tolerated in both healthy human volunteers as well as patients [40, 41]. In particular, formulation of siRNAs in lipid nanoparticles has led to significant breakthroughs in systemic delivery. Recently, two separate Phase I clinical trials have reported that LNP-formulated siRNA can safely and effectively inhibit the expression of two liver-derived genes, PCSK9 and TTR, in healthy volunteers and patients, respectively [42, 43]. While RNAi is still in the early phases of clinical investigation, the emerging safety and efficacy data suggest that RNAi will translate into meaningful therapeutics for human disease.

References

1. Nair H, Nokes DJ, Gessner BD, Dherani M, Madhi SA, Singleton RJ et al (2010) Global burden of acute lower respiratory infections due to respiratory syncytial virus in young children: a systematic review and meta-analysis. Lancet 375(9725):1545–1555
2. Falsey AR, Hennessey PA, Formica MA, Cox C, Walsh EE (2005) Respiratory syncytial virus infection in elderly and high-risk adults. N Engl J Med 352:1749–1759
3. Hall CB, Powell KR, MacDonald NE, Gala CL, Menegus ME, Suffin SC et al (1986) Respiratory syncytial viral infection in children with compromised immune function. N Engl J Med 315(2):77–81
4. Englund JA, Sullivan CJ, Jordan MC, Dehner LP, Vercellotti GM, Balfour HH Jr (1988) Respiratory syncytial virus infection in immunocompromised adults. Ann Intern Med 109(3):203–208
5. McConnochie KM, Hall CB, Walsh EE, Roghmann KJ (1990) Variation in severity of respiratory syncytial virus infections with subtype. J Pediatr 117(1 Pt 1):52–62
6. Olszewska W, Openshaw P (2009) Emerging drugs for respiratory syncytial virus infection. Expert Opin Emerg Drugs 14(2):207–217
7. Storey S (2010) Respiratory syncytial virus market. Nat Rev 9:15–16

8. Willson DF, Landrigan CP, Horn SD, Smout RJ (2003) Complications in infants hospitalized for bronchiolitis or respiratory syncytial virus pneumonia. J Pediatr 143(5 Suppl):S142–S149
9. American Academy of Pediatrics Committee on Infectious Diseases (1996) Reassessment of the indications for ribavirin therapy in respiratory syncytial virus infections. Pediatrics 97(1):137–140
10. Shah JN, Chemaly RF (2011) Management of RSV infections in adult recipients of hematopoietic stem cell transplantation. Blood 117(10):2755–2763
11. Kotloff RM, Ahya VN, Crawford SW (2004) Pulmonary complications of solid organ and hematopoietic stem cell transplantation. Am J Respir Crit Care Med 170(1):22–48
12. (1998) Palivizumab, a humanized respiratory syncytial virus monoclonal antibody, reduces hospitalization from respiratory syncytial virus infection in high-risk infants. The IMpact-RSV Study Group. Pediatrics 102(3 Pt 1):531–537. http://www.ncbi.nlm.nih.gov/pubmed/9738173
13. Crotty S, Andino R (2002) Implications of high RNA virus mutation rates: lethal mutagenesis and the antiviral drug ribavirin. Microbes Infect 4(13):1301–1307
14. Ventre K, Randolph AG (2007) Ribavirin for respiratory syncytial virus infection of the lower respiratory tract in infants and young children. Cochrane Database Syst Rev (1):CD000181. http://www.ncbi.nlm.nih.gov/pubmed/17253446
15. Sullender WM (2000) Respiratory syncytial virus genetic and antigenic diversity. Clin Microbiol Rev 13(1):1–15
16. Saksela K (2003) Human viruses under attack by small inhibitory RNA. Trends Microbiol 11(8):345–347
17. Wang Z, Ren L, Zhao X, Hung T, Meng A, Wang J et al (2004) Inhibition of severe acute respiratory syndrome virus replication by small interfering RNAs in mammalian cells. J Virol 78(14):7523–7527
18. Taylor JA, Naoumov NV (2005) The potential of RNA interference as a tool in the management of viral hepatitis. J Hepatol 42(1):139–144
19. Ge Q, McManus MT, Nguyen T, Shen CH, Sharp PA, Eisen HN et al (2003) RNA interference of influenza virus production by directly targeting mRNA for degradation and indirectly inhibiting all viral RNA transcription. Proc Natl Acad Sci USA 100(5):2718–2723
20. Bitko V, Musiyenko A, Shulyayeva O, Barik S (2005) Inhibition of respiratory viruses by nasally administered siRNA. Nat Med 11(1):50–55
21. Zhang W, Yang H, Kong X, Mohapatra S, San Juan-Vergara H, Hellermann G et al (2005) Inhibition of respiratory syncytial virus infection with intranasal siRNA nanoparticles targeting the viral NS1 gene. Nat Med 11(1):56–62
22. Alvarez R, Elbashir S, Borland T, Toudjarska I, Hadwiger P, John M et al (2009) RNA interference-mediated silencing of the respiratory syncytial virus nucleocapsid defines a potent antiviral strategy. Antimicrob Agents Chemother 53(9):3952–3962
23. Robbins M, Judge A, Liang L, McClintock K, Yaworski E, MacLachlan I (2007) 2′-O-methyl-modified RNAs act as TLR7 antagonists. Mol Ther 15(9):1663–1669
24. Schlee M, Barchet W, Hornung V, Hartmann G (2007) Beyond double-stranded RNA-type I IFN induction by 3pRNA and other viral nucleic acids. Curr Top Microbiol Immunol 316:207–230
25. Hornung V, Guenthner-Biller M, Bourquin C, Ablasser A, Schlee M, Uematsu S et al (2005) Sequence-specific potent induction of IFN-alpha by short interfering RNA in plasmacytoid dendritic cells through TLR7. Nat Med 11(3):263–270
26. Judge AD, Sood V, Shaw JR, Fang D, McClintock K, MacLachlan I (2005) Sequence-dependent stimulation of the mammalian innate immune response by synthetic siRNA. Nat Biotechnol 23(4):457–462
27. DeVincenzo J, Cehelsky JE, Alvarez R, Elbashir S, Harborth J, Toudjarska I et al (2008) Evaluation of the safety, tolerability and pharmacokinetics of ALN-RSV01, a novel RNAi antiviral therapeutic directed against respiratory syncytial virus (RSV). Antiviral Res 77(3):225–231
28. DeVincenzo J, Lanbkin-Williams R et al (2010) A randomized, double-blind, placebo-controlled study of an RNA-based therapy directed against respiratory syncytial virus. PNAS Early Edition: 1–6

29. Estenne M, Hertz MI (2002) Bronchiolitis obliterans after human lung transplantation. Am J Respir Crit Care Med 166(4):440–444
30. Trulock EP, Edwards LB, Taylor DO, Boucek MM, Mohacsi PJ, Keck BM et al (2003) The registry of the International Society for Heart and Lung Transplantation: twentieth official adult lung and heart-lung transplant report–2003. J Heart Lung Transplant 22(6):625–635
31. Estenne M, Egan JJ, Mallory GB, Yousem S (2002) Bronchiolitis Obliterans Syndrome 2001: an update of the diagnostic criteria. J Heart Lung Transplant 21(3):297–310
32. Kumar D, Erdman D, Keshavjee S, Peret T, Tellier R, Hadjiliadis D et al (2005) Clinical impact of community-acquired respiratory viruses on bronchiolitis obliterans after lung transplant. Am J Transplant 5:2031–2036
33. Gottlieb J, Schulz TF, Welte T, Fuehner T, Dierich M, Simon AR et al (2009) Community-acquired respiratory viral infections in lung transplant recipients: a single season cohort study. Transplantation 87:1530–1537
34. Milstone AP, Brumble LM, Barnes J, Estes W, Loyd JE, Pierson RNI et al (2006) A single-season prospective study of respiratory viral infections in lung transplant recipients. Eur Respir J 28:131–137
35. Khalifah AP, Hachem RR, Chakinala MM, Schechtman KB, Patterson GA, Schuster DP et al (2004) Respiratory viral infections are a distinct risk for bronchiolitis obliterans syndrome and death. Am J Respir Crit Care Med 170:181–187
36. Hopkins P, McNeil K, Kermeen F, Musk M, McQueen E, MacKay I et al (2008) Human metapneumovirus in lung transplant recipients and comparison to respiratory syncytial virus. Am J Respir Crit Care Med 178:876–881
37. Zamora MR, Budev M, Rolfe M, Gottlieb J, Humar A, DeVincenzo J et al (2011) RNA interference therapy in lung transplant patients infected with respiratory syncytial virus. Am J Respir Crit Care Med 183(4):531–538
38. Hall WJ, Hall CB, Speers DM (1978) Respiratory syncytial virus infection in adults: clinical, virologic, and serial pulmonary function studies. Ann Intern Med 88(2):203–205
39. Duncan CB, Walsh EE, Peterson DR, Lee FE, Falsey AR (2009) Risk factors for respiratory failure associated with respiratory syncytial virus infection in adults. J Infect Dis 200(8):1242–1246
40. Whitehead KA, Langer R, Anderson DG (2009) Knocking down barriers: advances in siRNA delivery. Nat Rev Drug Discov 8(2):129–138
41. Vaishnaw AK, Gollob J, Gamba-Vitalo C, Hutabarat R, Sah D, Meyers R et al (2010) A status report on RNAi therapeutics. Silence 1(1):14
42. Fitzgerald K, Frank-Kamenetsky M, Mant T, Ritter J, Chiesa J, Munasamy M, Hutabarat R, Clausen V, Watkins D, Smith K, Sutherland J, Cehelsky J, Nochur S, Binne L, Nechev L, Kretschmer M, Karsten V, Vaishnaw A, Simon A (2012) Phase I safety, pharmacokinetic and pharmacodynamic results for ALN-PCS, a novel RNAi therapeutic for the treatment of hypercholesterolemia. American Heart Association's Arteriosclerosis, Thrombosis and Vascular Biology [Abstract]; Scientific Sessions
43. Coelho T, Silva A, Suhr OB, Adams D, Lozeron P, Hawkins P, Mant T, Falzone R, Figeruoa Y, Hutabarat R, Butler J, Fishman S, Chen Q, Costelha S, Saraiva MJ, Cehelsky J, Vaishnaw A, Gollob J, Sah DWY (2012) Final Phase I safety, pharmacokinetic and pharmacodynamic results for ALN-TTR01, a novel RNAi therapeutic for the treatment of transthyretin amyloidosis. XIII international symposium on amyloidosis

About the Editor

Kenneth A. Howard is an Associate Professor at the Department of Molecular Biology and Genetics at the Interdisciplinary Nanoscience Center (iNANO), Aarhus University, Denmark. His group is based in a highly interdisciplinary setting with research and teaching activities focused on advanced drug delivery, nanomedicine, RNA interference, and inflammation. Dr Howard is CEO of an Aarhus University spin-out company that provides delivery solutions for RNA interference-based therapeutics. Kenneth A. Howard received a PhD in Pharmaceutical Science from the University of Nottingham, UK, in 1995, and has held postdoctoral positions at the CRC Institute for Cancer Studies, University of Birmingham, UK, and the School of Pharmacy, University of London. Kenneth A. Howard is an active member of the Controlled Release Society serving on the Board of Scientific Advisors (chair), Nominations Committee, and organizer of a CRS Educational workshop "RNA Interference Biology and Therapeutics".

K.A. Howard (ed.), *RNA Interference from Biology to Therapeutics*,
Advances in Delivery Science and Technology, DOI 10.1007/978-1-4614-4744-3,

Index

K.A. Howard (ed.), *RNA Interference from Biology to Therapeutics*, Advances in Delivery Science and Technology, DOI 10.1007/978-1-4614-4744-3,

GPSR Compliance

The European Union's (EU) General Product Safety Regulation (GPSR) is a set of rules that requires consumer products to be safe and our obligations to ensure this.

If you have any concerns about our products, you can contact us on

ProductSafety@springernature.com

In case Publisher is established outside the EU, the EU authorized representative is:

Springer Nature Customer Service Center GmbH
Europaplatz 3
69115 Heidelberg, Germany

www.ingramcontent.com/pod-product-compliance
Lightning Source LLC
Chambersburg PA
CBHW060743150926
51501CB00003B/80
* 9 7 8 1 4 6 1 4 4 7 4 3 6 *